Cyborg

This book provides in-depth information about the technical, legal, and policy issues that are raised when humans and artificially intelligent machines are enhanced by technology.

Cyborg: Human and Machine Communication Paradigm helps readers to understand cyborgs, bionic humans, and machines with increasing levels of intelligence by linking a chain of fascinating subjects together, such as the technology of cognitive, motor, and sensory prosthetics; biological and technological enhancements to humans; body hacking; and brain-computer interfaces. It also covers the existing role of the cyborg in real-world applications and offers a thorough introduction to cybernetic organisms, an exciting emerging field at the interface of the computer, engineering, mathematical, and physical sciences.

Academicians, researchers, advanced-level students, and engineers that are interested in the advancements in artificial intelligence, brain-computer interfaces, and applications of human-computer in the real world will find this book very interesting.

PROSPECTS IN SMART TECHNOLOGIES

Series Editor
Mohammad M. Banat
Mohammad M. Banat, Jordan University of Science
and Technology, Irbid, Jordan
banat@just.edu.jo
Sara Paiva, Instituto Politécnico de Viana do Castelo, Viana do Castelo, Portugal
sara.paiva@estg.ipvc.pt

Published Titles

Emerging Technologies for Sustainable and Smart Energy
Edited by Anirbid Sircar, Gautami Tripathi, Namrata Bist, Kashish Ara Shakil, and Mithileysh Sathiyanarayanan

Cyborg: Human and Machine Communication Paradigm
Kuldeep Singh Kaswan, Jagjit Singh Dhatterwal, Anupam Baliyan, and Shalli Rani

For more information on this series, please visit: www.routledge.com/Prospects-in-Smart-Technologies/book-series/CRCPST

Cyborg
Human and Machine Communication Paradigm

Kuldeep Singh Kaswan, Jagjit Singh Dhatterwal,
Anupam Baliyan, and Shalli Rani

CRC Press is an imprint of the
Taylor & Francis Group, an **informa** business

Designed cover image: Shutterstock

First edition published 2024
by CRC Press
2385 Executive Center Drive, Suite 320, Boca Raton, FL 33431

and by CRC Press
4 Park Square, Milton Park, Abingdon, Oxon, OX14 4RN

CRC Press is an imprint of Taylor & Francis Group, LLC

© 2024 Kuldeep Singh Kaswan, Jagjit Singh Dhatterwal, Anupam Baliyan, Shalli Rani

Reasonable efforts have been made to publish reliable data and information, but the author and publisher cannot assume responsibility for the validity of all materials or the consequences of their use. The authors and publishers have attempted to trace the copyright holders of all material reproduced in this publication and apologize to copyright holders if permission to publish in this form has not been obtained. If any copyright material has not been acknowledged, please write and let us know so we may rectify in any future reprint.

Except as permitted under U.S. Copyright Law, no part of this book may be reprinted, reproduced, transmitted, or utilized in any form by any electronic, mechanical, or other means, now known or hereafter invented, including photocopying, microfilming, and recording, or in any information storage or retrieval system, without written permission from the publishers.

For permission to photocopy or use material electronically from this work, access www.copyright.com or contact the Copyright Clearance Center, Inc. (CCC), 222 Rosewood Drive, Danvers, MA 01923, 978–750–8400. For works that are not available on CCC please contact mpkbookspermissions@tandf.co.uk

Trademark notice: Product or corporate names may be trademarks or registered trademarks and are used only for identification and explanation without intent to infringe.

ISBN: 978-1-032-47967-5 (hbk)
ISBN: 978-1-032-49222-3 (pbk)
ISBN: 978-1-003-39269-9 (ebk)

DOI: 10.1201/9781003392699

Typeset in Times
by Apex CoVantage, LLC

Contents

Authors xiii

Chapter 1 Cyborg: An Introduction 1

    1.1 Introduction 1
        1.1.1 Cyborg Body Control 1
        1.1.2 What Is Cyborg 4
        1.1.3 Types of Cyborgs 5
        1.1.4 What Is Cyberization? 7
    1.2 Defining the Cyborg Era 8
    1.3 Cyborg History 8
    1.4 Conceptual Schema of Cyborg 9
        1.4.1 Textual Transition 9
        1.4.2 Pedagogical Evolution 11
    1.5 Learn Cyborg Skill Literacy 14
        1.5.1 Posthuman Promises 15
        1.5.2 Posthuman Identity as Cyborg Identity 17
        1.5.3 Theorizing Racialized Humans as Cyborgs 18
    1.6 Cyborg Pedagogy 19
        1.6.1 It's Cyborg Pedagogy 20
        1.6.2 Traditional Teaching 21
        1.6.3 Ontologies of Cyborg 21
    1.7 Cyborg Responsibility 23
    1.8 Conclusion 24
    References 25

Chapter 2 Cyborg Ontology 26

    2.1 Cyborg Technology 26
        2.1.1 Define Technology? 26
        2.1.2 The Technological Desire for the Presence 27
        2.1.3 Innovative Cyborg Technology 28
        2.1.4 Communications Technology Media. Transcending Human Conscious Purposes 28
        2.1.5 Basic Analogous Protocols: Metaphysically Reminiscent Longings 30
        2.1.6 Preception Vision 31
        2.1.7 Modernity Humanity: The World's Largest Analog Depiction 32
        2.1.8 Cyborg Authenticity Representation 33
        2.1.9 Description of Body Parts 34
        2.1.10 Beauty as a Mediator 34

		2.1.11	Anthropological of Lacan: Beyond Requirements 34
		2.1.12	Cognitive Imagination 35
		2.1.13	Fantasy and Excess Pleasure: Between Gratification and Ecstasy 36
		2.1.14	The Technologization of Human Virtuality 36
		2.1.15	Basics of Virtual Technology 36
		2.1.16	Virtual Reality 38
		2.1.17	Four Concepts: Physical and Technological in Digitalization 39
		2.1.18	Virtualization 39
		2.1.19	Language: The Virtualization of the Real 40
		2.1.20	The Retroaction of "Real Time" 40
		2.1.21	"Law": The Virtualization of "Natural Forces" 41
		2.1.22	Technology for the Interfacing and Virtualization of Genuine Innovation Imagination 41
		2.1.23	The Technological Age: Conventional to Captured Pictures, Material to Interaction 42
		2.1.24	Digitalization and Representational Systems of the Intellect 42
		2.1.25	Augmented World of the Virtual Machine: from Knowledge to Release 43
		2.1.26	Interaction and Visualization Technology Representational Regions or Spaces Designed to Simulate 43
		2.1.27	It Appears Impossible to Connect the Indication with the Reference on the Interaction of It 44
		2.1.28	A Media Eradication of Subjective in Humanity? From the Performance to the Presentation? 44
		2.1.29	Cyber-Subjectivism or Cyber-Objectivism? 44
		2.1.30	From Semiotics to the Subject as a Mediating Window 45
	2.2	Imagination Shows the Important Significance of the Interaction Design 45	
		2.2.1	Representation of Digital World 45
		2.2.2	Imagination: Naturally or Artificially? Narrative? Imagination Seems an "Organic" Mediating Psychoanalytic Theories 46
		2.2.3	Fantasy as Imitation: Hallucinatory Wish Fulfilments 46
		2.2.4	Concept of Enjoyment Perception 47
		2.2.5	Philosophy of Categorical Imperatives: Perceptions of Mediation and Actuality 47
		2.2.6	Freudo-Lacanian Theory: Desirable Reality 48
		2.2.7	Behavioral Therapy 48

2.3		Technology and the Fantastic Relation to the Real 48	
	2.3.1	Technology and the Real Pleasure Principle and Its Beyond...	48
	2.3.2	Truth of Cyberspace	49
	2.3.3	A Historical Outline of the Real in Lacan's Work	49
	2.3.4	The Real as the Object of Lost Gratifications	49
	2.3.5	Concept of Reality	49
	2.3.6	The Fantasy Interface as a Screen	50
	2.3.7	The Screen as Principally Defensive: Phobia and its Computerized Treatment	50
2.4		The Hypothetical Virtual Environment....................................	50
	2.4.1	Cyborg Information	52
	2.4.2	Cybernetics	52
	2.4.3	Principle of Simulated World	53
	2.4.4	Embodied Space	53
	2.4.5	The Personality as Artificial Togetherness: The Mirrored Region	54
	2.4.6	The Aesthetics of Reflecting Official Numbers: The Impact of Cognition	54
	2.4.7	Virtual World of Human	54
	2.4.8	Pace: Projection of feelings to/from the Mucous Membrane	55
	2.4.9	Avatars: Engaging the Body in Space	55
	2.4.10	Imagination: The Dual Binding of the Virtual Environment	55
	2.4.11	Cyborg Subjectivity	56
	2.4.12	The Interminable Promises of the Cyborg: Among Insufficiency and Excellence	56
	2.4.13	The Cyborg Abundance: The Challenge to Real Annihilation	56
	2.4.14	Between Exploring and Dominating Space	56
2.5		Fantasy and Subjectivation: Excess Pleasure Good Things Innovation	57
	2.5.1	Principle of Mental Development	57
	2.5.2	The Sick and Twisted Media Pleasure: Not the Deed but the Setting	57
	2.5.3	The Vital Disavowal	58
	2.5.4	Interactivity and Technological Belief	58
	2.5.5	Narrative Visual Digital Media	58
	2.5.6	The Sinkhole: The Mark's Splendor	58
	2.5.7	The Partial Objects and the Cut	59
	2.5.8	Imagination Subjectivized: Self-Identification as an Objective Subjects: The Topic as an Entity	59

		2.5.9 Symbols, Facts, Engagement.. 60
	2.6	Conclusion 60
	References 61

Chapter 3 Cyborg Communication 62

    3.1 Enlightenment Cyborg.. 62
    3.2 The Communication with Anatomy Brain 65
    3.3 Extension of Soul Communication... 68
    3.4 Literacy Communications in Mechanical Operation............... 72
    3.5 Man-Machine and Intellectual Electricity 77
    3.6 Conclusion 82
    References 82

Chapter 4 Evolution of Woman Cyborg 84

    4.1 The Intellectual Female Cyborg 84
    4.2 Cyborg Reproduction Technologies in Modern Era................ 88
    4.3 Female Cyborg Origin Stories.. 90
    4.4 Innovation of Woman-Machine 93
    4.5 Female Vanity 95
    4.6 Domestic Woman Cyborg... 97
    4.7 Cyborg Mechanical Operation.. 98
    4.8 Reproductive Mechanical Accumulation Process 100
    4.9 Conclusion 102
    References 102

Chapter 5 The Cyborg Interdiscipline.. 104

    5.1 Introduction 104
    5.2 Enhancing Humans.. 105
    5.3 Humans, Bionics, and Cyborgs 107
        5.3.1 Humans, Bionics, and Cyborgs 109
        5.3.2 Brain-Computer Interfaces 110
    5.4 Biological Enhancements 111
    5.5 New Opportunities in the 21st Century 114
    5.6 Cyborgs and Virtual Reality 116
    5.7 Cyborg Disputes 118
    5.8 Two Technologically Driven Revolutions 120
    5.9 Merging with Machines.. 123
    5.10 Questions for Our Cyborg Future... 125
    5.11 The Reemergence of Luddites 127
    5.12 Enter the Horse 129
    5.13 Concluding Thoughts 129
    5.14 Conclusion 131
    References 131

Contents ix

Chapter 6	Cyborg Sensors	133
	6.1 Introduction	133
	6.2 A World of Sensors	136
	6.3 Our Reliance on Sensors	139
	6.4 The Network of Sensors	141
	6.5 Telepresence and Sensors	143
	6.6 Characteristics of Sensors	144
	6.7 Regulating Sensors and Being Forgotten	149
	6.8 Remotely Sensed Data	151
	6.9 Sensors and Intellectual Property Law	153
	6.10 Surveillance, Sensors, and Body Scans	155
	6.11 Using Sensor Data in Trials	157
	6.12 Conclusion	158
	References	159

Chapter 7	Telepathy Signals in Cyborg	160
	7.1 Introduction	160
	7.2 Questions of Law and Policy	161
	7.3 Towards Machine Sentience	164
	7.4 Telepathy, Brain Nets, and Cyborgs	167
	7.5 Bodily Integrity	171
	7.6 Singularity and Concerns for the Future	174
	7.7 Introducing Super Computer (Watson)	175
	7.8 Who's Getting Smarter?	177
	7.9 Returning to Law and Regulations	179
	7.10 Conclusion	181
	References	182

Chapter 8	Intelligent Cyborg Brains	183
	8.1 Placing an Exponent on Intelligence	183
	8.2 The Numbers behind Brains	185
	8.3 Law and Brains	187
	8.4 More about Artificially Intelligent Brains	189
	8.5 Machine Learning and Brain Architectures	194
	8.6 Brain Architecture	197
	8.7 Hardware Protection for Artificially Intelligent Brains	199
	8.8 Our Competition against Better Brains	202
	8.9 Conclusion	204
	References	204

Chapter 9	Neuroprosthesis in Cyborgs	205
	9.1 Introduction	205
	9.2 Medical Necessity and Beyond	206

	9.3	Third-Party Access to Our Minds 209
	9.4	Concerns and Roadblocks 211
	9.5	A Focus on Cognitive Liberty 213
	9.6	Reading the Brain, Lie Detection, and Cognitive Liberty 217
	9.7	Towards Telepathy 219
	9.8	Creating Artificial Memories 220
	9.9	Litigating Cognitive Liberty 221
	9.10	Implanting a Software Virus in the Mind 223
	9.11	Conclusion 225
	References 225

Chapter 10 Body Sketch for Cyborg 227

10.1	Making, Modifying, and Replacing Bodies 227
10.2	The Shape of Things to Come 229
10.3	The Androids Are Coming 232
10.4	Culture Is Important 235
10.5	Our Reaction to Cyborgs and Androids 236
10.6	The Uncanny Valley 238
10.7	Observations about Discrimination and the "Ugly Laws" 240
10.8	Mind Uploads and Replacement Bodies 244
10.9	Copyright Law and Appearance 247
10.10	Derivative Works, Androids, and Mind Uploads 249
10.11	First Sale Doctrine 252
10.12	Right of Publicity for Androids 253
10.13	Androids and Trade Dress Law 255
10.14	Gender, Androids, and Discrimination 256
10.15	Our Changing Faces 257
10.16	Concluding Examples of Lookism Discrimination 259
10.17	Conclusion 261
References 262

Chapter 11 Cyborg Body 263

11.1	Introduction 263
11.2	Hacking the Body 265
11.3	The Risks of Body Hacking and Cyborg Technology 267
11.4	Prosthesis, Implants, and Law 270
11.5	Body Hacking in the Digital Age 274
11.6	Sensors and Implantable Devices 275
11.7	Issues of Software 278
11.8	Machines Hacking Machines 279
11.9	Hacking the Brain 280
11.10	Hacking Memory 282
11.11	Implanting False Memories 283

Contents

11.12 Hacking the Skin	285
11.13 Hacking the Eyes	287
11.14 Hacking the Body with Sensors	289
11.15 Sensory Substitution and a Sixth Sense	290
11.16 Hacking the Ear	291
11.17 Sensing Electromagnetic Fields	292
11.18 Cybersecurity and the Cyborg Network	292
11.19 Conclusion	293
References	293

Chapter 12 Cyborg Futures on AI Robotics 295

12.1 Cyborg as an Intelligent Machine	295
12.1.1 Universal Machine Intelligence	295
12.1.2 Generating Robotic Progeny	297
12.1.3 Framing of Robot Code	297
12.1.4 Bodies in Relation	297
12.1.5 Slow Robots and Slippery Rhetoric	299
12.1.6 Challenges of Robots	299
12.1.7 The Labors of Violence	300
12.2 Autonomy of AI	301
12.2.1 Existential Danger Debate in Artificial Intelligence	301
12.2.2 AI and the Illusion of Transcendent Rationality	302
12.2.3 Capitalist Context of Rationality	302
12.2.4 Knowledge of Transcendent Rationality	303
12.2.5 Bostrom's Occult Motivations of AI Machines	303
12.2.6 Colonization of Cyborg	303
12.3 Visions of Swarming Robots	304
12.3.1 Autonomous Robotic Systems	305
12.3.2 Swarming as Natural Inspiration for Military Futures	305
12.3.3 Emergent Stupidity	306
12.4 Business of Ethics, Robo, and AI	306
12.4.1 Cyborg Ethics?	307
12.4.2 Why Ethics Now?	308
12.4.3 Why Robots and AI and Ethics?	308
12.4.4 Corporate Robot	309
12.4.5 Why We Need a Feminist Ethics of Robots and AI	310
12.5 Fiction Meets Science	310
12.6 Racing of Robotics	313
12.7 Conclusion	315
References	315

Index 317

Authors

Kuldeep Singh Kaswan is presently working as a professor at the School of Computing Science and Engineering, Galgotias University, Uttar Pradesh, India. His contributions focus on brain computer interface (BCI), cyborgs and Data Sciences. His academic degrees and thirteen years of experience working with global Universities like Amity University, Noida, Gautam Buddha University, Greater Noida, and PDM University, Bahadurgarh have made him more receptive and prominent in his domain. He received a Doctorate in Computer Science from Banasthali Vidyapith, Rajasthan. He received a Doctor of Engineering (D. Eng.) from Dana Brain Health Institute, Iran and he obtained a master's degree in Computer Science and Engineering from Choudhary Devi Lal University, Sirsa (Haryana). He has a number of publications in international/national journals and conferences and is an editor/author and review editor of many journals and books.

Jagjit Singh Dhatterwal is presently working as an associate professor, at the Department of Artificial Intelligence and Data Science at Koneru Lakshmaiah Education Foundation, Vaddeswaram, AP, India. He completed a Doctorate in Computer Science from Mewar University, Rajasthan, India and received a Master of Computer Application from Maharshi Dayanand University, Rohtak (Haryana). He has also worked with Maharishi Dayanand University, Rohtak, Haryana and is a member of the Computer Science Teacher Association (CSTA), New York, USA; the International Association of Engineers (IAENG), Hong Kong) International Association of Computer Science and Information Technology (IACSIT), USA; professional member Association of Computing Machinery, USA; Institute of Electrical and Electronics Engineers; and life member, Computer Society of India, India. His areas of interest include artificial intelligence, BCI, and Multi-Agents Technology. He has a number of publications in international/national journals and conferences.

Anupam Baliyan is working as an additional director of engineering at the University Institute of Engineering, Chandigarh University, Punjab. He has more than twenty-two years of experience in academics and has an MCA from Gurukul Kangari University, an M.Tech (Computer Science and Engineering [CSE]), and a PhD (CSE) from Banasthali University. He has published more than 30 research papers in various international journals and is a lifetime member of Computer Society of India and Indian Society for Technical Education. He has chaired many sessions at various conferences across India and has also published some edited books and chapters. His research area includes algorithms, machine learning, wireless networks, and artificial intelligence (AI).

Dr. Shalli Rani is pursuing postdoctoral from Manchester Metropolitan University, UK, from July 2022. She is a professor at the Chitkara University Institute of Engineering and Technology, Chitkara University, Rajpura, Punjab, India. She has 18+ years teaching experience. She received an MCA degree from Maharishi

Dyanand University, Rohtak, in 2004, an M.Tech degree in Computer Science from Janardan Rai Nagar Vidyapeeth University, Udaipur, in 2007, and a PhD degree in Computer Applications from Punjab Technical University, Jalandhar, in 2017. Her main area of interest and research are Wireless Sensor Networks, Underwater Sensor networks, and Machine Learning and Internet of Things. She has published/accepted/presented more than 100+ papers in international journals/conferences (SCI+Scopus) and edited/authored five books with international publishers. She is serving as the associate editor of IEEE Future Directions Letters. She served as a guest editor in IEEE Transaction on Industrial Informatics, Hindawi WCMC and Elsevier IoT Journals. She has also served as reviewer in many repudiated journals of IEEE, Springer, Elsevier, IET, Hindawi, and Wiley. She has worked on Big Data, Underwater Acoustic Sensors, and IoT to show the importance of WSN in IoT applications. She received a young scientist award in February 2014 from Punjab Science Congress, Lifetime Achievement Award and Supervisor of the Year Award from Global Innovation and Excellence, 2021.

1 Cyborg
An Introduction

1.1 INTRODUCTION

A cyborg (/ˈsaɪbɔːrg/) body is a body of both organic and biomechatronicidal elements and is a portmanteau of "cybernetic organism." Manfred Clynes and Nathan S. Kline coined the term in 1960. The terminology cyborg is not the same as that used in Figure 1.1 but refers to an organism having re-established function or increased capabilities due to the incorporation of such artificial components or technologies that are based on the feedback of some kind. Although cyborgs are generally considered mammals, like humans, they may also be some type of creature. D. S. Halacy's *Cyborg: The Superhero Innovation* of 1965 was an introduction to the idea that advanced cybernetic enhancements, combining human and machine elements to possess extraordinary abilities—the connection between mind and matter might exist [1].

A human with visible artificial components, such as the superhero cyborg from comic books or the Borg from *Star Trek*, is the most common representation in the evolutionary biology of a cyborg. But cyborgs may also be seen as robots or as regular people. Cyborgs may be seen as prosthetic devices like the DC doom patrol robot or Doctor Who humanoid aliens or they may be seen as non-humanoid robots, like the Doctor Who Daleks or other Battle Angel Alita Motor ball players. More human cyborgs, including Darth Vader of *Star Wars* or Misty Knight from Marvel comics, will cover their mechanical components with armor or garments. Figure 1.2 shows that cyborgs can also be artificial or human appearing. The man and the bionic woman of *Six Million Dollars* had bionic elements that appeared like their body parts. Motoko Kusanagi is a full-body, shell-like cyborg with a human-like appearance. In the previous examples as well as in many others, the physical and mental capacities of cyborgs are common outside the capacity of people. Super powerful, augmented senses; machine brains; or built-in weapons may be used.

1.1.1 CYBORG BODY CONTROL

According to certain terms, perhaps the most primitive inventions rendered cyborgs by humanity's tangible connections. A common example of this will be a cyborg human with an electronic heart pacemaker, since these machines calculate the potential voltage in the bloodstream, do signal processing, and can provide power to the body using this synthetic feedback system to keep the person alive. This may be one kind of cyborg. Cyborg improvements are also observed in Figure 1.3, particularly cochlear implants that integrate mechanical changes and feedback. Many researchers mention modifications such as contact lenses, hearing instruments, smart telephones,

DOI: 10.1201/9781003392699-1

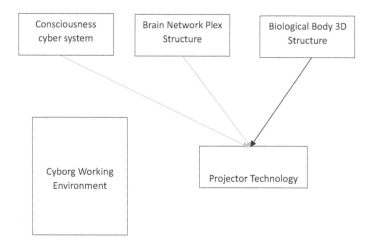

FIGURE 1.1 Structure of cyborg and its environment.

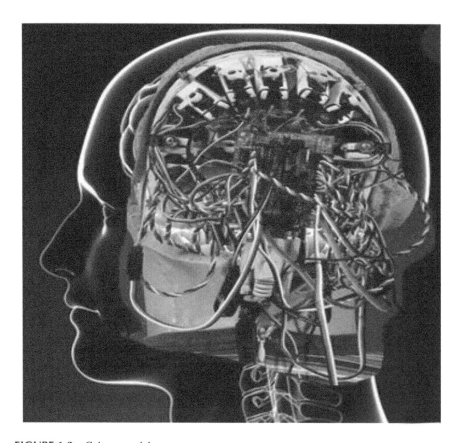

FIGURE 1.2 Cyborg model.

Cyborg

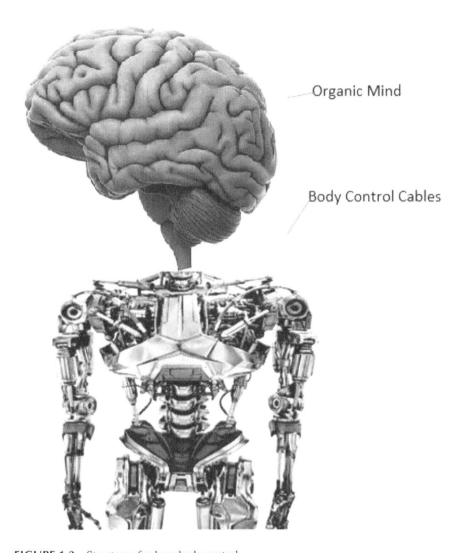

FIGURE 1.3 Structure of cyborg body control.

or intraocular lenses as indicators of how people are equipped with technologies to improve their biology. Because cyborgs are on the increase, some theoreticians contend that new conceptions of aging need to be developed and, for example, a biotech no-social definition of aging is proposed.

The concept is often used in the abstract to deal with mixtures of human technologies. This includes, for instance, plum and paper, language, and talk—not just widely used technological items such as telephones, laptops, the Internet, and other artifacts that might not be considered popularly. Increased with these innovations and linked to people elsewhere, an individual would be able to do a lot more than ever. One example is a device that can be connected to other computers using Internet protocols. An additional and increasingly important example is a human-assisted bot that aims at social networks with likes and actions. The cybernetic technologies include roads, pipelines, cables, homes, power stations, libraries, and other infrastructures, which we are barely aware of but are vital aspects of the cybernetics in which we operate. Bruce Sterling proposed a concept of an alternate cyborg named lobster in his shaper/mechanic world, which would not use an internal implant but rather an external shell (e.g., a Powered Exoskeleton). In contrast to humans are cyborgs who physically seem human, though fundamentally synthesized. Lobster appears beyond barbaric but retains an inner human (e.g., Elysium, Robocop). The *Deus Ex: Invisible War* video game prominently features cyborgs called Oman, where the pronunciation of the term Lobster by Russian is "Omar" (since the Omar are of Russian origin in the game).

1.1.2 WHAT IS CYBORG

In Wiener's Book of 1948, *Cybernetics, or Control, or Communication in the Animal and Machinery*, cybernetics is a term invented by a group of researchers headed by Norbert Wiener. Cybernetics is a theory or study of control or regulatory processes in human and software systems, including devices, based on the Greek "cybernetics," meaning steersman ship and governors. Cybernetics may be seen as a technology that has been recently created, but in some way, it crosses current sciences. If physics, chemistry, biology, etc. are standard studies, the classification of cybernetics is that everything is in Figure 1.4. The science of communication and computing in organisms, man, and machinery is formally known as cybernetics. It removes information retrieval and controls from any context. One of cybernetics' main characteristics is its concern about model design, which overlaps operating science. Cybernetic structures are normally characterized by a hierarchy that is flexible and feedback loops are permanently used. In several cases, cybernetics is like organizational science, focusing especially on the complex existence of the structure.

Everything in a cyborg creates an additional prosthesis system. The concept of a mobile phone being a technological social object that activates an actor (user) to

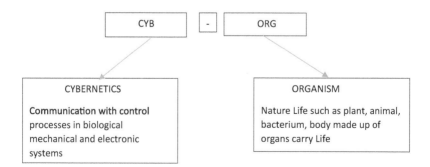

FIGURE 1.4 Meaning of cyborg.

Cyborg 5

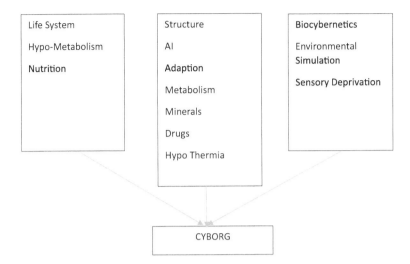

FIGURE 1.5 Cyborg diagram.

interact in the network with other actors (users) is what David Hess calls low technology cyborgs, as in *Informational Sharing and Interconnection*:

> I learn about how nearly everyone might be seen as a cyborg in modern environments when they invest most of the day in vehicles, telephones, laptops and, of course, TV machines. I ask the cyborg anthropologist if there could be a cyborg device with someone watching a TV. (I feel like an unconscious homeostatic machine when I watch TV.) I assume there's even often a convergence of personalities with the black box.
>
> *(http://cyborganthropology.com/Low-Tech_Cyborgs)*

Displayed in Figure 1.5.

1.1.3 Types of Cyborgs

Four different types of Cyborg technology are restorative, standardization, reconfiguration, and improvement. Cyborg interpreters are generally seen almost entirely as strengthening, by speeding up, improving, and increasing the reliability and cost-effectiveness of current translations. And there is no reason why cyborg transcription should only improve.

(https://cyberartsweb.org/cpace/cyborg/4categ.html)

- **Proto cyborg:** a Chris Gray proto-cyborg describing a cyborg that "mistakes its full execution." The prefix "proto" comes from Greek form of protos, or "first," meaning that protocyborg as a term would describe an early or first cyborg.

The early prosthesis of Steve Mann was an effort at an advanced cyborg paradigm from a contemporary perspective. Today, many of the capabilities of Mann are included in ordinary cellular phones and networks that make cyborgs completely integrated into real industrial humans.

- **Neo cyborg**: "neo cyborg has the form of a cyborg, like an artifactual limb, but does not fully integrate the prosthesis in a homeostatic way" (Gray, 14); the translator of neo cyborg consists of an on-computer human translator but only serves the computer as a typewriter, without full use of word processing, word management, e-mail, or Internet-browsing functions.
- **Hyper cyborg**: the hyper cyborg defines a cybernetic translator that is layered or paved into a greater whole of cyborgs; the translator of the hyper cyborg consists of a network of many smaller cyborg translators when a semi cyborg translator-editor team is connected via listserv or web board, and its collective results are fed into a shared database, a term management tool, or another system (helped).
- **Retro cyborg**: retro cyborg explains a transformation in cyborgs designed to recreate some form lost, and a retro cyborg translator might have a session on a computer like an old pre-electric typewriter for the sake of historical illustration at a translator fair, for instance, where a human translator can make translation decisions typical of proto cyborgs when hitting the screens
- **Pseudoretrocyborg**: pseudoretrocyborg is a term that describes a cyborg conversion intended to recreate a lost shape that never emerged; well, here, we are rather far away from science fiction but, say, we can assume a translator made of retro cyborg, someone having received the words of the spirit world target text may be an attraction in the main LDS museum in Salt Lake City.

In fact, cyborgs exist; approximately 10% of the U.S. population is thought to be technically cyborgs, which include persons with automated pacemakers, artificial joints, drug implant structures, corneal lenses implants, and artificial skin. There are a considerably higher proportion of participants working in the process of making them metaphoric cyborgs, including a computer keyboard with the computer monitor in a cybernetic circuit, a fiber optical microscopic neurochip, and a local video game player. Author of *Terminal Identity* Scott Bukatman calls this condition an "unmistakably twofold joint," which indicates the end of traditional models of authenticity even as it points toward the cyborg loop which generates a new type of objective reality.

This combining of industrialized and advanced cells has been called many things: bionics, vital machinations, cyborgs, this integration of the builder, the constructed ones, those systems of dying, the flesh, and undead, living, and artificial cells. They are the main characters of the late 20th century. But the cybernetic narrative is not only a tale about the glow of the TV fire. Among us in society are many actual cyborgs. Technically, a cyborg is anyone who is reprogrammed to resist (immunized) diseases or intoxicated to feel and think/behave/feel better (psychopharmacology) or has an artificial organ, limb, or supplement (like a pacemaker). Anyone is a cyborg. The spectrum of these close relationships between men and machines is stunning.

Cyborg

It's not just Robocop, it's a pacemaker for our grandmother. The establishment of these four cyborg classes is characterized in "Cyborgology: building knowledge in cybernetic organisms":

> cyborg technology can restore operations and substitute for lost organs and limbs; cyborg innovations they can be normalized, by restoring certain creatures to barely distinguishable norm.

- They can be reconfigured ambiguously, giving rise to posthuman creatures equal to but different from humans, such as what one is now when communicating with other creatures in cyberspace or, in the future, the types of alterations proto-humans will undergo to live in space or under the sea after forsaking all the conveniences of terrestrial existence;
- They can improve the objective of most industrial and technological investigation and the fantasizing of others with Cyborg envy or Cyborg philia.

The latter category seeks to construct everything from plants controlled in mind-controversial exoskeletons by a few "worker pilots" and infantrymen with the imagination that many informatics have downloaded their conscience into immortal computers.

1.1.4 What Is Cyberization?

The process of human development into a new species does not simply take place at night, as Transhumanism examined. Increased technology does not reach a tipping point where we immediately claim that we are "posthuman"; that sounds absurd. We need an additional, fluid concept to illustrate technological processes that alter the conception we have of humanity as a culturally constructed perspective. This chapter will call this process cyberization, the "cybernetic organism," the sliding scale between the human and cyborgs (cybernetic means mechanical processes supplementing the biological functions). In 1960 Clynes and Kline (1965), developed cyborg, but this is more properly explained in the *Cyborg Manifesto* (1991) and *Animals-Human Type Cybernetic Cyborg* (1965). The autonomy prototype among autonomous self-controlled machinery and organizations, particularly human ones. The blurred boundary lines are of great importance here to help us understand how humankind's structure is progressively eroded over time. The diagram shows a growing technological increase that is accompanied by increased improvements. It is to be pointed out that every phase is open to significant debate, but that should not dampen the overall subject of the diagram [2]. A smartphone wearable device offers the same benefits, plus more of the time. Bionic limbs promise a better use of humanoid form, and health advances in nano-bots and other such technologies can lead to unsearchable life expectancy improvements that can lead people to use these techniques for much longer periods of production. The ultimate computer technology classification of transcendence technologies promises to bring the human consciousness to inconceivable efficacy levels, maybe even outside of the form of the humanoid; human meaning is no longer confined to the biological mechanism but is rather like the light velocity and processing capabilities of super quantum mechanics in Figure 1.6.

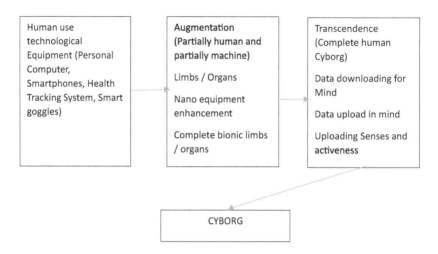

FIGURE 1.6 Human stages to build cyborgization.

1.2 DEFINING THE CYBORG ERA

There are opportunities for living creatures with in-body technology to enhance their natural capacities to be cyborgs with superior robotic capacities. Cyborg types include people with mass-produced biomedical implants; mass-imagined people with rust-colored assaults; and people with mass-custom insider capabilities [3].

Most cyborgs are growing with the company introducing new technologies in the body while individuals are aiming to increase cultural capital through corporate projects. They are also more advanced embodied cognition.

1.3 CYBORG HISTORY

Before the Second World War, the concept of a man-machine blend was widely spread in science fiction. In the short story *The Man Who Used Up*, Edgar Allan Poe had already characterized a man of enormous prostheses in 1843. In 1911 in *Le mystère des XV* Jean de la Hire introduced the science fiction hero Nyctalopia, who may have been his first literary cyborg (later translated as *The Nyctalopia on Mars*). In his novel *The Comet Doom* of 1928, Edmond Hamilton presented a blend of organic and machine components to space explorers. He was later used in a straightforward scenario, in all the exploits of his iconic hero, Captain Future, in the talk and living mind of an elderly scientist, Simon Wright. He specifically described "the mechanic analogs" named "Charlies" in his short story of 1962 "After the day of judgment," stating that "cyborgs were called in the 1960s . . . cybernetic species." They are a kind of "cyborg." C.L. Moore wrote about Deirdre, a performer whose body was burnt and whose head was put in a facial yet exquisite and artificial body in the short story *No Woman Born* in 1944.

The idea of an improved person that may live in the extraterrestrial world was coined in 1960 by the artists Manfred E. Clynes and Nathan S. Kline.

> We suggest the word 'Cyborg' for the exogenous institutional complex, which unconsciously functions as an interconnected homeostatic device.
>
> **(E. Clynes and Nathan S. Kline Manfred)**

Their idea came about when the modern frontier of discovery of space began to open, and there was the desire for a romantic partnership between humans and machines. Clynes was Chief Scientific Officer at Dynamic Simulation Laboratory of Rockland State Hospital in New York, the manufacturer of physiological instruments and computer data processing systems. The concept first appeared in print five months ago when the *New York Times* announced that Clynes and Kline were first presenting their papers on the psychophysiological aspects of the Space Flight Symposium.

A cyborg is a human-control system in which medications or regulating systems modify the control mechanisms of the human component externally to allow the person to survive in a separate world [4].

1.4 CONCEPTUAL SCHEMA OF CYBORG

A concept of cyborg narrative can be built on this conceptual basis; see Figure 1.7. First, the reframing of accessible computer science and the culture of writing from 1979 until 2000, since cybernetic literature, makes the political resources of this collection of information visible, thus highlighting how professional knowledge can politically assist individuals in different contexts. Themes, directions of academic growth, emerging from the closely and holistically considered experience of scholars may be found in the body of information [5].

In the following pages, I investigate two cyborg stories directly from the notable machine and writing scholarship and show how they guarantee legitimate work in the world, current or future. The stories are as follows:

1.4.1 TEXTUAL TRANSITION

The advancement of text, ranging from early experiments investigating on-screen web browsing and syntax to more recent analyses of the social and cultural ramifications

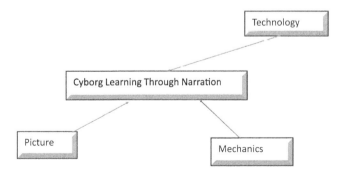

FIGURE 1.7 Cyborg narratives.

of hypertext and multimedia, can be found in one of the most multidimensional cyborg stories in the computing and writing community. This framework would allow a broad variety of scholarly research in the present and future to be legitimized due to its range and reach.

The story starts in the early and mid-1980s when students first suggested that computers replace existing authors, revisers, and editors via computer programs. WANDAH, founded by Michael E. Cohen, Lisa Gerrard, Andrew Magpantay, and Ruth Von Blum at UCLA, was among the earliest applications. This application had a strong focus on creating ideas, drafting, revising, and editing, which meant that its users were subject to the process model of writing. The model of the method is still very common, and today and tomorrow designers can learn more about WANDAH to understand and overcome the difficulties faced by Gerrard, Cohen, Magpantay, and Von Blum. Gail Hawisher said that word processing influences writing across fields in "Studies in Word Processing," published in *Computers and Composition* in 1986, not only in the analysis of composition. Hawisher's article shows successfully that programmer and writer study can influence multiple fields in crucial ways and that this experience resonates directly with current and prospective researchers, who want their projects to be accepted in and through different contexts. However, there were still concerns, including the ever more common use of word processors at the time [6].

A literary transformation narrative begins with the advent of an electronic hypertext in the early and mid-1990s, which triggers a discussion about whether it should be seen as a distinctive and modern kind of text. In the future, researchers will undoubtedly present more new texts to be hopefully original so that they can predict the field's reaction, through recognizing the controversy surrounding hypertext. In addition, the hyper-list scholarly action reveals that such an argument inevitably activates a dynamic and thorough procedure. In this paragraph, electronic hypertexts must be identified first, not just hypertext. The term hypertext was coined as far as it applied to news, short stories, and electronic media in 1945 in "As We Can Think," an *Atlantic Monthly* paper by Vannevar Bush. Electronic hypertext is therefore particularly referenced to nodal and multilinear texts activated by electronic media, such as story space, CD-ROMs, and the World Wide Web. Jay David Bolter's word, the electronic writing room, described the ability of hypertext to give new write-and-read opportunities through connecting and several nodes, perhaps the first commonly accepted representation of electronic hypertext on computer and writing communities.

In the early to mid-1990s, researchers started to learn of special features such as email and multi-user dungeons (MUDs) in other electronic media. The implementation and acceptance of these measures show how digital media are viewed, enabling scientists both to expect and help with challenges.

In the early and mid-1990s, MUDs appeared prominently in the machine and writing world, allowing current and prospective researchers to see if a second form was often adopted. If you look simultaneously at the advent of MUD and email technology, you will be able to get a more holistic feel of how newspapers are received and introduced into the field by the students of technology and the reading culture. MUD is a multi-customer or multi-user dimension, as the readers know, and it acts as a simulated universe, which enables in-depth communication and in-depth interaction. It allows players to create their environments, as they are called. A specialist MUD, MOO, has become most popular in computers and authors, highlighting artifacts that players can create along with the interaction and world-building of MUDs. Many and quick scrolling threads of text appear in MOO environments and what happens first is the feeling of the new player being overcome by any opposition to MOO textuality. Such reluctance doesn't mean that the players don't try or feel safe; it just means that some people find it difficult to adapt to MOO traditions. The way opposition arises as part of the emergence of digital media in the field should be considered by current and future computers and writing scholars. Even people who do not feel certain about a medium can express questions and concerns that are particularly helpful to the industry and designers. The several text threads can be read more and more quickly as players MOO. Last, the players are not reading faster; they are developing what can be called a human heuristic meta-reader, scanning screens for interesting words or sentences, or even the messages of other players they're acquainted with. Multimedia stresses the next level of literary transformation, applying in this case to the graphical display of various visual platforms, as well as graphics and audio clips and more conventional print training, such as phrases and words [6].

As multimedia is such a large concept, it is particularly instructive for current and future scholars as to how it can be embedded into the machine and writing culture because it may rely on more advanced versions from at least a few of the same media for its invention. The increase in the popularity of multimedia, as with the individual press presented earlier in this chapter, entails both opposition and cautious development. Multimedia technology advocates today argue that its introduction into education means a new and special generation of "connected children," capable of thinking and interacting with multiple media over some time but unwilling to concentrate on educational contexts more singularly. The wired children are a glimpse of the future, so the thought goes. Critics of multimedia are concerned these wired children, on the other hand, believe their schooling leads them away from the diligence and dedication that the future requires. Print and electronic texts will remain popular in the future, as shown by the story just outlined. Students need to recognize that their preference is not between all computing technologies, and Figure 1.8 contains no such technology. Rather, all current and future challenges would be to investigate what are the most relevant activities in textual processing and use, from rich people with computing technologies to those without. This vision means a broad variety of research projects will lead to the growing awareness of print and electronic textual materials in the computer and writing communities.

1.4.2 Pedagogical Evolution

When the cyborg story of teaching and learning innovation started in the 1970s, most of the written software was written directly by teachers, essentially because there were no other alternatives. While it is unclear if machine and writing educators can do such work again in the future, this training is vital if such work is undertaken. One of the most significant aspects of educational technology in the 1970s and 1980s is the way a wide range of pedagogical ideas were integrated. Early apps, such as those

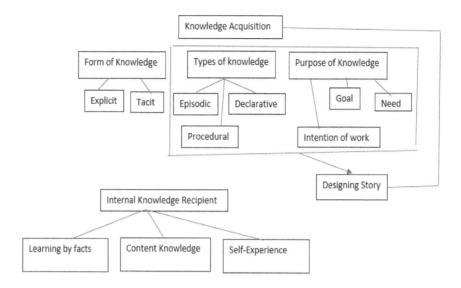

FIGURE 1.8 Cyborg textual transition.

in the Write Well series, have made students autonomous. Following this example, the student will sit at the computer terminal and participate in a sequence of punching exercises, speech, or another subject that is taught during the day. Other tools, such as the Workbench and the Epistle, have helped the students edit their documents, which they have closely examined before they can apply to graduation. Perhaps the most well-known early computer-writing program developments are those of BURKE, TAGI, and TOPOI, each intended to help students reflect on paper topics by posing hypothetical strategies. In early word processors, such as WANDAH, a more pedagogical change shown by software can be seen in allowing the students to insert their ideas in a database file, which can then be stored and found if a writer wishes to make corrections or edits, as well as creative ideas for discussion and expansion. Current and future researchers should also recognize that applications from the 1970s and early 1980s have not today been properly developed. Early word processors, for example, were not like today's text processors with sophisticated user interface designs and command systems that make service users comfortable. They needed special commands and complex interfaces.

Pedagogy and tech were also strongly linked during the early and late 1980s, with companies taking part in the mix. It makes sense for today's and future academics to get a feel for the past as a way of shape their decision-making and appraisal of tech choices with too much teaching and training software by companies and more exciting of the future. In the early 1980s, companies made major computing improvements in the form of greater budgets and more comprehensive resources, especially to make software easier to use. This transformation is well shown by applications like Apple Writer and WordStar, which worked far more easily than previous text processors. Moreover, they illustrate the distinction between corporate finance and sponsorship in a short period. Now, companies range from giants such as Microsoft,

Apple, Compaq, and Hewlett-Packard, to tiny, local organizations and, everywhere goods are bought and incorporated, they inevitably influence the way scholars function in computers and in writing culture.

The future is going to depend on companies for better or bad tech solutions, so scientists can consider carefully before choosing. The shift towards design and manufacturing companies have started to pose crucial issues on pedagogical principles in the early to mid-1980s. Should learning schools, critics ask, educate students that technology goods are "happy little consumers?" How do conspicuous consumption and the increasing need for trained technicians affect curricula? Richard Othman's *Literacy, Technology and Monopoly Capitalism* (1985) may be the most often cited early publication of such important issues in the computing and writing world. By the late 1980s, the advancement of the pedagogical processes of computers and writing within the Culture started to increase in diversity. Intellectuals who function in the present competitive environment can be particularly attentive to these early encounters as they demonstrate how interdisciplinary relations can be built by teaching and how the group can develop significantly by diversifying the contributions. Few machine-rich pedagogy places were more important than the classroom for literature and George Landow was one of the first scholars to apply computers to literary studies. He and the students at Brown University created the Victorian Web as a hyper-textual guide to Victorian literature.

The World Wide Web hadn't been launched yet, so the view of Landor was much more incredible. Classes used UNIX-based Intermedia design tools and its purpose was to allow files on a server to be connected conceptually and physically so that many users can access them and manipulate them on several computers. In the following screenshot, the intermediate Victorian Web is shown:

> A configuration of the Victorian Web files or networks is shown in the left-hand screen window of the capture, and users have chosen themes of interest, which allow them to view and choose increasingly detailed subjects, ultimately to find the desired content. The right window showed the required material, in this case, a Victorian disease conversation. While viewing the Victorian Site, readers cannot doubtless see how Intermedia's previous employment interacts with today's web-based pedagogies. Current and prospective academics should be aware of some early pedagogical creativity to help them understand the way they learn and learn. The experiences of learning are linked to previous experiences [7].

In the late 1980s and early 1990s, networking technology was introduced into the schools and a major event in computing and writing was the genomes of the culture. At that time, channels had evolved as technology to educate and study, making knowledge of past encounters particularly relevant to the future of pedagogy of computers and writers as well. First implemented in the late 1980s, computer networking was called a local region network or LAN, and it included mostly many functional requirements: a source computer modem that has required information to be converted into a transferable format, a wire between all computers that have to be involved in such transactions, a central data transport router, and an objective computer modem that has enabled the transaction in readable and accessible formats of the received data. A LAN does not allow users of external machines to take part in transfers of information; unique computers are chosen, and persons must have access to the LAN on those computers. Electronic Interaction Networks (ENFI) are some of the first programs that used network technology to promote creative teaching and learning. They were first set up at the University of Gallaudet but then quickly expanded to other universities. The artist, Trent Batson, worked on ENFI to promote teamwork and the allocation of the jurisdiction for classrooms. Thomas Barker and Fred Kemp (1990) wrote in the 21st century, continuing the use of LANs as a teaching tool in their schools, *Network theory: Postmodern pedagogy for the writing classroom*. Their use of network technology also represents exciting developments in interactive pedagogy. Kemp was also a writer of the Daedalus Integrated Writing Environment (DIWE), LAN-based applications enabling users to exchange and communicate documents. DIWE is a web-based company with no more LANs in its new update, thereby demonstrating some pedagogical development in connection with network technology.

Present and prospective academics who learn about the development of DIWE should consider how their teaching can contribute to the program reshaping. And innovations such as Batson, Kemp, and Barker's earliest contributions promise to continue to educate pedagogy into the future, as machine and writing culture educators are increasingly considering how networks will open new opportunities. The World Wide Web, including e-mail, opens several pedagogical possibilities and today a powerful foundation to integrate into machines, as does the writing culture. The web provides resources, first and foremost, for teachers and students for electronic publication, and this aspect has a major influence on the teaching-learning process.

For example, teachers may involve students in rich and important public dialogs and then encourage students to adapt their deeper understanding to the design and production of a browser window. It is also interesting to note that web writers return their computers and the writing world to software creation debates—if teachers develop software, how much technical expertise do teachers want to have? The key to this controversy is the hyper-text mark-up language or HTML, which is not an arduous language. However, file system information and programming tags are essential. Software solutions have been available such as the Composer of Netscape, Dreamweaver of Macromedia, and FrontPage of Microsoft, which allow users to create web pages without learning HTML. In general, advocates of HTML-level expertise also think that web authoring software like Composer and FrontPage adds extra or operationally meaningless tags as well as arranging HTML code, and advocates of the software conclude that the more user-friendly alternative should be used by teachers and pupils, that software is an evolutionary stage beyond HTML.

1.5 LEARN CYBORG SKILL LITERACY

Several words have been introduced for the development of the current meaning: literacy in technology. These viewpoints, however, cannot be considered equivalent; they reflect distinct beliefs and viewpoints. In Figure 1.9 computer literacy, for example, is also described as a collection of skills; it means that you can perform a limited variety of skills effectively, including writing a paragraph in a Word document or saving a file on a floppy disk or hard drive. Information literacy refers most

Cyborg 15

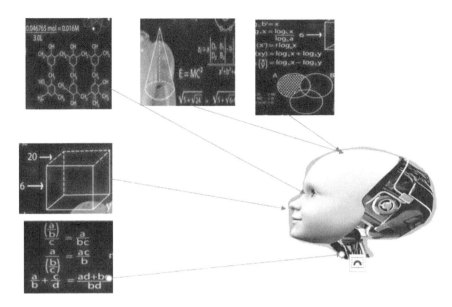

FIGURE 1.9 Cyborg literacy.

frequently to the ability of a person to effectively communicate with information systems such as libraries and communication messages. The platform is different. Literacy is usually characterized as the ability to use and learn about various devices such as the computer; it is less about the abilities that someone can do with innovations and more about their function generally. The use of the word "alphabetization" in both terms indicates a relationship to the printing tradition and not a distinction. The defining characteristic of their work was vocabulary or words combined with literacy for academics who had adopted different terminology [7].

1.5.1 Posthuman Promises

The concept of man as a sovereign and independent awareness is not so much in line with lived human experiences. In subtle and unconscious ways, environmental influences, whether social or normal, affect our minds. Our bodies are complex, and our body also reveals disparities in agency capability, such as the strength of male bodies and the abiding or accessible only at high expense. Because of this imbalance, posthumanism suggests that we should understand ourselves as posthuman instead of just individuals.

N. Katherine considers posthumous as a definition of the cluster. According to Hayles, posthumously, Figure 1.10 gives priority to shared knowledge about body biology, epiphenomenal rather than agency consciousness, mechanical prostheses rather than privileged locus of action, and humans' continuous artificial and cybernetic modes of life. The following is written in Figure 1.10. Hayles compares the posthuman with a notion of humanity, so known, which comes from early European philosophy: a soul that naturally possesses its corporeal body, which employs

FIGURE 1.10 Humanity after transhumanism.

conscience to enforce its will and regulate its corporeal behavior and which, due to its potential for autonomy, stands apart from all animals and machines alike [8].

A similar contrast is given by Braidotti. For Braidotti, man has an inexhaustible potential for authenticity and understanding, an awareness that can overcome spatial and geographical characteristics and gain access to temporal and universal realities and a logical mind that dominates the body while imposing on the world sense and significance. The posthuman is, on the other hand, fluid and relationship-based rather than unifying and self-sufficient, affective, and observable rather than logical and intangible. There is something in the posthuman that avoids categorizing in terms of these identical conceptions. Braidotti points to the root of resistance: [the posthuman] is thus not bound to pursue an appropriate portrayal of his life within a regime that is fundamentally incapable of giving proper respect, since it is not framed by the inescapable forces of meaning. The same statement (Posthuman 188) characterizes the notion of posthuman as a means of identifying, (maybe) potential, the uncertain, altered personality of human beings, by integrating diverse techniques into our bodies and our self. It is therefore a paragon word that covers several similar concepts: GM people, artificialism or androids, uploaded consciousnesses, cyborgs, and chimeras (mechanical or genetic hybrids).

Accordingly, these distinctions do not eliminate the sense of posthumanism because the posthuman is different from the human. The posthuman is ineffable, rejecting the rigor of conceptual framework, unlimited by fixed categories that are mutually incompatible with the present and accessible to unconceived futures.

Along with its inefficiency, posthumanism seems to provide a basis for understanding our socioeconomic injustice, negativity, and injustice of all kinds and better imagination of the future. The adverse effects of ethnic hierarchies on representatives of the marginalized races were overwhelmingly demonstrated. There is also an increasing social desire to improve or avoid those harms represented by the Black

Lives Matter and We Are Here protests in the USA. Therefore, the notion of posthumous people should be expected to promote critical theorization regarding the basis and effects of racial hierarchies and successful change and transition policies to racial justice.

1.5.2 Posthuman Identity as Cyborg Identity

The cyborg is one of the most powerful and suggestive visions accessible for theorization through a posthumanist orientation. The perception of the cyborg comes from Donna, who characterizes it as a "kind of dismantled and reconstructed community, post-modern and private identity." According to Haraway, the cyborg is "a cybernetic body, an organ and machine hybrid, a socially active creature and a fictional creature." Cyborgs are hybrids and their machine and body are inseparable and cannot live as separate from each other. The other elements are indistinguishable. These hybrids are emotional creatures, because current social networks give priority to synthesis methods over another method and because different hybridizations are experienced in different societies. Cyborgs are, however, often fictional beings, and they serve as social and political structures that provide alternate hybrid capacity in the amount of additional [9].

In Figure 1.11, Braidotti describes Haraway's view of cyborgs as "high posthumanism," subject to the precautions that Haraway disavowed her posthumanism designation and that Haraway understood her work on compliant organisms in earlier cyborg works. However, it is very far from Haraway's philosophical preferences that cyborgs give a view of the posthuman. For the idea of the cyborg, posthumous theorization lends itself. Cyborg anthropology believes for example that human objects and subjective experience are as essential as they are computer manufacturers and operators of computers, machinery, and knowledge exchanges.

In the sense of systemic inequality, Chela Sandoval outlines how to realize the theoretical potential of the cyborg idea. She concludes that suppressed races are adopting five methods to achieve self-government within harmful racial hierarchies: read cultural symbols, challenge signs from domineering ideologies by separating the sign from the intended meaning, take these symbols by placing them in a new

FIGURE 1.11 Posthuman identity as cyborg identity.

sense, direct such deconstructive efforts to build equal social relations, and choose the right way to act. Sandoval says that we can better appreciate how cyborgs contradict classical significance due to the expansion of human abilities. These techniques, used by racialized people, are used by theoretical cyborgs.

The essence of the study of Sandoval, without the ethnic background, is shown by Eyeborg Neil Harbison. Harbison is born with black and white only; it carries an Eyeborg gadget that transforms colors into sounds. Over time, the Eyeborg has modified its neural paths to give it new visualization capabilities. Harbison can "interpret" visual signals, but his unorthodox readings allow him to engage in new meaning in traditional practices. For example, through "viewing" mosaics he can compose music, and he paints speech presentations. Harbison's cyborg color ability contradicts traditional sound and visual separations; in which he was historically placed. Similarly, the Sandoval initiative to characterize the methods used by marginalized races as cyborg inventions suggests, rather than as individual practices, efforts to challenge harmful racial hierarchies [10].

1.5.3 Theorizing Racialized Humans as Cyborgs

Haraway's conception of the cyborg is constructed by destabilization of frontiers among itself, the independent entity, and the deterministic system, to subvert "the legacy of racist and male-dominant Capitalism." Haraway claims that the success of gender, ethnicity, or class is guided by the horrific historical history of the conflicting social realities of patriarchy, colonization, and globalization.

These inconsistent perceptions restrict the probability of phenomenology. For example, if colonial ideology reserves personality for white people to the exclusion of Black people, while patriarchal ideology preserves personality for women, black women only inherited "A waterfall of derogatory identities," dismissed not only by patriarchal women but also by colonialism-resistant anti-imperialists.

The perception of African Americans as cyborgs prevents these negative events. "The woman of color," as a cyborg personality, is "a powerful subjectivity synthesized infusions of external identities." In so doing, Black women achieve identification as composite human body integrations, and the "woman" and "blacks" parallel preconceived notions of the other. The cyborgs are the product of their life at the edges of capitalist and imperialist philosophies. This power enables them to survive and prosper by seizing the myths, languages, and instruments used to identify them as others and eradicate them from sight. As cyborgs, black women problematize "man or woman, person, artifact, race, individual object or body position." This undermines basic gender and racial inequality assistance.

This, roughly speaking, means that posthuman beings are seen as cyborgs for racial rights and activism. Haraway says that "Cyborg imagery may mean a way out of this labyrinth of dualisms, through which we have expounded to ourselves our bodies and instruments." We recognize that seeing racialized people as cyborgs is an incentive for the innovative discovery of political and social replacements. We also note the ability of this vision to clarify racialized human living realities. For instance, it helps expound the unfair or oppressive classification of individuals of color as cyborgs, because cyborgs themselves are frequently de-humanized, regarded as less

Cyborg 19

than completely human, or conceived as threats to existing social order. Theoretical race as a key component of cyborg identification also tends to understand why ethnic colorblindness is a myth: since the fundamental parts of cyborg identity are inseparable, a human who does not appear as a single person should not be a separate individual from ethnicity.

1.6 CYBORG PEDAGOGY

This education aims to provide an understanding of how teachers inspire their students to take the subjects they learn individually and to reflect and write about how their personalities and daily life are inseparable from the problems they discuss in the school of geography. This paper covers the organization of undergraduate courses on geographies of cultural heritage, including three concepts—Situational awareness, cybronic ontology, and Boundary Pedagogy. It tries to persuade students through easy consumption and obligations to look at their contacts with the lives of remote others. This pedagogy shows the type of teaching practice that can be derived from a course of this kind and the forms of this question of accountability.

In his magazine, for our Materials Culture course, Geoff was making the point of using this annotated polaroid. Instantly he was moving in Figure 1.12 above the coffee granules from the Nescafé jar, with water from a close tap connected by miles of tubing with a reservoir and who knows what else to cook with the help of a heating element connected to the national grid, its wires, pylons, transformers, power plants, and its fuels, the ceramic flavors with a few pieces of milk from the plastic container in the white "Day Match" ceramic mug that rests on the formal edge of the beige kitchen and uses the stainless steel spoon with the blue plastic handle held in that hand to lift the arm to pour out its content into a mouth recently used as an important dental device. It could have all gone wrong too easily had there not been any correct connection if the order had been broken; if the power had been shut off; if the milk had been cut off; if plumbers had not adequately connected the joint causing

FIGURE 1.12 Cyborg pedagogy patterns.

draining of the water; if the drivers had been a member of the recent fuel protest; had he decided not to buy this brand of coffee for the last time he went to supermarkets; the list could go on. He tripped over his flatmate's shoes and spilled his coffee onto the hall carpet. And that is the matter: it is a "cybernetic body," a cyborg, a network node. He writes a newspaper on the ties to where the "cyborg shows." There are fleshy relationships, interactions that explain the sorts of things that would always have to happen in order for him to be who he is today (although he can only speak about a small part of that!): networks, tankers, product exchange plants, petrol pits, oil wells, bodies, and so on (organic, tracts, ships, vein synapses, etc.) and body networks external to farms, warehouses, etc. . . . all over the world.

These risks combined are the framework of his material-semiotic self. It is difficult in one of those distinctions and structures as he or she thinks to locate the boundaries between inside and outside, between him or herself and others. And so it is; as soon as he begins to search for and make associations and blur those borders, this should usher about additional duties: to those that are closely tied up to his life, to his body, to his self, to the public, to the animals, to the atmosphere, to mechanical equipment, etc. We're not working alone. It's a big cooperation endeavor to be ourselves. However, this cooperation is never on an even footing. You could not claim that slaughterhouse employees "collaborate with" cows for beef cutting or that "in cooperation" sweatshop worker multinationals make clothes. We're talking, then, about associations, relationships of power, and obligations.

1.6.1 It's Cyborg Pedagogy

Our Materials Geography curriculum seeks to inspire students to dig their teeth into the highly contentious yet basic problems in Figure 1.13.

- When their interactions with commodities are considered, they follow a cyborg ontology.
- Think about "commodity chains," "cultural circuits," and/or "actor networks" through their links to other people.
- Establish such a piece of knowledge by reading and discussing comprehensive scientific analyses of intake, output, and flow in classes.
- Work on community presentations that establish more core problems resulting from those discussions.
- Continually place their experience in the worldly conditions of their daily lives.
- Maintain a report that describes how such knowledge can be based and how it evolves throughout these circumstances.

In the beginning, we give only one lesson. We then organize the course: preparing comprehensive handouts, ensuring that correct readings are readily made accessible, organizing class meetings, scheduling additional coursework time for smaller debates, evaluating newspapers following well-defined criteria, and as far as possible decentralizing ourselves. In 1999, there were 64 students, and in 2000 there were 30

Cyborg 21

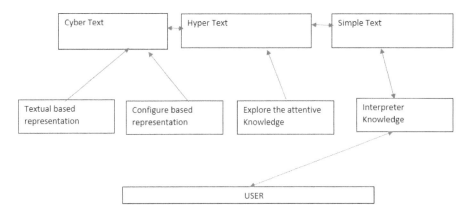

FIGURE 1.13 Cyborg learning capabilities.

students. We didn't offer the "correct" responses or want them back. In the specific situations of their lives, the readings shared, and the topics brought up in the class, we looked for compelling, insightful, creative, and knowledgeable solutions. You may wonder, though, why should this progressive pedagogical approach be considered exactly?

1.6.2 TRADITIONAL TEACHING

The fundamental pillars of this cyborg pedagogy must be briefly described in the reply to this question: "situated intelligence." Donna Hardaway's work on the first and second is best known, and the third has attracted a lot of attention. The basic criticisms of conventional methods of knowledge, being, and teaching are established. Their subject of "situated knowledge" (1996) criticizes both absolute knowledge in empirical objectivity and social constructionism as relativizing knowledge. She criticizes researchers who act in these ways to "promise fair and complete vision from all over and north" for making knowledge claims which, as a result of this positioning, are "unloadable, and so irresponsible unable to be called into account."

She concludes that the "reasonable" and "intellectually honest" science of the universe is created, expressed, and placed in "the awareness of self that is part of all its forms, unfinished, whole, simple, and original; often built, stitched together imperfections, and thus able to unite with another, and seeing together without pretending to be another.

1.6.3 ONTOLOGIES OF CYBORG

Neither of us corresponds to the conventional "charter person's character as an arbitrary vessel of atomistic prosocial reason." This is the second foundation in which the epistemology of placed expertise is moved to cyborg ontology, which is not so far from what Haraway (1991) argued in her "Cyborg Manifesto"—that advances in digital technologies, surgical treatments, genetic manipulation, and other fields of technoscience made the environment an increasingly mixed environment. She said now "both of us are machine and species hybrids, theoretically and manufactured." The consequence of this is disruption of how people live their lives, for example through IT: binary opposition and rigid categorizations of Enlightenment thinking—"self/other, mind, culture/nature, male/female, civilized/primitive, real/aperitive, all/part, agent/resource, maker/made, active/passive, right/erroneous, truth/illusion, whole/part, God/Man." Due to these binary oppositions and strict categorizations "the logic of dominance by women, persons of colour, nature, laborers, and animals have been systemic—in summary, the dominance by those who constitute themselves as others, whose duty is to represent one." Haraway argues that "progressive people" should encompass and exploit "cross-border, powerful fusions and hazardous possibility" if they want a "more adequate, rich, better world account to live well and with a critical, reflexive relationship between ourselves and others" and "unfair privilege and oppression" practices that are the basis of any position. Haraway argues "We live in a world of links—and it's important what's done."

Finally, we have "boundary pedagogy": the path towards schooling and learning that takes the "banking method of education as a nemesis, in which students, to sum up crudely, are urged by evaluations that decide their academic success to acquire a dominant understanding of the world and replicate these dominant understandings. Not unexpectedly, these interpretations are structured by Haraway's binary resistance to such exclusions as being central. Critics suggested that many pupils became excluded from their education by this mix, since they would not participate in the hegemonic project, or they just couldn't. Border pedagogy, by comparison, believes that students have valuable information and perceptions about society, culture, and economy and that they are critical of this knowledge and draw on it.

Please provide pedagogical approaches that allow students to recognize, in action criticize, and/or locate spaces between these binary logics to disrupt the modes of dominance that are the product of their use.

All those alternatives are to have students write newspapers with a placed epistemology and a cyborg ontology. It should make it possible to see, make, reflect, and articulate these relations in line with the logic and boundaries of that mechanism. However, it cannot split the universe into good and evil, good and bad, legal, dishonest, responsible, and reckless. Things here aren't white or black. If anything, they're gray colors. It's a progressive educational initiative in theory. But (how) does it succeed if there can be no straightforward answers?

What is this kind of fundamentalism? What might be the effects? Where did Geoff get it? Well, he has been dragged around the town, to slaughterhouses, coffee fields, young people in Southall, televised soap, meat, sugar, spoonsful. Phil (philosophical Geoff) in Matt was assisted by the course (material Geoff). It informed him how much better problems can be understood when related to personal experiences. Yet Geoff's brain has been struck by the course. As a cyborg, it is obvious that he might not be anything if he wasn't connected; he can't be entirely unrelated, but his understanding of the position makes him unable to see his actions clearly in the corner of the network. The network's inconsistency is that a successful deal does not affect each aspect of the network positively.

Cyborg 23

The acquisition of fair trade coffee will harm farmers operating for commercial businesses in the Third World profits. Moreover, what good would he be doing in such huge networks as such a small actor? Was his behavior now more accountable? Or less than that? Indeed, are others not responsible to him as a semi-material being? His interpretation was scrawled by cyborg ontology! He has spent hours about what kind of café we should drink at and what kind of trainers we should wear and how our decisions affect people worldwide. It's impossible ever to meet and it is never met. What's happening; how does he even know? Will he have something he doesn't know about? These matters made Geoff's life difficult. It was a hassle at first for cyborg shows, but it has cyborg eyes now. He challenges his personality, his responsibilities, and his networks—connections exist all over the world and he feels like he has total understanding and confusion, on the one side. With assistance he made a deliberate decision from the first few months of the program to ignore the plight of coffee planters. The coffee aisle is longer than average in the store. It takes years to shop. The path has taken over his life in certain ways, so he pushed himself to find things. The breakdown in his analysis style, which we could not lose sight of, was one of the implications of this reconfiguration process for many years. The loosening of style was considered quite liberating for Geoff—to bring in his journal entries of what he wants and thinks. Even while he realizes that this course works only with a different set of rules than usual, he cares to return to what he now considers blind education. Even if Geoff knows that he can't survive without any of his brands in his genuine times and is most pleased to ignore some of the networks in which he is involved, he says it's one of the only courses he recalls for over five minutes after it ends. Indeed, this lesson isn't completely done because Geoff is now a messy cyborg who always finds objects, loves them all at once, and hates them.

1.7 CYBORG RESPONSIBILITY

I'll next move to what I'll call cyborg responsibility, building on the questions discussed in the expanded excerpt that has just been introduced, and thus the influence of many computers and the several voices of writing culture. Cyborg accountability is a special computer/writing definition that is an acceptable criterion for measuring actions, rather than a guideline for those actions. Few scientists will describe a cyborg figure as having a major obligation if the feminist activist Donna Haraway (1985) is concerned about the Case in William Gibson's *Necromancer* (1984) or the "Cyborg Citizen" of Chris Hables Gray (2001). Fewer will also recognize any cyborg duty linked, including computers and publishing, to a specific culture. What is cyborg duty, then, is anti-responsibility? To be responsible for cyborg terminology is to avoid traditional accountability explanations or meanings. It is to ask important questions instead about the machine and write group concerns and experiences, irrespective of the degree to which a person is investing in something asked. It is Important to see meaning instead of one or the other in reasons for and against any project or initiative and it's to consider who's quiet and who's spoken in any discussion. The responsibility of cyborg is to generate conflicts by compromise and to override problems. It offers places for inclusion and diversity, even though it is not easy to open, and it depends on people who can make technical and personal sacrifices for the sake of the future. However, the responsibility of cyborgs won't raise prominence.

On the other hand, certain persons in the machine and writing culture may dislike the many questions raised and the extra time needed to make decisions now and in some upcoming future. However, anti- or reputation in many ways fits into the anti- or liability that appears in cyborgs. Of note, change agents hardly win competitions for popularity.

It is crucial to recognize that cyborg accountability is an important danger in computers and the written and other cultures. Any person taking this approach must take account of his or her behavior in different contexts. As seen in Figure 1.14, having said that, cyborg duty often must be changed or tangled into contexts where the risk is highest. I mean this by suggesting that the concept is a dream rather than a path map or a practical one. Evaluation of these methods in different scenarios can be adopted. I may not have the ability to fully take responsibility for cyborgs if I am a supplementary or part-time lecturer at a university, but I must adjust to them. I could doubt decisions taken about technologies and teaching, for example, but do so in a way that does not convey the more confrontational style that the word and its ideal shape also implies. These changes do not deteriorate the definition of cyborg liability because the changes enable it to embody some of the perfect concept's essence, related to cyborg duty. Ultimately, cyborg duty must represent the behavior of several people rather than one or two, as in the agenda in this chapter for computers and writing. Much like someone who can't carry out cyborg liability, no matter what standing they hold, as they have been ideally identified, some probably won't be able to carry out all the elements on the agenda, if a decision is made or whatever your contexts entail. However, it's a term like a cyborg accountability that makes a difference in helping to steer efforts for diversity and inclusion, if necessary.

1.8 CONCLUSION

The dynamic of mass frameworks, domestic infrastructure, and cultural capital are increasingly affected by cyborgs. More businesses introduce mass paradigms of in-the-body innovation, and much more innovations are prevalent through

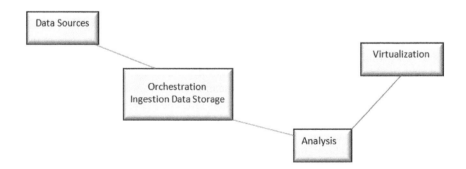

FIGURE 1.14 Cyborg responsible for the improvement.

domesticating technology. Around the same time, more people are trying to raise cultural capital using corporate ventures. Consequently, discussions on the future of robots-centered culture are faulty, as the increasing numbers of cyborgs with higher robot capacities are not taken into consideration. Discussions concerning humanity in the future should also consider the possibility both of cyborgs and robotics replacing people. The hypothetical cyborgs should be included in discussions over society's future, which include concerns as to whether human beings' substitution leads to opportunity or exploitation, dystopia or utopia, liberation, or extinction.

REFERENCES

[1] Cybernetics. *Wikipedia*. Available on-line at https://en.wikipedia.org/wiki/Cybernetics.
[2] Giselbrecht, S., Rapp, B. E., & Niemeyer, C. M. (2013). Chemistry of cyborgs-to link technical systems with living things. *Applied Chemistry*, 125(52), 14190–14206.
[3] Karlsruhe Institute of Technology. (2014). The cyborg era has started. *ScienceDaily*.
[4] Morimoto, Y., Onoe, H., & Takeuchi, S. (2018). Biohybrid robot powered by an antagonistic pair of skeletal muscle tissues. *Science Robotics*, 3(18), eaat4440.
[5] Institute of Industrial Science, The University of Tokyo. (2018). Cometh the cyborg: Improved integration of living muscles into robots. *ScienceDaily*.
[6] Tian, B., Liu, J., Dvir, T., Jin, L., & Jonathan, H. (2012). Macroporous nanowire nano electronic scaffolds for synthetic tissues. *Nature Materials*, 11, 1–19. doi:10.1038/nmat3404
[7] Harvard University. (2012). Cyborg tissues: Merging engineered human tissues with bio-compatible nanoscale wires. *ScienceDaily*.
[8] Song, C., Gehlbach, P. L., & Kang, J. U. (2012). Active tremor cancellation by a smart handheld vitreoretinal microsurgical tool using swept source optical coherence tomography. *Optics Express*, 20(21), 23414–23421.
[9] Optical Society of America. (2012). Cyborg surgeon: Hand and technology combine in new surgical tool that enables superhuman precision. *ScienceDaily*.
[10] North Carolina State University. (2014). Paving the way for cyborg moth Biobots. *ScienceDaily*.

2 Cyborg Ontology

2.1 CYBORG TECHNOLOGY

What is the effect of technology on human life? How does it affect our conscious awareness? I would first attempt to characterize what technology is to shed some light on those basic issues. Not an easy job since the definition of the word technology is not widely accepted. There is, moreover, an understanding of what innovation is like—if not entirely homogeneous.

2.1.1 Define Technology?

First, there is the "personal responsibility" perspective, which associates technology with specific devices such as instruments, equipment, computers, etc. The second principle highlights the notion that, instead of making things, the central problem of innovation is the manufacture and use of those items. It concentrates on technological invention, architecture, and public use. A third interpretation sees technology as a kind of intelligence consisting of the talents, codes, laws, and hypotheses we want to accomplish. Figure 2.1 shows how the fourth technology design applies to man's goals, intentions, wishes, and preferences as the "consumer" of technology: technology as will. I will concentrate on this very complicated and "secret" part of technology [1].

Can a man use technology autonomously, or does technology in our environment represent a self-supporting mover? The strong view that innovation has propagated through its first generation, the existing people such as Neil Gagman and Don Hide, Wittgenstein, Mumford, and Ellul, are transforming our understanding and our consciousness, policy, civilization, and community. The technology's vision has a major impact on the future of society and politics. Technology then interferes deeply with subjectivity: it changes the impression we have of ourselves by being a powerful "mediator." This perspective firmly opposed the view that innovation is an impartial instrument we should employ for all sorts of purposes, but it is not very popular among theories. This does not indicate that the large production of culinary creations is substantial, but the other way round. The theory of technology, which is mostly American, is that technology is a (social) structure.

The "soft" variant of technical decisively has a deciding effect on technology too, but it is not the only one. In several influences (economical, military, social), technology is the product of the form of civilization, culture, and subjectivity. Figure 2.1 displays this. Although this focus on multiple triggers entails an "over-determination" of an impact, in a system of cause and effect, the "soft" version of the technical determinism always believes. Social constructivism is trying to break this system down. While social constructivism has various methods, the vision of

Cyborg Ontology

FIGURE 2.1 Conception of technology in cyborg.

technical progress as a contingent mechanism involving heterogeneous variables is a popular element. In technical transition, different players or social classes play a deciding role. To mold technologies in their planes, they are employed in various techniques. Therefore, the paths and objectives of technology rely on the decisions and perceptions of the various social classes that plan and adopt them. Social constructivism, emphasizing the significance of actor and party decisions and its methodological methodology, attempts to dissociate that the dismissal of "enlightenment ideals" is much more consistent with efforts to separate "homogeneous" concept of technological future.

2.1.2 THE TECHNOLOGICAL DESIRE FOR THE PRESENCE

The digital world is the "electronic domain" created by the combination of many computer networks between the 1960s and the early 1990s, and it became a general social phenomenon. "Digital communications networks are digital technology. The new room acted as the location of correspondence, socialization, organization and exchange and a new market for information and knowledge sharing."

In keeping with the many thinkers of modern technology, who see cyberspace as a new country to satisfy our wildest dreams, as the basis for our intervention in cyberspace, Michael Heim sets the old Platonic Eros. He discovers that what is missing, what is beyond our bounds—or our opposite half—motivates people to use technologies. This is the ancient need for appearance, the new "material" embodiments of which are the Singularity technology. In her important essay on the phenomenology of electronic presence, Vivian Sobchak communicates that information technology seems to design or build the second parallel universe. This is the technical design of being and existence conceptually speaking. However, the question is whether many views (utopian or idealistic) regard this parallel universe in a Platonic sense: as a world of substance separate from human matter. Cyberspace is a place in which knowledge is already accessible, and it just waits before we can show it. It turns cyberspace into a field of immaterial information, which does not rely on hardware, software, or human neural implants and is independent of the machine and networks. Like Plato, concerning how the ideas are expressed, the material is neutral: concepts occur regardless of the human user's awareness, experience, or imagination. Cyberspace also propagates the Platonic dualism between mind and body (as seen in Figure 2.2) and these uncritical viewpoints. Or do they think of the Cybernet as an immaterial consciousness unimpeded by its physical constraints by the cyberspace information flows? ICTs seduce or invite the viewer to believe that there is a constant point of touch between the feature extraction and its interpretation [2].

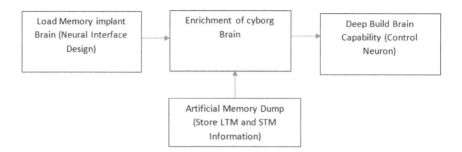

FIGURE 2.2 Enhance cyborg's brain.

2.1.3 Innovative Cyborg Technology

The digital world is the "electronic domain" created by the combination of many computer networks between the 1960s and the early 1990s, and it became a general social phenomenon. "Digital communications networks are digital technology. The new room acted as the location of correspondence, socialization, organization and exchange and a new market for information and knowledge sharing in Figure 2.3."

Conforming to many psychologists of modern knowledge, who see cyberspace as a new country to satisfy our wildest dreams, like Heidegger, machines and language systems were seen by Ivan Illich as a huge challenge. They take away the essence of man with their natural language. The language is the heart of the human being for Heidegger and Illich. The wish is the core of existence; Lacan says the same. From the research, "technical eros," the Postmodern viewpoint, appears to touch upon the existence of technology in the field of intellectual archaeology.

To illustrate technical eros, Carl Mitcham uses the differentiation Paul Ricoeur makes among three stages of human will. Science as voluntarily may also be categorized as technological appetite, inspiration or technic action and agreement to technology within at least these three volitional senses. The dialectic of "Technical eros" (Jakob Hommes 1955) may be elucidated by such a study (Michael, 1994, 255). Technology and Eros' partnership is only one of four "traditional" wisdoms of the technology. My research on the "technological eros" does not, however, concern the entire "computer technology" realm but is limited to that part that intimately affects us and is, therefore, the most difficult to understand. This declaration should also help to escape the pit of mismatch.

2.1.4 Communications Technology Media: Transcending Human Conscious Purposes

The "traditional" approach maintains that man is a faulty species, which needs science to live. When the human being characterizes biological flaws and weaknesses, technology is a way of substituting these weaknesses. It thus can offset or accommodate biological or natural requirements that are central to the innovation.

This predominant definition of technology positions its social constructionist level of importance. Technologies and their applications change or manipulate the

Cyborg Ontology

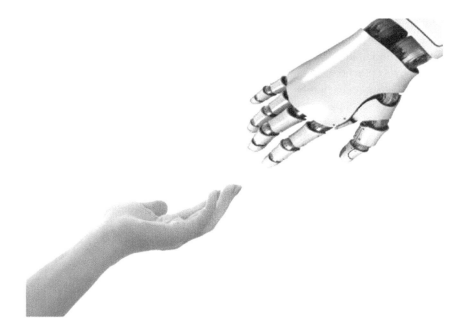

FIGURE 2.3 Exchanging information man and technology.

environment to satisfy modern goals. It is a type of teleological or deliberate intervention that meets utilitarian or functional purposes and objectives.

For the first time, the graphical user interface provided a spatial dimension to data structures, allowing the device to show up as a user-passing environment. This concept of the machine as a medium became very popular when the Internet boomed in the 1990s. As this machine conceptualization as a tool closely links with data object displays on all kinds of displays, a valuable analogy for my relationship to cyber warfare is possible. The reality that the digital world is presumably a combination of numerous analogies—both in terms of production—should not be ignored. And at the consumer level, in architecture and receipt, a special interest is given to the metaphor of the medium if you dwell on the "volitional" element of creating a reality by the machine—unintentionally. In the general field of information technology, the key distinction is between computers as instrument or medium.

They raise the major question of whether certain devices are a human being's extension or a means by which he creates new "forms" of himself. We ought to wonder whether technology (instrumentally) contributes to this society or whether it produces (substantially) a world: it is just medium to solve bionic problem. Are we using only technology to protect our biodiversity or are we using technology to change our environment and ourselves as we wish? To underline my willing control construction, I note here (again) that several philosophers are aware of this notion of technology as a willingness to turn. A philosophical exploration of the Ortega Y Gusset method for achieving one's goals by one's own decision-making. The technology of Hannah Arendt is seen as a response to the ancient tribal dreams of realizing the wish to abandon the world and its circumstances. Technologies to simulate true pleasure and to recover "True ecstasy is simulation." Technology is also used as a means of eliminating (the sense of) depletion. Therefore, Peter Weibel explains all technology as psycho technology: "technology tends to fill, bridge, and transcend" the lack of absence. Each type of technology is tele- technology and amplifies space and time. This triumph over distance travelled, though, only constitutes a phenomenological component. Mental disruptions are the actual result of the media (fears, control mechanisms, castration complexes, etc.) caused by all modes of absence, leaving, breakup, vacancy, delay, cancellation, or lack due to distance and time. The technical media are turned into care and appearance technology by resolving or shutting down the derogatory horizon of lack. The media often transform the damaging effects of the absence into fun ones by visualizing the absence, rendering it symbolically present.

We continue to violate, confront, alter, or reposition our limits with (psycho) innovations. In a Lacan sense in which the truth is precisely what is absent from our understanding and thus limits itself (the real core between and without us), the real is what we know to be. What should be confronted or achieved except for a moderate? Technologies are media, as Peter Weibel shows, that make a connection between tele-technology to cross distances and interactive technologies to close the gap between simulated and physical environments.

The purpose of technical media is therefore to hide and assert a real existence as a medium and to admire the presentation of the objects that seem to be fact on opaque windows. What are the "twin concerns of modern media," according to Bolter and Grusin? "The straightforward portrayal of reality and the exhilaration of media opacity themselves."

This study is shown by two cases, electromechanical technologies and digital technologies, which aims to replace the actual with its "simulated form," the feeling of speed most of all. The futuristic car can provide us with this insight; the car, in its driving version, provides a framework for the reality that helps us to alter our understanding of reality. This gives a hyper-realistic view of actuality: as in the elevation perspective but also the existing world order with the usage of vehicles. The automobile was consequently a contemporary vehicle to take it away. And, as the microcomputer and scientific changes in 1993 illustrated by Kaufmann and Smart, the cvomputer system radicalizes this push into the digital sphere: they can reproduce items none have seen—biomolecules or the beginnings of the universe—or they may construct locations that cannot be reached by man and thereby almost separate our perspective from our physical role [3].

2.1.5 Basic Analogous Protocols: Metaphysically Reminiscent Longings

All know the nostalgic syndrome. For days that have passed, we will be full of longing. Or we can be nostalgic about a place that interests us. The target of nostalgia, however, typically has a particular place in time and space. At a certain moment, or in a long span of our lives, our minds return to a certain world in a certain location. We assume we belong home in these repeated feelings. Or at least we feel as if we belong to the world that fulfils our wishes. There appears to be a mirror universe

Cyborg Ontology

in our imagination: an envisioned environment that is analogous to our aspirations is experienced. The major consideration is what this world is like: is it real or just a delusion? Is it a true reflection of the "original" incidents that have arisen in our portrayal or is it rather an imitation of our wish for a moment of achievement? Briefly, was there a "house" like that?

The concept of nostalgia in Figure 2.4 reveals that a "house" has a psychological memory. At least there is a possibility that the psychic entity is related to the "true" object and this connection could mean an equivalent need for this longing. The suggestion or knowledge of a relationship of resemblance distinguishes analogy in its broadest sense. The case in the past, to which we are referring, is quite obviously long gone in nostalgia. However, the issue at hand relates to the relevance of a case at the representational level. Is my interpretation of the event consistent with the "real" event; is there a consistency between these events? Is the psychic universe analogous to "true" reality? Is that what the original condition means by forming a parallel relationship? If so, the aspect of interpretation and the "initial" event is similar, imitated, or even united. A connection between the objective of representations and the topic appears to exist in remembrance.

2.1.6 Preception Vision

The basis of this notion of "vision" (theory) states that the source of knowledge in its true nature resembles the object of view. Then suddenly, without mediation, the subject and object shall be together. This is a clear perception, imagination, and vision framework. The information thus represents or represents "actual" truth and consists of sufficient images or representations on the "mental screen." The philosophical belief is that a real understanding is based on an Initial Fullness guaranteeing the essence of representations.

This Platonic foundation in Western thinking is often referred to as "metaphysics of presence." The initial existence is combined with logos, God, and teleology by Jacques Derrida, who continues the critique of Heidegger's metaphysics. The idea of a cohesive self that can create genuine awareness of experience is made possible by this original presence.

In Figure 2.5, Platonism explains the argument that real experience plays a key role in memory or reminiscence. The precise memories of the still-existing knowledge

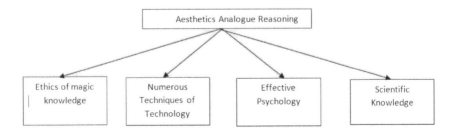

FIGURE 2.4 Nostalgia: an analog desire.

FIGURE 2.5 Platonic knowledge tree.

(that may have been "forgotten" or turbid) enable people to get a real glimpse into themselves and the world. Therefore, it is a nostalgic desire that leads the individual to (re)find the missing presence. The crucial point of Platonism is its development into an entity that can be "seen" by the "eye of the spirit" (which overcomes our intelligence and vision, and this is what our wishes are addressed to). Likewise, Plato transforms into an entity of light, which gives all else visibility, a term never used previously by the ancients. This platonic concept of mimesis binds the world of the real objects (ideas) to the sensual world, driving this metaphysics. Sensitive universe representations hold the only reality, whether they so truthfully as practicably mimic or resemble the concepts. Plato thus maintains a correspondence theory of truth: an assertion is valid only when it corresponds to a non-linguistic fact.

2.1.7 MODERNITY HUMANITY: THE WORLD'S LARGEST ANALOG DEPICTION

The Perfect Wisdom Platonism is a kind of act of aggression. It isn't active participation or advancement, and the words "development" or "output" can also be seen as such characteristic of modern thinking, according to Heidegger. The transforming of man into a subject that is based on reality is decisive for the technological society: everything is real only if it is represented by the subject. The "being represented" is synonymous with being. Therefore, the meaning of the new era is the representation of the future (the world image is "the age"). This definitive transformation in the West ties Heidegger with a fundamental dreaming recognition differential. The Greek thinking (pre-Socratic; phantasies), the "expression" of being, also concerns how it exposes the secrecy of life. However, the modern man who made the universe into a picture fantasizes: he thinks all things are nothing but objects he uses: "fantasy occurs unconcealed": the coming into being, as a specific aspect of what presences—for the human being, who presences himself of what seems. Man as the object that represents, however, "fantasizes," that is, in imagination, goes into the world as an image, in his representation, pictures what is and is the objective. There was a mistake.

Cyborg Ontology

In Hegel's philosophy, where Spirit seeks its fulfilment (will find its fulfilment in itself) a tradition of comparison of highlights were the mind and the world, interpretations, and experience in the field and honesty: "there's absolute knowledge," as expressed in the world.

2.1.8 Cyborg Authenticity Representation

Mental images are a manifestation of a (transcendent or immanent) "material truth" in the first and last phase in the intellectual heritage of addition to implementing Plato and Hegel's philosophy. This identity of representation and truth is supported by the comparison theorem and a certain memory notion as well.

However, Freud's "discovery" of consciousness was an important influence in blocking the pretentious authenticity of representations. Figure 2.6 further indicates that memory may be an unintentional procedure; a painful event may be recalled at an unaware stage. While Freud felt in his early work that these painful experiences should be remembered and released, he quickly realized that psychic experience cannot be resolved into "actual reality." There is a "void" in fact and thus (of the past) never (completely) comparable representation. This isn't an exact similarity to (historical) reality; representation is often determined by codes rather than an analogy. The unconscious includes an image difference, which makes the object's similarity short. Thus, the discovery of Freudianism is a break between the similitude of experience and reality or even (complete) personality. Conscious consciousness never correlates entirely to our own (unconscious) reality.

Hypertexts, therefore, seem to lead from the interpretive research hermeneutical undertaking to the discovery and exploration of the (hidden) context of a text or a narrative. Thus, as the text has become a maze, it becomes a matter of exploration and interpretation.

The need for it is not mainly inconceivable, something that happens. In contemplating the archetype of what is perfect, the kind of beauty, people will understand his fundamental, "nostalgic" wish, which continues to make him seek more and further. A (re) seeing attractiveness is used as a guide to beauty in general in a single

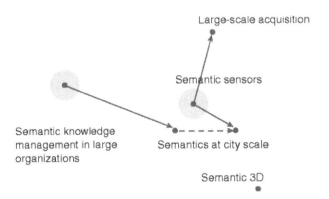

FIGURE 2.6 Gap in analogical mind.

object. Physical beauty is divine beauty and social institutions' beauty corresponds to intelligence and science, the beauty of the government groups. Platonic Eros is an appetite for magnificence and reflects its fundamental goal. It is in a higher ontological field than beauty, in which all other entities participate [4].

2.1.9 Description of Body Parts

In the *Dialogues*, Plato gives a great explanation of passion. At dinner at Agathon, guests speak of Eros and explain what is considered the principle of Platonic friendship. Plato establishes, by Aristophanes, a story concerning the original human body (with four wings, arms, and so on) that Zeus has cut in half to weaken humankind as revenge for human efforts to attack the gods. Love is the separated person's longing for his Other Quarter: the want to go back to one. Aristotelian Eros is the internal turbulence in which one longs for the sublime and it is perfection that love seeks in all its forms. He finds what he is after in elegance. We will be glad only if a fully satisfying love can so re-establish us to our pristine configuration.

2.1.10 Beauty as a Mediator

Lacan overcomes existentialist philosophy, with the focus on the other real thing in the phase of differentiation following this transition. He now holds that the importance of the interpretation—but also the (idiosyncratically) satisfaction they may give—is not only determined by the sense of signifiers. As a libidinal body, Lacan emphasizes the subject. This (bodily) subject is not simply alienated by the inter-Subjective existence of the human world but is isolated from the subjects of its partial movements. Segregation then presents an intrinsic boundary for "pleasant" preference considerations by identification. The drives shown in Figure 2.7 do not worry about identification. Despite our loss, the urge remains linked to the illusive notion of the goal of total fulfilment that can meet our requirements. By imagination they are looking for its "simulation." Plato motivates the need for "self-realization" through the recollection of beauty that the psyche of man "saw" before he saw the first daylight as a physical being. Because our physical life now binds us to the natural world and binds our minds, the objective is to liberate us from the physical bounds. In reminiscence, Lacan dismisses any speculation about larger dimensions than our reality on the planet. This recollection hesitates to direct our interest in what it called the true, actual, and early fulfillment of our past. As the satisfactions that guide our desire no longer exist and cannot be represented as they were, they are in the order of the actual. Only consider trying, as much as possible, to rebuild a happy event in the past: it should not be the same. So, says Lacan, we refer to the real by imagination. Eros, not the Platonic ideal beauty, is motivated by lost pleasure with its drives.

2.1.11 Anthropological of Lacan: Beyond Requirements

In reality, one must be more precise because, according to Lacan, the word "man" cannot be preserved in a religious structure that draws man away from the center of the universe. Then he proposes that "desire is the essence of man" be substituted for

Cyborg Ontology

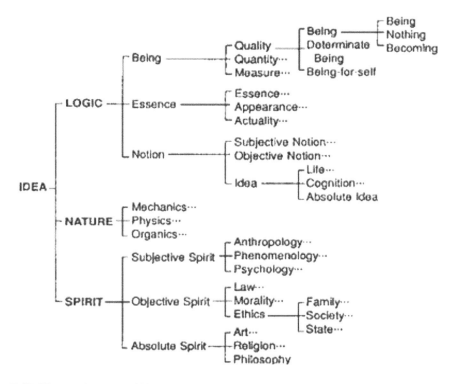

FIGURE 2.7 Structure of idea.

"desire is the embodiment of truth." Religious anthropology, in its connections to the gods or God, recognizes a man as the "homo religious." It combines the appetite of men with a supernatural component, which can remedy human weaknesses. As discovered by Ancient Greeks, logical anthropology places an independent explanation on the forefront and finds it as the exciting example of humanity ("animal rational"). Kant argues from a pragmatic viewpoint in his anthropology (1798) that humanity can realize the process of enlightenment. Human rationality cannot be achieved by the individual but by humanity by understanding the goals it is given [5].

2.1.12 Cognitive Imagination

Our regular status is controlled by statute. His "never too much" axiom normalizes our satisfaction—interpreted as pleasure—into the boundaries of the concept of happiness. However, there are "points" within this precept of the place principle where the remains of deeper gratification emerge in a masked form: sublimated, imaginative. This is the fantasy domain. The word *entity* introduces Lacan for the rest of the elusive fulfilment that the imagination attempts to frame within the boundaries of the concept of enjoyment. The object is a remnant of the pleasure of the "perdue heaven," that is, the excess of pleasure after the statute. The appetite and, as such, the intended subject is causing this "object." It is beyond meaning in its cognitive component: we can't tell what it is precisely. The realm of language subjectivity is transgressed, and the meaning of the body is made clear to us.

2.1.13 Fantasy and Excess Pleasure: Between Gratification and Ecstasy

An appreciate the pleasure of fantasy "at work," one must see Lacan's understanding of pleasure first. The most widespread misunderstanding is that happiness is equivalent to feeling good. To get this misapprehension started, however, you should perform whatever Lacan characterizes as comfortably pleasurable. Pleasure isn't fun! It can be seen only where psyche control hits its limits through the concepts of gratification and unpleasure. Anyone who runs his car as quickly as possible is not happy with the thrill. The thrill of a peeping tom that can be trapped with his looks every moment is not just pleasure. "Let us be clear: enjoyment is not fun, but rather an over the playable state; it is undue anxiety, maximum tension, to use Freud's words, while pleasure is a decline in stress." Our boundaries are at stake with pleasure. With that one can discern three main types of pleasure: pleasure, which acknowledges the law and thereby attempts to benefit from the imagination in the realms in which it counts not; pleasure, which transgresses the law; and pleasure, as "playing" with the rule. The concern is how the statute is related to (pre-subjective) pleasure—as a standard for personal identity (it is pleasure that cannot be reconciled with working as a competent person). An area hangs on to it in the first-pleasure attachment to a subject and at the same time recognizes the laws and practices that forbid (the "childish") ways of pleasure. This tension condition would then cause (neurotic) symptoms to develop. Such signs provide the clearest expression of a repressive type of pleasure.

2.1.14 The Technologization of Human Virtuality

It is confronted by a generally agreed conception of virtuality. This opposition in Figure 2.10 brings us to the first sense of the term "virtual" places where there appears to be a unique. It is "a picture or space which is not true but appears to be" like telephone storage or electronic currency. The virtual still has a significant metaphysical interpretation, apart from this daily significance. We will take account of its technical significance later in the next paragraph.

2.1.15 Basics of Virtual Technology

We can see that it comes from the Latin *virtual*, that is, strength and quality, in addition to understanding the philosophic sense of the word "virtual." The term virtuous can be traced back to virtue: man, or manhood (as in virility). So, one comes to the concept of virtue in terms of a more physical meaning, where fitness and sexual purity are equated. Virtue refers to "virtue" in the moral sense, which reflects bravery, competence, and virtue. In this sense of force, Latin philosophical vocabulary includes *virtual*; the idea of perceived reality was not understood by Ancient Greek philosophers.

The transcending use of the word "virtual" refers to the positive relationships between things. Thomas of Aquinas tries to present a collaborative or "virtual

Cyborg Ontology 37

consequences or reclusion" as a synonym for Plato and Aristotle's capabilities, indicating that the result has always been prevalent in the cause: the tree is almost prevalent that is in the grain. The virtual provides an impetus in this grand dimension that leaders understand, which was developed in the Aristotelian theory of potential and reality.

Duns Scotus expands this principle from the philosophical to the epistemological realm on virtual material, capacities, or substances ("essence") by arguing that the inference already exists in the assumptions. So, because it's real that machines don't has feelings because I'm a machine, it is almost true that I don't have emotions. Although the thesis on interactive material was controversial, Leibniz was persistent in the contemporary era when he took Scotus to a new role with his theory that the subject either directly or virtually includes it in all true phrases.

Scholastic vocabulary incorporates virtuality, effectiveness, and performance in the 14th- century. "Virtuality" gained a sense other than that which would be expected from semantics in its scholastic description. The digital amongst scholastics takes on a sense of as-if or of "virtual distinction." It is possible to consider what is unlikely to be differentiated (what is "virtualized") as if it had been distinguished similar to how a cyclist in the Tour de France may nearly be wearing. Perhaps, he was that far behind the winners during one competition. A sense of virtuality here becomes fictitious. There is a second, imaginary order in which situations are arranged differently from "as they are." It's not just as it seems: it's "digital multiple." The classical concept of "virtuality" corresponds to potential. Virtuality came to distinguish man as a being still capable of realizing its powers beyond those limits. This misunderstanding between the virtual and the potential is criticized by Charles Sanders Peirce. The virtual is associated with a divergence in instructions. That's not the same order as the possibility, because as a potential, it simply has not yet been understood.

All have the same essence as everything is like the future is from the seed tree.

> A virtual X (in which X is a typical substantive) is anything, not an X, that has the effectiveness (virtue) of X That's the correct sense of the word; but it was confused with 'potential' that is nearly its opposite. For potential X, but without real performance, is X's existence.

Therefore, the questionnaire that was administered is usually an artificial feature. It is like Gilles Deleuze's thought, as expressed in differences and repetitions. Deleuze makes a capital difference between what is real and what is virtual. Possible beings are already formed and static; only they lack life and must realize themselves for that purpose. Deleuze says that this understanding is very different from the update of the interactive, which is a "development." "The feasible is the wrong idea, the cause of false issues" (Deleuze, Bergsonism, in his book). "All have been offered already entirely: all the truth in the shot, the pseudo-actuality of the conceivable." So, what Peirce and Deleuze tell us is that the many forms in which the digital can make itself fashionable ("What somebody is depending on the circumstances to him") is fundamentally different. That seeks to be rendered trendy by the consequentialist endeavors of the possibility [6].

2.1.16 Virtual Reality

These days, most people think of "virtual" as meaning "computer-generated." The name given to this augmented reality often merely references to one of the two basic meanings of "virtual," which is that something just seems to exist; it is not "actual." This was also the approach used to explain immersive computing networks in the vocabulary. In 1980, in the following terms, Theodore Nelson, the very first man to employ virtual environment in software applications, created the name "hypertext": "by the virtuality of an item," I mean its appearance, as distinguish from the more tangible "reality." In its conventional meaning I use the word "virtual," a contrary to "physical." The truth of a film comprises the painting and playing of the actors among scenes, so who cares? The virtuality of a film seems to be what it contains." The film is not only virtual, because it is not true, but more so because it produces "truth effects," which lead us to think the illusion is true. This has potential to induce results, "The virtual is not imaginary, is the most significant significance of simulated reality. The results are generated." Interface with augmented worlds by the machine differentiates itself from previous concepts of virtuality in two essential respects. The following Figure 2.8 particularly applies to interactive Virtual Reality (VR). The user can be connected via the detector interaction with a simulation environment via a head-mount device with an impressive sense of "reality" in the virtual environment.

For example, the user may alter the visibility of the details or the circumstances of the virtual environment he is in. This sensor-motor feedback interactivity produces an unfounded phenomenon in media such as film or TV, which gives people a certain knowledge of how their bodies move.

The second characteristic of the Singularity generated by the machine is the interactive nature already described. The actual sense of being submerged in a computer-generated world comes from stereo glasses and data gloves. The most modern version is now a direct representation of the pictures on the retina. Innovations often utilize helmets that display images on the user's machine (head mounted display; CAVE). The most powerful effect of virtuality is interactive digital technology. But even an interactive augmented world can be felt on the two-dimensional interface of your personal computer. Virtual reality like World3D is seen to access a virtual

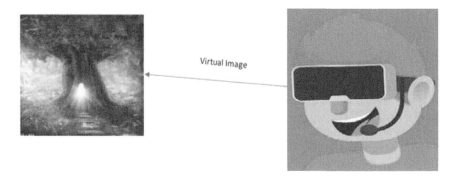

FIGURE 2.8 Creation of virtual image.

Cyborg Ontology 39

world in which others live and thus a sense of entry or insertion is felt. An overall feature of imaginary spaces is the continuous presence of connectivity, quest, and recuperation resources. So augmented digital realities (in the worst meaning augmented worlds) means both software things and their contents, for instance, computer programs and databases.

2.1.17 Four Concepts: Physical and Technological in Digitalization

The first iteration of Singularity, emulation, recognizes the simulated as a copy of the physical. Actual things are considered as original and self-identical in the principle of correspondence of representations that drive this debate (A statement is legitimate only if it pertains to additional amounts of factual elements). The virtual here is a risky addition, as the picture is in the writings of Aristotle.

The second iteration of augmented reality, suppleness, is subject to the same approximation discourse, but this is a transposition. The actual is impartial, insufficient, and incomplete in this case. That can be supplemented by the interactive. The virtual relates to the actual only as perfectly relates to the imperfect; the real errors may be corrected.

The third instalment of flirtation, or education and this "fetish virtual ideal" will be the existence of a complete realization of the possibilities of the future. This results in complete essence extinction. With a good quote from Baudrillard, you add this. The IBM mechanics take up the mission of translating the 130 million names of God from the congregation of Tibetan monks: in a few months, their machine will accomplish the task of achieving and ending the goal of the universe according to their teachings. The whole thing is "a real drag" that can be left behind in this iteration of the relationship between real and virtual.

The fourth edition of augmented reality that the writers adhere to is around the idea of Deleuze-simulacrums. Guattari's idea avoids the error to confuse the virtual with what is conceivable in ideologies of assassinations and hyper realization (both first forms; the third).

2.1.18 Virtualization

In no way does virtualization or the transformation to a challenge imply the absence of the distortion or dematerialization shown in Figure 2.8. Instead, the term "desubstantialization" can be understood. This decrease has been divided into several significant improvements: deterritorialization; the Effects Moebius, which coordinates the endless loop between within and outside, is to swap private elements and subjectively include public things. Conceptual frameworks include the psychological and somatic processes by entities of technological, semiotic, and reform movements. Objectivities shall be described as the collective involvement of subjective actions in building a common environment. There are also two important dimensions of virtualization, subjectivities, and objectivities. Indeed, neither subject nor object is quantities but oscillating networks of events that interact with each other and envelop each other. The thesis by Lévy emphasizes that virtual machine and institutionalization are simultaneous processes in their conceptual and archaeological elements.

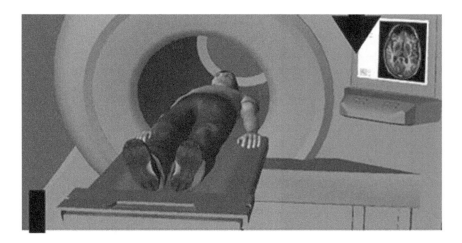

FIGURE 2.9 Virtualization is the mediation of language.

His thoughts teach us that a synthetic component often pervaded the reality of the human being. This is just to suggest that "desubstantialization" characterizes the truth of the human being. This "destroying" is divided into various divisions by Lévy. First is the deterritorialization process as a separation between here and now. He explains this process, pointing to the work of one of his previous contemporaries in the visual representation, Jacques Serre's, who describes the digital as a movement of abandonment in his novel *Geography*:

> The virtualization vectors of creativity, memory, intelligence, and faith have allowed us to abandon this 'there' long before the advent of computers. Objectivation shall be described as the collective involvement of subjective actions in building a common environment. There are also two important dimensions of virtualization, subjectivation, and objectivation. Indeed, hardly subject nor object are quantities but oscillating networks of events which interact with each other and envelop each other.

The thesis by Lévy emphasizes that virtual machine and institutionalization are simultaneous processes in their conceptual and archaeological element.

2.1.19 Language: The Virtualization of the Real

I shall now move on to a Lacanian concept on virtualization, following Levy's keen insights into the virtual. In Lacanian words, language mediation opens the way to "the time and location." The concept of language as a symbolic other is based on the basic assumption of Lacanian thesis that man is an interpretative connection as a participant of communications in Figure 2.9.

2.1.20 The Retroaction of "Real Time"

By inscribing "events" at the other's locus, they gain meaning that the recipient does not understand and cannot predict. A jihadist artist once again reveals

Cyborg Ontology 41

that the subconscious is "out there" in a symbolic structure; it only occurs in a debate as Lacan says on his television. Therefore, the understanding by Lacan of Psychoanalytic concepts of "alternative drama" as ignorance is the focus of the other individual. Another one is also the location where the enthusiastic aspects of time are installed: history, current, and potential. Together with the Levy definition, terminology opens a space of potential possibilities for virtualization of "actual environments [7]."

2.1.21 "Law": The Virtualization of "Natural Forces"

We are inherent "subject" to the rules governing these discourses, representing truth in all kinds of dialectical constructs. This may be illustrated by games. When I enjoy a sport, that is when I am a certain player; I'll follow the rules that decide how to play that game: how to communicate, etc. In its most basic form, Lacan considers fact even as a "game" (shown by his saying that the principle of reality is the principle of collective fantasy). The cloth in Figure 2.10 is composed of the language. The rule that recognizes language is also fundamental: the law of the meaningful person. "Law is the compilation of common rules making possible social life, mechanisms governing all types of social exchanges."

2.1.22 Technology for the Interfacing and Virtualization of Genuine Innovation Imagination

The expansion of speech by man makes it possible to distinguish intelligence from the far broader truth. Without language, Bergson says, the consciousness of human beings must remain fully interested in its artifacts . . . Bergson believes that even cognition is a spread of man, diming the bliss of union in the communal unconscious.

(In Creative Evolution*)*

FIGURE 2.10 Virtualization as a natural game.

As an expression of the human being, language is the first invention that allows people to grasp the universe consciously beyond the subjects of their interest. Language means world mediation. All kinds of technology further expand the possibilities of people in matter and energy (mediatization). So generally, technology provides in "space-time distance" much as language as the first technology: it takes us further from the here and now. In his mediation theory, John Thompson emphasizes this complexity, since all types of media are concerned with symbolic modes (the name of this symbolic communication is not so problematical for him, because its study addresses the modernity of communications technology and the post-modernity of IT that puts symbolic—yet another—under pressure): "transmission of a symbolic object inevitably entails a separation of this form to some degree from its original areas of communications: separated spatially and temporally from this context and integrated into different situations, located at various times and locations."

2.1.23 THE TECHNOLOGICAL AGE: CONVENTIONAL TO CAPTURED PICTURES, MATERIAL TO INTERACTION

The analogy to the modern transfer called digitalization is the first key feature of the "technological revolution." Consistent data is converted into a numerical image. In other words, all kinds of objects are stored in the "language" of zeros and the language of digital knowledge. This "language" has several units for a variety of newspapers. Photographs are represented as pixels, voxel tones, number and letter messages, disturbing images as multi-gyms, and the script or software sets are acceleration units. As the objects thereby become a machine digital language, they can be quickly manipulated, transported, and replicated at the speed of light. The object that is meant to be the unit of various updates on the computer screen is transformed into a strict target environment. Although updates address the coded object in a certain way, they are never the same. In representation, the encoded object lacks its true form. Why should a "box" of zeros and ones be the appropriate form? The attributes and features that are perceptible or phenomenal are not available in the encrypted entity as such. The aspect of the organization relies on the user's software, their settings, and the operation of the user (do you use Google Chrome or Outlook Express?).

2.1.24 DIGITALIZATION AND REPRESENTATIONAL SYSTEMS OF THE INTELLECT

I can immediately observe how the digitalization operation transforms the two basic founders of the subconscious mind's sense of truth: temporality and location. The way we envision, explain, and take time, according to Jeremy Rifkin in his book *Moment Wars*, modulates all our views of the world, thus, our identity and the society in which we exist. The way we view time, how we deal with the past and the future, will improve by digitalizing time. The programming will decide the order, length, and speed of an operation in advance: programmed machines instantly tell how to manufacture or serve commodities. Even if human bargaining breaks down (modification, fault, caprice), the design of time is essentially separate from the schedule; the arrangement is our "Traditional scheme" for our relationship with the future.

Programs often remove users from their contextual knowledge of the past that he normally utilizes as a resource (a "system") for potential activities. Thus, time digitization implies a (further) detachment from the subject of our "natural" or "immediate" memory. This time scale digitalization is typical of the changing relationship that the machine causes to the world around it. Whereas the clock corresponds to the circular time determined by the earth's orbit as an analogy reflection of time, a digital temperature increase is no longer bound by this circle. We are not so much constrained by the machine as we may be available in numerous time zones in the reference frame of our direct surroundings The time with its (analogy) representations is less perceived as a temporality than pace ("virtual immediacy") [8].

2.1.25 AUGMENTED WORLD OF THE VIRTUAL MACHINE: FROM KNOWLEDGE TO RELEASE

The robot does not bring about a change in paradigms. This represents an objection to techniques altogether inappropriate to adopt the human relationship with the outside world—but not generally. The method of self-construction appears to provide one with a better view. The "Came to fruition" section of this Graphical User Interface (GUI) shows off the virtualized aspects of our "physical world" that have been implemented in advance by various code frameworks. In keeping with Vivian Sobchak's understanding that cinema shows us the subjective framework of our perception for the first time, machine Singularity often reveals our subjectivity—but more of that.

> The film projections mechanically in its pre-electronic state and original materiality and makes the objective universe apparent for the first time, not just the very form and mechanism of subjective articulated vision—so far the only thing we can perceive as 'mean' until now is the unseen and private structure that has been directly accessible to human beings. In other words, the materiality of film offers a real and analytical perspective and makes the reversible, dialectical and social character of our subjective view logically clear.

Virtual reality will carry one to the very core of "the intangible and personal framework which we each experienced as my own." The mechanism of subjectivation by legislation will become explicit on the computer screen. So Sobchak's thesis that technology can make more noticeable the mechanism of our subjectivity seems to be supported by the programming interface. Or as McLuhan expresses it: modern technologies could lead to an "unconsciousness."

2.1.26 INTERACTION AND VISUALIZATION TECHNOLOGY REPRESENTATIONAL REGIONS OR SPACES DESIGNED TO SIMULATE

The body is a simple mechanical reasonable process in a cartesian conception, defined by implication in the physical environment. The mind is, therefore, the example of truth. So does IT create a "transformative" expression space? Can it distort online and offline space? Do they blur? In both cases, cyberspace was represented. First, the user will walk around as a bodiless intellect, surrounded by highly technological rhetoric as an (non-physical) emptiness, with intangible informational items in the population. It will be liberated with a literal definition of all the restrictions placed by the external surroundings in its purest thought. This will provide Platonic access to the ethereal world of thoughts and realize philosophers' uncompromising understanding goal.

2.1.27 It Appears Impossible to Connect the Indication with the Reference on the Interaction of It

Do any signals always apply to a fact in the "Information Age"? Does the modern revolution push us into modern architecture, a "new profundity," in which the vast impact of online technology makes it difficult or impossible to differentiate between a meaning and a meaning, according to philosophers such as Eco, Jameson, and Baudrillard?

> Eco . . . reflects on formal characteristics as the major facets of popular society and considers the aims of creativity and the metaphor (defining modernity). . . . In both cases, Jameson and Baudrillard view advances in reproducibility technology as a key factor, possibly eventually the core factor in the process of the emergence of an overall dominance of forms over the material.

2.1.28 A Media Eradication of Subjective in Humanity? From the Performance to the Presentation?

Do modern innovations take us into a post-Cartesian age: a posthuman period where technological influence is so convincing that the participant would seem to be erased who thinks he can depict the cosmos in his actual shape prudently? That kind of elimination will take us to the simulation paradigm for the human topic of representation. The Eros, the subject of my research, is the need for a "complete copy."

2.1.29 Cyber-Subjectivism or Cyber-Objectivism?

When digital transformation loses clarity about the correlation of signs with the truth of the logical subject, a concerning problem could emerge. Losing the rigid distinction between the indications and the reference may be thought of as freeing us from the patriarchal narratives that have defined this interaction based on a particular "ratio." Then digital transformation reveals that the signs convey no "actual truth," and in effect, everyone constructs his reality; cyberspace is a realm of freedom, in which everyone can be, without any restrictions on conventional narratives, who he or she is. It is a domain of emancipation. This framework contributes to a kind of subjectivism that takes individual awareness as a starting point for considering human thoughts and behaviors. Digitization will allow us to see the universe regardless of human subject weaknesses. This is the statement of moral relativism that often drives scientific thinking: empirical observation of the universe, which is free of the human conscious mind's subjective preferences.

Cyborg Ontology

2.1.30 FROM SEMIOTICS TO THE SUBJECT AS A MEDIATING WINDOW

In the space of imagination, the world of knowledge as interdependent in the natural world is unavoidably focused. The potential relationship between facts and context must also be considered. It might be a helpful trip into the past. In conversation about the essence of knowledge at the Macy Conferences, the connection between knowledge and context was also the general issue. Not all preferred knowledge is mathematical philosophy. Donald MacKay, a British academic, tried very hard to make the connection between study and life. He held said this is about a changing of mind in the recipient, and he concluded that "subjectivity, far from being a dull thing, is exactly what makes it possible to connect knowledge and context."

2.2 IMAGINATION SHOWS THE IMPORTANT SIGNIFICANCE OF THE INTERACTION DESIGN

Architecture is a subject for artists and engineers alike. "Efficiency" ideal engineering architecture is opposed to the aesthetic ideal of elegance. Beauty is not so much about energy and resources as about shape. In the study of electronics, Derrick de Kerkhof provides this intermediary design with an even wider reach. As an expansion of our mental and physical functions, He sees technology as an outward manifestation of our inner self. The design gives these mechanical extensions of us a shape and hence lies at the intersection between the human body and mind, the material and spiritual, the inner and outer. "Design," in my view, is a modulation, as changed by technology, of the relationship between the human body and the environment. Technology comes from the human body and architecture is meaningful. . . . Mind and body are so interwoven that separating them is meaningless.

2.2.1 REPRESENTATION OF DIGITAL WORLD

In designing the shape of digital content, symbolism plays a significant role. They allow us to display, communicate, and interpret data (graphic representation for the folders on the computer; speech-based symbolism such as "the super-road detail" as a Network description). The analogies are ways to form what is not (yet) what is "as we know" in fact. They also connect the unrepresentative as such with known interpretations; thus, it can be assumed that all people talking about God are figurative. Therefore in "how virtual worlds are objectivated representations and arguments as cognitive models," for example in Lakoff and Johnson's theory, we should be careful to regard cybersecurity as a premise or an impartial piece of knowledge. The concept of representations is therefore vital for interpreting virtual realities. The product of human creativity is a modern field of knowledge and communications, with established metaphors. Multiple perceptual chains are socially constructed thus. A dream picture and another "shape of the consciousness" are made up of various associative components that often will not have a singular basis of comparison (for instance: a dreaming component is a "construction" of separate characteristics). But these "literal and figurative groupings" connect an individual's everyday existence to his subconscious [9].

2.2.2 Imagination: Naturally or Artificially? Narrative? Imagination Seems an "Organic" Mediating Psychoanalytic Theories

The discovery by Freud's immaterial sensation as the correct field for an investigation into psychoanalysis indicates that a detailed psychiatric examination of a monitor that connects humans and technology should necessarily turn away from objective reality. The concept of psychic truth in Figure 2.11 shows a reality of desire that operates surrounding drivers and their psychological representations. In "objective experience" the drives don't have a natural object; they have a psychic structure. The unadulterated difference between fact and fantasy then lacks its significance: drive images are as much fictional as true fiction. Building on the impressive projection of the things that appear to "naturally" correspond to the disks, Man continually lives "lifestyle on televisions" as a pleasure-seeking entity more often than merely channels contributing to the operations of the laptop screen. And like innovation is a practical way of achieving a clear aim at first glance, human drives are normal ways of meeting biological requirements. In both cases, though, the fantastic aspect that works as a monitor between man and world "is forgotten." Lacan emphasizes Freud's assertion that the drivers are simply our myths to justify this intervention.

The dimensions of our "natural" connection to the universe are given by fantasy. For starters, the woman who draws me physically is not my (sexual) desire's natural target. It only seems normal that it blends into the wonderful coordinates that control my drivers. In the thesis that "no sexual association exists," Lacan follows. With Freud he shares the notion that our darkest dreams are the focus of the discipline:

> The experience of the basic fantasy becomes the driving force after mapping the subject concerning [objective].

2.2.3 Fantasy as Imitation: Hallucinatory Wish Fulfilments

The engineering interfaces that bring us into the world demonstrate why it is impossible to disconnect technologies from emotion. "This is the now the standard of computer graphics towards software technology-opened new worlds": new areas that can

FIGURE 2.11 Fantasy as like nature.

overcome all the old boundaries. Digital technology aims to conquer truth and connect us back to the heaven we have been given by existence. Return to the alternate routes of life and delays: immediate reward! We may link to porno websites that fulfil sexual desires, be the hero of our own (game) universe, etc. with the machine. The iconic term "consensus hallucination" in Cyberspace (William Gibson) refers intimately to lust. By simply imagining our wishes as completed in cyberspace, our wishes are accomplished in hallucinating fashion. As an electronic realization of the dreams that surround and sustain the technical business, cyberspace demonstrates the loving relationship between human beings and technology.

In certain cases, these dreams include the idea that development gives us ways to go beyond the boundaries imposed on us by reality: it allows us to free ourselves from actual burdens. This satisfying feature of technology works as realized visions of hallucination from the Freudian perspective. What we can't have can be achieved across the dream screen (of the computer). A modern dream universe or, on the contrary, little more than an imaginative delusion will be a "consensual hallucinating" cyberspace. To properly explain a clear distinction between fact and perception from a Freudian viewpoint, we must look closely at the (first) Freudian fantasy paradigm to look at cyberspace as either rationality that overruns or is lost (which relies on one assessment of its transcendental capabilities).

2.2.4 Concept of Enjoyment Perception

He believes that a proper classification is not inherently a product of this between existence and the digital communications dream space. She applies to Erikson to make it possible for a sense of identity to retreat from reality. Relatively non-impactful experimenting helps to create a "heart self," a personal understanding of what Erikson calls "identity" and provides a sense of existence. Turkle's remarks in Freudo-Lacania mean that cybersecurity is not only a reserve for the sole processing of the concept of enjoyment. It is also a glimpse into what the initial attraction is. It may deliver a predictably desirous object as an inspection of the drives, a vision of what it desires and which object answers. This is the prerequisite for imagination: you can only arrange the items in imaginative scenarios if you know what you want [10].

2.2.5 Philosophy of Categorical Imperatives: Perceptions of Mediation and Actuality

This is the aim of the contemporary revolutionary. I then showed that information technology, along with Rifkin and Chester, provides another awareness of the spatial and temporal, changing these key "schemes," with which we reflect the facts. Within a Kantian world, we may conclude that "informational creativity" contributes to various world views: more multifunction, less subjective, extremely expandable. Since a desktop computer (new) gives a structure to the dimensions of spacetime of objects, it is indeed a (Kantian) imagining. The augmented world may underline this. Based on (the "actual") characteristics, the computer turns the stimulus of a virtual environment into a reality-like "bodies detailed analysis."

2.2.6 Freudo-Lacanian Theory: Desirable Reality

In addition to operating the virtual keyboard as a screen that converts different visual experiences into such manifestations (the transcendental Kantian imagination), which we consider effectively or otherwise as a "true presence," there is also a (psychoanalytic) problem of a "desirable presence." In the creation and use of objects, the subject of desire poses unconscious illusions ("schemes").

2.2.7 Behavioral Therapy

Lacan's problem with Freud's imagination governing the world's understanding itself indicates that the validity of a dream is unconnected to an empirical fact. Evaluation of fantasy is not as factual as it looks real mode. A psychotherapist does not try "unreal illusions" to liberate his patients. Moreover, denying an issue of appetite ends in pathology and not self-sufficiency. Let us assume that in composing this work my conscious ambition is to build with myself a famous philosopher.

2.3 TECHNOLOGY AND THE FANTASTIC RELATION TO THE REAL

What I would like to reassess, in connection with the ("ancient") relationship between techno and touché, is whether we exist in a "technological world," where the natural does not serve the paradigm of the idea but creates it employing technical processes (distilling). For Aristotle's technos, what is not feasible is created and is, therefore, a constructive mediation between nature and mankind. The term technology, which may mean either craft or the arts, originates in Greek philosophy.

2.3.1 Technology and the Real Pleasure Principle and Its Beyond

The artist and the craftsman both use scientific expertise and instruments to produce their works. Apart from the artistic element that Aristotle emphasizes, techno also represents a tool for controlling and shaping the environment (at the same time). It seeks to monitor the complexity of the universe expressed by the Greeks in touché: chance. This is the strange duplication of technique: as an artistic method, it is indeed "a way to mold the world in line with our will," while at the same time it is creative as a scientific creation of objects. The mechanical processing of objects is concerned while technèto is limited to the aspect of "mechanism," which is generally considered a subheading of technical technics. The algorithm and the robot were among the most renowned and sophisticated types of machine-to-machine processing. The simulated worlds of cyberspace may be the latest stage of the techno, as art and technology are combined in a particular world disclosure. Computational productions would suggest that self-representations are controlled by the concept of enjoyment in cyberspace. This is because cybernetics bridges the gap between automated control and artificial intelligence. Lacan's great interest in the philosophy of cybernetics illustrates the use of the autonomous and touché concepts. Lacan is asked if the (unconscious) chain of signifier and signified works as an algorithm.

Cyborg Ontology

2.3.2 Truth of Cyberspace

When machines resemble expression or the workings of the meaningful, they create an imaginary magical universe. The notion of touché by Lacan points out that the symbol is not accountable for performances by themselves. Life is not the imagination of technology. And the psychoanalytic theory is not an idealist type. To not plummet into delusions—here Lacan's ontology is compatible with Kant's—appearances must have connection to the truth. This disruption is made by Lacan as the interaction with the true universe, the perfect world. Thus, a view of the true and the truth of cyberspace comes into being. So (American) troops come back traumatized and infected from the Gulf War (Gulf War Syndrome), despite the striking fact that electronic warfare has taken place. An accidental or useless word from another person might let the neurotic portraying an angry macho online realizes that its self-image is not independent. The obsessed man who continues to surf the Net to try to satisfy his wishes may eventually discover that "it's hard to locate."

2.3.3 A Historical Outline of the Real in Lacan's Work

What is Lacan's true meaning? To uncover the concept that plays a very important role in his (later) work and is so difficult to understand, one should look first at the history of that concept in his work. Ellie Ragland refers in her summary of the real to the recurrence of Lacan's lessons by Jacques-Alain Miller to Lacan's creation of the real. In the 1950s, Lacan called actual a brutal and pre-symbolic truth that returned to the same position "concretely and completely already." Lacan pointed out that it was the stars that were still in the same position in the first place (Lacan, 1988, 238). When he makes it clear that it is not entirely explainable that people see a Big Bear or Orion in some star constellations, he illuminates the dialectic between actual and symbolic. Lacan compares the actual with the Freudian thing during the second phase of his teaching in the 1960s and the early 1970s. Freud says that "das ding" is the part of the person or neighbor we are unable to comprehend.

2.3.4 The Real as the Object of Lost Gratifications

This brief analysis of the creation of the real in Lacan's search process is the ongoing construction of the real type. This is not so unexpected because the idea of the real is key in what the psychoanalytic theory is all about, and it is the status of the source of drive and therefore of desire. Man wishes to reverse the loss that constitutes his life as a wanting being. That means he needs to restore satisfaction at the drive stage.

2.3.5 Concept of Reality

For instance, in the fifth chapter of *The Four Foundational Principles of Psychoanalysis*, Lacan's theory of trauma is defined as such a limited concept and describes the theory of reality as trauma. Freud tells us of the unrememberable suffering, against which the subject builds protective organizations. It's not easily triggered or reminded of consequently interpreted. Another feature of the conception of the actual by Lacan is this aspect of resistance. "When the subject says" his tale, something latently works that rule and condenses this grammar. Touché animates techno: a dual relationship to the truth of security and revelation If this true plurality cannot be assimilated, so there is a dialectic between exploitation and dispossession. The vocabulary of English has a curious equivalent: to arrogate. It sheds light on the notion of arrogance, which we can now characterize as something actual (a real hero, a real writer, etc.). Around the same moment, it is still important to help the fantastic image that the real danger of disrupting it. Fantasy is a screen that defends against the intervention of the real too directly. The "missing thing" is crucial.

2.3.6 The Fantasy Interface as a Screen

Freud was taken up by the topic of memory and its inconsistencies in the years following his understanding of the value of fantasies in psychic life (through fantasies). In 1899 his observations on the protective mechanisms of some memories ("Screen Memories") were published. He argues that certain experiences mix elements of memory with elements of imagination—while they are mostly separated from their backgrounds. In Figure 17 they protect this topic against the remembrance of a traumatic event, but by focusing it on a defined and objective form, they articulate this childhood aspect that is so highly emotional. Since memories are not a flawless display device, Freud compared it with a children's toy called the Mystic Writing Pad in a later document.

2.3.7 The Screen as Principally Defensive: Phobia and its Computerized Treatment

The computer screen will function similarly as a definition against anxiety. The fetish and the phobic entity are like Lacan in that they are abstract and fictional replacements to a threat of some presence. Then one becomes much more interested in the case of Virtual Reality treatment of phobias. The usual psychological application in Virtual Reality is immersion therapy: the patient is increasingly exposed to the object or condition being dreaded. Research reveals that gradually the anxiety of phobic patients is reduced, as they get used to, for example, spiders and social situations as in Figure 2.12, and they are relaxed. This in vivo cognitive therapy is a mixture of cognitive-behavioral therapy, which enables people to think positively (about spiders), and behavioral therapy, which attempts to reassess the phobia sensitivity to a feared object via the unlearned sensory reaction (desensitizing the subject).

2.4 THE HYPOTHETICAL VIRTUAL ENVIRONMENT

Through discussing the underlying similarities of the communication and regulation of animals and machinery Norbert Wiener has changed the understanding of the human transmission of signals with his work *Cybernetic enhancements: Communications and manipulation of mammals and technology* (1948). Both humans and computers are cybernetic devices that collect signals through sensory organs or receptors

Cyborg Ontology 51

FIGURE 2.12 Fantasy interface with screen.

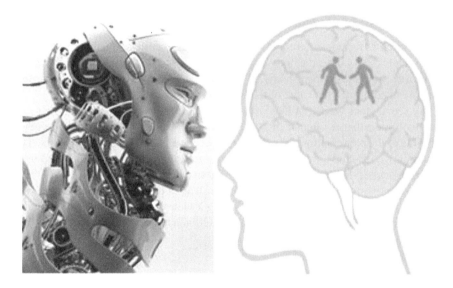

FIGURE 2.13 Human control cyborg.

from outside the environment and that control contact with the world through feedback bands. The operator ("steersman in Greek: cybernetics") of a car controls the engine via the gas pedal, just as the computers can be regulated via signals so that a system-world constancy or regularity can be achieved at homeostasis. The entropy regulation in Figure 2.13 of the cybernetic systems is a thermo-dynamic term that defines the inclination of an organic system to an increasingly chaotic environment. The explanation of cybernetic systems by Wiener was especially interesting in his attempt to remove the distinction between men and machines. Bruce Mellish's book The Fourth Discontinuity is defined in his cyborgology outlined as Gray, Mentor, and Figueroa-Carrier: human and machinery's co-evolution. "In Mazlish's story,"

West philosophical history is seen as a surmounting series of great, discontinuity, illusions, which posed four natural artificial distinctions, i.e., 1) human and cosmos (Copernicus overcome); 2) human-life (Darwin overcome); 3) human-unconscious (filled by Freud); and 4) human-induced human-life (Darwin overcome); Wherever we see the fourth interruption that dissolves, cyborgs flourish "There was a mistake." In Lacan's job the cyborg even prospers. He explains this perfectly in his 1955 seminar by his reply to Octave Mannoni, who is concerned that a computer could produce language and therefore no longer human:

> Don't be soft. Don't be soft. Do not say the computer is evil, and it overwhelms our lives. This is not the point. This computer is merely the progression of little 0's and 1's because it is completely resolved whether it is human—it does not. And that there is still the issue of whether a man is as human as anything in the way you perceive it.

There was a mistake.

2.4.1 Cyborg Information

Lacan addresses the contemporary matter of cyborgs' incarnation in his examination of cybernetics. How does the body connect to the cybernetic circuit and incorporate it? The dilemma is described clearly by Frank Bicocca, and it is linked to Descartes' mistake of "outliving" the value of the body and dividing it from the mind.

"We see another variant of Descartes" mistake from the perspective of the robot as a giant brain that is often seen in the 40s and 50s. This link was from mind to core. One of the conversations was contact between the person and the computer. The discussion was around an important artificial brain that was disembodied and used as either a peer, a slave, or a foe. We have only two incarnate conversations, a sterile link between Abstract Symbolic Generators, instead of the mind speaking via a body to another body. It is the emblem of exploiting a view of early artificial intelligence and not the placed incarnation of increased intelligence.

2.4.2 Cybernetics

The subconscious is considered by Lacan to be an independent cyborg network. The natural progression of signs via interactions of recurrence and free association. There is an (unconscious) mechanism that determines the meaning that will occur at any stage. When used "as the discourse of my circuit," the language of the unconscious operates. For instance, my dad's lecture on how I'm doomed to repeat his mistakes while building couches. Lacan spans this topic of determinism in his subsequent lecture on psychoanalysis and cybernetics, which I would attempt to demonstrate hinges on imagination. He utilizes cybernetic enhancements to elucidate the analyzing context in which the investigation is made possible to talk unintentionally (free association); this means that "he should deliberately approach the possibility as close as practicable." Yet this very speech shows a certain degree of ontology as shown by errors being repeated. Cybernetic enhancements think of a binary scheme of 1s and 0s (as a few other basic symbols and operators such as a, B, x, y, and +—will form a "system") that reduces language to its barest bones. The elements 0 and 1 can be a composition of a circuit that transmits a particular message. It can clarify whether the analysis once said something or repeated the same feature or behavior—to pull it into the area of psychoanalytic theory properly. Repetition may then prove that the element of fixation exists outside the "pure" codified text.

2.4.3 Principle of Simulated World

In this "imaginary" part, the incarnation component of the virtual subject is embedded. This is also the lesson of Sandy Stone's incarnation principle in simulated worlds. For example, she renders the libidinous aspect in sign communication very plain in her study of phone sex. In this purely verbal correspondence, the libidinal component of expectation of a certain picture or scenario includes the embodiment factor. The worker verbally codes and exemplifies gesticulation, presence, and neighborhood, often in no more than one word. The customer reconstructs the symbols and builds a complicated and dense communication picture. . . . In the modes not articulated in the token, the client mobilizes desires and current corporal codes: i.e., phone tokens are solely verbal. The client uses verbal tokens to create a multimodal object of desire with form, tactility, fragrance, etc. Meaningful knowledge is comprised of an embedded relationship with the subjects of our everyday life, instead of context-dependent symbols and laws. Lacan's study of cybernetics can also be associated with the contemporary discussion of the driving metaphor in ICT. The symbolism of conduct presupposes ideas as objects that can be placed into words and transmitted to the recipient through a conduit that extracts ideas from the words. It assumes that language exists critically and analytically in people.

2.4.4 Embodied Space

It is not a matter of open and shut that comprises space; there are different definitions of space. After the emergence of modern technology, there is a spatial perception of space, in which space is an unlimited expansion that includes everything, including the space on my bookshelf, which is an aspect of the universe that is outside of bodies—whether it is filled with books—the logically and ontologically separate component of this angular acceleration of the Newtonian physical space. This understanding is so well established in daily use that we generally consider it the primary sense of space through which all others come. Space is a container that includes anything without any relation to context. Leibnizian relationism in the contemporary understanding of space does not regard space as a complete and limitless substance, just as history culminates in Newtonian mechanics. The condition, distance, or connection between one body and another determines one thing; it has no fixed position in a scientific coordinating structure. Space is what all these areas have, according to Leibniz. It has no true truth. It has no reality. Leibnizian rationalism is intense according to George Berkeley's epistemological idealism. If there is no room without bodies, there is "just nothing" as such. Because Berkeley rejects the presence of objects beyond our experience, space derives from the projection of images that God fascinates upon our consciousness, rather than from connections between things.

Though we can believe the images. They're thought from God, coming from real things from outside ourselves. It probably leads to the conclusion that God is the environment: we simply experience "the Author of our being" in perceiving objects in space.

2.4.5 The Personality as Artificial Togetherness: The Mirrored Region

"Ego is a perfect ego simply because it's a picture. . . . The subject would still learn that this self-image is the very frame of its definitions, of its understanding of the universe—of the object."

We can see and see a branch in the branch. "All innovation is based on the ability to manipulate and double realities, on the heterogeneous of the real," which he believes to be the foundation of technology. The interchange of us with the "actual" entities, which we feel, virtualizes the real by its duplication. "The dialectic function is the foundation of the interactive since it still provides a second environment in various ways." Lévy emphasizes the competition between the ability of man to create mechanical objects and his ability to create a second world: no creative creation. Creativity itself is a component of (technology) knowledge. We still are a hypothetical duplication, but this medium is according to Lacan's inventiveness. In his renowned mirror hypothesis for maybe the first period, he establishes this constituting bond between the organism and its double. His key notion involves the recognition of the specular picture as a conceptual unity for the self.

2.4.6 The Aesthetics of Reflecting Official Numbers: The Impact of Cognition

The context of the linked connections between resistances and facilitation is the only lens through which ideas, emotions, aspects of the indigenous neurological network, and individual characteristics can be understood. All that can be an aspect in our inner vision is virtual, as is the impression that is created by movement by light rays inside a telescope, although we should assume that there are systems like the telescopes blades that throw pictures (these are in no way psychic beings themselves and they can never be accessed through our psychic perception). And by using this comparison, we can equate the repression of two mechanisms with the refraction as a beam of light enters a new concept.

2.4.7 Virtual World of Human

The ego expands the virtual world as an association of photographs. The alter ego is the perpetual partner of (ideal) pictures. Thus, we live not only where our own body is biologically but also where we perceive ourselves to be. This picture order parallels the working of the imago that Jung established in psychoanalytical theory. The imago paints how we connect to others: through the light of the different images, we view the other beings. Although Lacan's view of the concept of universal templates in Jung's theory is overwhelmingly negative, the picture often acts as a mechanism of insight. This is seen by his concept of the mirror point. "I am therefore guided to consider the reflective surface role as an imagery specific case that is intended to link the individual to its experience—or, as they suggest, between the inner world and the environment."

2.4.8 Pace: Projection of feelings to/from the Mucous Membrane

The envisioned role of the situation opens numerous possibilities. The environmental representation of the client differs from the "actual." The mechanism of naming that makes us "inhabit the world" thus has an initial physical expense. He mentions a well-debated phrase of Freud's in his debate over the association with a surface or imagination. "The ego is primarily a corporal ego; it isn't simply a surface being but itself the projection of a surface." It can also be considered a mental projection of the body surface. This Freudian conception of an ego, which comes about by identification with the mental representation of herself or body sensations, is upheld by Lacan's principle of Spiegel. The imaginary role is an empirical psychic fact.

2.4.9 Avatars: Engaging the Body in Space

The disappearance, embodiment, or manifestation of the goddess was in Hinduism avatars. The concept extends to shifting states in which someone works. Both online ("personal") and (ancient) "self" modes in Lacanian theory can be seen as "Avatars" (I think . . ., I think I perceive myself as . . ., I think I think I idealize myself as. . . .). We will experiment with these shapes, restructure them, and reconstruct them in the physical world (the realm of—abstract—representations). Therefore, at a psychoanalysis conference, Sherry Turkle should link the online person and the ego: each of them is substantially interactive, either built in the analytical space or in an online community role-playing virtual space. An avatar in a virtual environment will unify otherwise perceived trends as discordant and unsettling, much like Lacan's theory of the Spiegel phase's association with the virtual picture. I may formalize those traits that otherwise stay dark and mysterious by selecting an avatar (erotic, violent, animal-like, etc.).

2.4.10 Imagination: The Dual Binding of the Virtual Environment

The child's jubilee in the mirror stage where he sees his appearance is a recognition of his own body's power. Similarly, in some temperaments or states of mind, emotions (foreseen) of superiority or power, in general, are conveyed. Fixation is one of the definitive arrangements that man builds. He absorbs or interacts in the film as a virtual duplicate.

Therefore, vital insight into the formation of a biological, virtual environment that combines the new feeling with a new self-sense is provided by these experiences of fixation and hypnotization. The digital world is crucial to the captivating impact of the picture and exhilaration. "To achieve an ego phenomenon, fascination is utterly necessary. Since it is amused, the uncoordinated, incoherent diversity in primitive fragmentation unifies. Reflection is curiosity, too."

2.4.11 Cyborg Subjectivity

"Cyberbodies . . . a modern social imagination where a complicated mixture of image, technology, and innovation reconfigures the body." "In the end, the way you express yourself and your behavior determine emotional levels"—distinction between both the individual as the individual and the body as the representation represents an alienation from the senses. That's the lesson that Lacan's imaginary theory will draw. The imaginative ego maintains strong illusion and lures components, yet its impact is powerful. It "virtualizes" our direct feelings by making it an outcome of imagination. Therefore, someone might feel an incredibly painful slap on his side. It can also understand why a fakir may withstand a strong feeling like sleeping on a clock bed by renouncing the ego. This consciousness issue is highly significant for the interaction study as it is used for emotional difficulty. The results of human experience is represented here via pictures.

2.4.12 The Interminable Promises of the Cyborg: Among Insufficiency and Excellence

Marshall McLuhan compares the Narcissa image to the concept of man's mirror-like extension in his chapter in the *Media of Comprehension* called "The Gadget Lover: Narcissus as Narcotic." Where Lacanian improve shareholders that anything should McLuhan works as a mirror, using technologies or what he calls "character." The notion of awareness as the immediate relief of tension on the central nerves readily spreads to the root of the medium to communicate between the speech and the machinery." Based on medical science, McLuhan describes the root of media from the central nervous system as neural inadequacy to cope with all sensations arriving from outside the globe. The media are thus essential tools to synthesize the various sensations, in a Kantian-Lacanian manner. For example, McLuhan regards the wheels as an expansion of the foot as the product of "the weight of fresh burdens caused by written and monetary media exchange rating."

2.4.13 The Cyborg Abundance: The Challenge to Real Annihilation

There is potential exposure to mechanisms of thinking and feelings, especially in cyborg forms of communications, as a question of pictures (in large measure). Affected avatars point up the potential for a cyborg person to entirely manage his feelings and affects (as a cause of concern) "rationalizing" their presence when surface images of feelings are decisive for emotional exploration. This is the case, for example, about the topics of logical emotional therapy (RET) and the topics of Diestel's lung, which are covered in the second chapter, which determine feelings and values. Lacan points out that since the sense of self is a question of representations of the self, it is all that can hurt the walls of the mind.

2.4.14 Between Exploring and Dominating Space

It is difficult and tight that the ego lives in the virtual (subjective) room opened by a mirror image. Technology, self-replication, and the separation it presupposes are required prerequisites for contemplation and autonomy, on the one hand. On the other hand, the ego could reach that distance, guided by the will to reconcile with its picture. The ego's autonomous imperative and its unrealistic and selfish expectations remain in sequence. Lacan believed that the psychological reality "distracts in the vertigo of dominating the room the so-called "instinct of consciousness." Lacan points out that since personal identity is a question of representations of the self, it is all that can hurt the walls of the mind.

2.5 FANTASY AND SUBJECTIVATION: EXCESS PLEASURE GOOD THINGS INNOVATION

Trucial reproduction of the real, ideally by another reproduction medium, including photography, is the creators of existence itself in hyperrealism. The physical is deprotonated, an emblem of death. From medium to medium in a way, however, it is still strengthened by its death. For its purposes, the knowledge emerges, the hypersexualization of the lost object: the joy of negating and of its ritual annihilation no longer belong to representations. Dream and creativity are not any falseness and beyond, but the use of electronics also is paralleled by fetishistic activities in the real. The real is a hallucinating approximation to itself. Such a view says that inventions disapprove of the boundaries of everyday lives by opening an almost infinite number of scenarios. The "hallucinating" interpretation of reality in digital technology synchronizes man with the principle of enjoyment. They produce wonderful encounters as "engines of enjoyment," which transcend the natural limitations of the human state.

2.5.1 PRINCIPLE OF MENTAL DEVELOPMENT

"There is a reality not just such that we bump our faces it against wrong roads through which the working of the principle of enjoyment takes us."

Reality is not saved from pleasure itself. Therefore, Freud's concept of imagination (his "central use") is undermined by Lacan, drawing a strict distinction between the two concepts of psychological capacity. In Freud's "Two Principles of Mental Development" formulations, Lacan recognizes that there is such a strong difference. But for Lacan there is more to it than just presenting anything of need, even in a minor delusion, being someone who feels hungry about food. There is not just a need but also a (sexual) attraction to be involved in any hallucination.

> One might disagree in both cases, but in any hallucination, it's utterly crucial to map the dimension of meaning if we want to comprehend what the concept of enjoyment entails. The connotation of truth is hallucinated from the stage at which the subject wants. And if Freud compares the concept of truth with the rule of enjoyment, it is to the extent that reality is untextualized.

2.5.2 THE SICK AND TWISTED MEDIA PLEASURE: NOT THE DEED BUT THE SETTING

The unacceptable ideology of reality and the notion of enjoyment offer one certain authority over "physical actuality" amazing delight. Such pleasure is notable as a violator of the law; it is not an act or mode of conduct that contrasts directly with social truth standards. Leisure within the legislation means an organization of facts that already reflects our ability to discover ways of leisure. That is, my subjective experience included a perversion aspect, not because of night, beyond my screen.

2.5.3 The Vital Disavowal

Now that I've highlighted the twisted essence of dreams, disavowal must follow. The basic operation of immorality is disavowal. In the world of pleasure—fetishism, castration, division, and perversity—the various words that I used so far come together. About the castration complex, Freud elucidates on denunciation. The word "disavowal" is used to cover the acknowledgment of traumatic perception by a particular defense system.

2.5.4 Interactivity and Technological Belief

We are converted to a (virtual) subject by our original replacement by the (speaking) meaningful individual. An interacting subject is a human being: the signifier substitutes him.

Thus, what I do not believe is placed within a sort of "absolute truth" of my inner self but by meaning (this I, therefore, live only as a reference point to which the means are referring; it is not a substance). Somehow, "think" signifies to me, acts to me, and "believes" to me (I believe by performing or executing the rituals of belief, etc.). This is the first understanding of the "belief" of another. We inevitably accept as linguistic topics that forms of representation are meaningful. Without representations, there is no "absolute truth." Such "truth" may have been his defining point and what Lacan calls the true one.

2.5.5 Narrative Visual Digital Media

A change from expressing narrative society to a culture based on the sounds and visuals of digital media may be identified by Lacanian thinking. Lacan examines the massive impact of mass media in his seminary's last section *Psychology four basic principles* (1964).

> It's possible that the so-called "mainstream media" is to blame for the disturbing nature of our relationship in this day and age. Considering the increasing penetration of science into our area of study, it's possible that the reference sheds light on all of them. Planetaries, even stratospheric zed, our engine—the look whose perpetual nature is not less expressive—since it's not so much our eyesight that's needed by so many spectacles, by so many imaginations, as it is our gaze that's awakened. Planetaries, even stratosphere zed, our engine.

2.5.6 The Sinkhole: The Mark's Splendor

The impossible (truly) to conclude an analysis means that latent wishes cannot be brought entirely to the stage of expression (meaning). A (libidinal) rest of

Cyborg Ontology   59

meaninglessness is not part of the story order. Thus, the study ends with the suggestion of the imagination and fact fusion into "not sensitive officials."

> It is, therefore, wrong to assume, as has been stated, that perception is accessible to any significance under the presumption that it only concerns the relation of a concept with meaning and hence an uncontrollable link. There is little sense of understanding. This would be to admit that all the explanations, obviously ludicrous, are possible for those who are against the character of ambiguity in their analytical analysis. It does not mean that interpreting itself is nonsense that I said that the result of interpretation is that a kernel, a kernel, isolated from the matter, Freud's own words, of nonsense uses. Perception . . . will produce an irreducible meaning."

2.5.7 THE PARTIAL OBJECTS AND THE CUT

Second, through reading Lacan's description of the object in his sixth seminar conducted in 1958 and 1959, we will elucidate the change from "phallic" theories to "theory of subjects." During the seminar desire and its perception, which Lacan aims at describing as fantasy (S.6, 18), three sources of a subject are given. He provides us an example. By describing man at a material exchange level as an animal, Lacan presents his first order, pregenital objects, as the organism with two different points of entry: one from which he enters, and one through which he enters. Pre-genital target plays a significant role in the dream, precisely this trait.

2.5.8 IMAGINATION SUBJECTIVIZED: SELF-IDENTIFICATION AS AN OBJECTIVE SUBJECTS: THE TOPIC AS AN ENTITY

Subjectivation and imagination go hand in hand, for imagination is the (non-pathological) need to give the transcendence framework of desire a real (empirical, "material," real, particular) substance. Notice that fetishization is a necessity to desire enjoyment but not pathology in Freud's observations. The object of Lacan reveals the absence of a natural privilege to find our "real wish" for patriarchal or phallic subjectivation. Another process of subjectivation is also given way to the object. The phallus is not (necessarily) the privileged meaning must be remembered. The argument is that this question of subjectivation still exists. Because of the conditional role it plays in imagining, the subject cannot avoid developing into an entity. The wonderful object does not have a set or stables double, thus the item cannot "release" it; instead, it is a manufactured thing, which means that it was produced, developed, or intended.

The historical and community history of birth, social heritage, and status is no longer unchangeable because of the 1960s. The most significant cultural movement of the day involves the idea of identity as an "eligible" building, which can be personalized from a variety of photographs, instances, roles, and all kinds of works (re)presentations on photographs, fashion, culture, films, news, and advertising industries. This means that the society is not directly linked to the neighborhood or village. It becomes a question of common desires and markings like clothes, hair (style), inclination to sexuality or ethics, and any other form of paraphernalia: tattooing. Brands and trademarks transform marketing products into identities that can express the charism of the consumer by buying them.

The physical form and the abode of sin constructing one's identity requires a collaborative effort between idiosyncratic, "material" and social models. Examples are transmitted, supplied, and manufactured more and more by the newspapers. We are no doubt the product of a grand (phallic) story that creates our name, but we produce stories. By interacting with very personal "stuff" (mementos, souvenirs, snapshots, and any kind of trivia—and in the light of digital media we can mention personal blogs, baby albums, gadgets, etch.), we have very personal material for our life stories. The "idiosyncratic things" are at the sin home stage. "The eye and ear passion: Again." The fantastic connection between body and environment is a fundamental type of intentionality, which lies at the limit of more conventional subjective types such as the (self-)individual, Scott Bukatman's Identity Terminal (1993) "Terminal identity" is at the corpus and computer terminal interface and is inscribed on an ancient selfhood surface. While his vision of hyper individuality and the modern socialization of the Internet need not be supported, his research does make clear that, in our interface with new technology, we need to search for new ways of subjectivation. These innovations build our world, a high-tech environment that focuses on the ear and the eye. Today experience moves primarily in an AVC, which "has become the mirrors to find an identity" in certain technical pictures.

2.5.9 Symbols, Facts, Engagement

Symbolic engagement places a cap on the subject's mere virtuality. If a symbolic (ego-ideal) regulation frames the want, the imaginary expectation that everything is possible is "forbidden." So, I don't need to eat as much as possible if I stick to, cling to, or promote the ideal of modesty. "Symbolic law" says, too, that we must have the implications of what we have to say or do (the rule implies the impossible). Thus, the virtual is not equivalent to the possible in a Lacanian context. Santomean's concept increasingly refers to identity as a matter of physical connection to items (with the inscription in or on the body, enjoyment of action, and use of things). We associate with consumer objects, devices, exclusive signs, and lifestyle tokens. Biometrics is another interesting development in technology. The biometric technologies are producing digital representations of the distinctive physical properties of a person, such as a hand form, fingerprint, iris, sound, face, or hand or retina vessels in the blood.

2.6 CONCLUSION

The objective of the chapter is to emphasize some of Donna Haraway's principal philosophical origins, a topic that is seldom discussed and often overlooked in the literature. However, it allows one not only to grasp her thoughts more easily but to place them even within the intellectual movement. It indicates in general that Haraway's thoughts come from a Cartesian and Heideggerian tradition, in which, indirectly, it stems from the devastation by Heidegger of metaphysical anthropocentrism to criticize the distinctions between humans, animals, and machines that Descartes insists upon in his Method Discourse. However, I propose that Haraway is implicitly affected by Heidegger's criticism of the binary logic, which is part of Descartes' anthropocentric. I argue first that their support for Heidegger's critical lectures by Jacques Derrida, Bruno Latour, and Giorgio Agamben leads him to the Heidegger suggestion that to overcome that logic calls for re-questioning what it means to be, instead.

REFERENCES

[1] Guattari, F. (1993). Machnic heterogenesis. In V. A. Conley (Ed.), *Rethinking technologies* (pp. 13–27). Minneapolis/London: University of Minneapolis Press.
[2] Gurevich, J. (1999). *The Jouissance of the other and the prohibition of incest: A Lacanian perspective*. Available on-line at http://dept.english.upenn.edu/~ov/1.3/jfg/other.html.
[3] Harpold, T. (1994). Conclusions. In G. P. Landow (Ed.), *Hyper/text/theory* (pp. 192–210). Baltimore: Johns Hopkins University Press.
[4] Havelock, E. (1963). *Preface to Plato*. Cambridge, MA: Harvard University Press.
[5] Hayles, K. (1999). *How we became posthuman. Virtual bodies in cybernetics, literature, and informatics*. Chicago/London: University of Chicago Press.
[6] Hegel, G. W. F. (1977). *Phenomenology of spirit*. Oxford: Oxford University Press.
[7] Heidegger, M. (1997). *The question concerning technology and other essays*. New York: Harper Torchbooks.
[8] Heim, M. (1993). *The metaphysics of virtual reality*. New York/Oxford: Oxford University Press.
[9] Heim, M. (1998). Virtual reality and the tea ceremony. In J. Beckmann (Ed.), *The virtual dimension* (pp. 157–177). New York: Princeton Architectural Press.
[10] Kennedy, B. (2000). Introduction to part six: Cyberbodies. In D. Bell & B. Kennedy (Eds.), *The cybercultures reader* (pp. 471–476). London/New York: Routledge.

3 Cyborg Communication

3.1 ENLIGHTENMENT CYBORG

In the 17th and 18th centuries, there was a transformation from an early machine culture to an early plant cultural culture; from spiritualized theories of there as the uncommon air of heaven and of the human spirit to subtle yet material flows subject to material properties and physiological and behavioral ether theories; from me, the rare air of age is the important component of capitalist and trademark culture; mind a mechanism like material. The telescope and magnification helped to change the understanding of humans and their connections to one another [1].

Once represented as ideal spheres inside perfectly concentrated celestial spheres, the planets were now seen as imperfect entities that journey through mathematically represented punctuation marks. A plethora of minute bodies in a drop of water was revealed in the microscope for the first time.

Micrographic or other physiological representations of Minute Bodies Made by Magnification Glasses started with a characterization of point and plane proto-cyborg-theorist Robert Hooke. What was the shortest immutable symmetry component—the optimal and optimum dot—or the perfect geometrical feature with a length but not width or thickness—may now be seen as skewed and scarred bodies with dimensions representing a needle's point and a written duration and the edge of a razor. It was no longer possible to imagine what was invisible as effective and useful or to see literally possible futures [2].

The assumption of Hooke was that the mechanisms of God were flawless on this magnification stage, whilst man's artifice rose significantly. Both, however, were material rendered accessible to expose previously unimagined alien frontiers at an insignificantly small size. Modern search technology, along with new chemical and mechanical expertise, was intended as a body sense for a multitude of infinitesimal mechanical bodies.

Hooke argued that human intelligence has been limited to the shortcomings of vision but that these "sicknesses" could be reversed by "instrumental" and "artificial organs" added to the natural. In recent years, he finished by inventing optical lenses that have already been achieved. That Earth itself, which yes, beneath our feet, reveals something very different to us, and we now look at a whole number of Creatures in every little molecule of its substance, like we were. Capable of calculating themselves in the whole universe. With the modern "mechanic intelligence," he wrote, "we can possibly discern all hidden works of Nature just as we do others, the productions of Art that have been designed by Human Wit, and are managed by Wheels and Engines and Springs" [3].

To comprehend the mechanism of the cosmos—and to perfect those produced by humans, the spread of restricted human senses for perceiving material bodies was a crucial element. Another was experiments and math showing what could not be perceived immediately. The microscope has shown that the universe has a different material dimension. Philosophers now tried to experimentally show the smallest particles of matter, bringing existence and movement to the universe through chemistry, optics, and later in the 18th century developments in the fields of magnetism and electricity.

Later the electricity discovery and its consequences on organic bodies forced the body mechanics and animation discourse to change the language in Figure 3.1. As a symbol of a groundswell, Samuel Taylor Coleridge could stand against the mechanistic worldview, which then opposed and accepted Hartley's mechanical physiology as the human body persisted in its material processes to investigate and clarify. In his unique way, the horror of a modern mechanistic mode of human imagination, which usurps God's natural order and threatens human spirituality, was dreamed of by Mary Godwin Shelley and William Blake. The annual Croonean lectures of the Royal Society, which were prepared by the legacy of Dr. Croune on his death in 1684, offer an instructive account of the evolving minds and research into the mechanics of muscle action. The invention of the input system James Watt would have become established as the administrator was also seen in the long centuries following. The developments both led in different ways to a human-machine dialogue marked by doubt and hope and attributed to an increasing body of medical, metaphysical, and

FIGURE 3.1 Enlightenment cyborg.

technological literature, gradually characterizing the artistic forces of God or nature and humanity in the same way [4].

This research concentrates on the works of the Franco-materialist philosopher but mainly on the works of English restoration and coronation authors, who have contributed importantly in the collective memory of modern medicine and information technology as well as in linguistic and literary studies. The English materialist (often simply anti-Cartesian) human consciousness interpretations and will I shall examine here had not the same right winged and aggressive objectives as the radical philosophical in their modern-day version.

The Royal Societies members whose activities lead to improvement in the English tradition of human machinery were more conservative than La Mettrie in their claims on the soul but a dialog can still be observed in a variety of genres from the methodological and empirical analysis of Isaac Newton and the mechanistic approach to evolutionary biology during the era. Locke, Hartley, and Swift, Pope, Arbuthnot, Sternen, Coleridge, Mary Shelley's ideologies and poems. Ontology is thus less important in this history than physical mechanics, iatrochemistry, or anatomy and the expression of poetry and fictional studies [4].

All these documents tell the tale that, as from the late 17th century, Westerners had been carting a single iteration of the "Cartesian ego," a spirit that was distinct from the substance of the body, a ghost in the machine.

The clear signs that the mental and physical function as mechanics never wane, but the soul issue was becoming ever more incidental to medical discourses that were intended to comprehend the physical phänomens of bodies. In eighteenth and nineteenth centuries, the research and experimental presentations by Royal Society in philosophical or literary philosophies, but in physics and medical science, the immaterial and eternal spirit—which always followed Descartes—has become a practice. The medical and technical interference of human organisms that produce cyborgs does not constitute a legacy of substance dualism but is strongly monistic where human nature is deeply physical and incarnated. The presentation of Willis in the animal and human engines of a fleshly type of intelligence and an embodied spirit/consciousness marks the beginning of a marked rejection of cognitive and physical cartesian dualism in the physiological sciences. The man-machine of Descartes with an immaterial spirit in the middle of the brain in the spine is just a part of a long tradition, even though it dominates contemporaneous cyborg fiction. In the anime film *Ghost in the Shell* (modeling Masamune Shirow on Manga Series), Mamoru Oshii's pictures of cyborg give an example: the memory and personality of Motoko Husanagi's cyborg are implanted into a totally artificial cybernetic body along with the fragment of brain left of her original body. In order to get the minds of human sufferers across the data network, Husanagi and fellow operators investigated an electronic sentient actor capable of infiltrating. She later finally leaves her body and remains in the cellular connection. The anime *Ghost in Shell Innocence* (2004), which is clearly modeled on the cyborg thinker Donna Haraway, refers to the relationships between early modern thought and current theory.

The cyborg investigator Batou commented to Coroner Haraway and Togusa police officer, "Descartes could not discern man from machine, animate from the inanimate. Extracted a doll after her, he was called Francine by his beloved five-year-old daughter. He's pointed at it. That's what they're saying, at least." Then the hacker, Kim, told Batou and Togusa: "Dolls confront us with the fear of being reduced to mere structures and materials." In other words, anxiety that all people belong to the vacuum essentially. In addition, science has caused this fear in trying to unlock the secret of creation.

3.2 THE COMMUNICATION WITH ANATOMY BRAIN

Willis carefully clarified that he ought not to view his mechanical understanding of the soul as anything resembling atheism. Archbishop Scheldon's letter *The Anatomy of the Brain* is as protective of a defense as the later one, "I am unaware," wrote the great work I carry out. It has long been considered to be a certain mystery of Atheism, as though we might derogate from religion some reason that we provide to philosophy.

But really, the name of philosophy is too abused, through this discovery, and considering machine made so much of a profit from him, he should not recognize the wheels, curious frames, settling together, small pins and all that a clock makes and provides for, can be known and evaluated exactly if he had discovered the course of time, orders of the months, or movements in the sea and other things of that sort [5].

His brain research and explanation are like one reading the "Pandects of Nature there is surely no page that does not show the author and his authority, his goodness, his trust, and his wisdom." He describes it with craft. Moreover, according to him, no right judge can blame him for studying "these natural roles because certain atheistic people can be created of them." At the completion of his directions on the dissection of the brain, Willis defended his many animal dissections as a way to uncover and show God's "wonderful artifice," whereby "the strongest and most invincible argument could be opposed to the most twisted atheist by showing the finger and divine workmanship of the Deity." In both books, Willis' restraint is wise, for he has dangerously nearly undermined the body monarchy, the everlasting and intangible spirit, as he portrays a human being as a machine and as a flesh. Indeed Willis' text takes and redefines the sun-king concept of William Harvey—not, as one would expect, for a rational and immortal soul to reign from the head over the whole body but for a sensitive soul that spans all over the body: "when a Head contains the head and power of the sensitive person, it shall be taken as a digestive system of a Luminary, as a Sun or as a Star," he wrote, "The brain and cerebellum then bring the animal spirits out to 'irradiate the respiratory system' from a 'double light.' " When Harvey actually endorsed a "normal" body system, instead of the heart, Willis strengthened this image of a representation of the spirits of animals spread over the whole of the body and lit by two sources of light.

Several Cartesian grounds were expressly refused. Whether spirits produced in the ventricles are concentrated in the hypothesis and travel through the hole nerves was shown to be otherwise by Willis' dissections. He dismissively states that the "moderns" estimated the blood vessels to be "so disgusting, that they affirmed that they were all sinks to work out the exceedingly important thing." In reality, "almost all Anatomists, later-ages, have given this more inner chamber of the brain this vile office of a Jakes or a drain."

Willis says that the neurons do not seem hollow, but instead they have "solid smooth bodies," whereby humor or spirits might slip into the fine cloth or sponge-like liquids. He established that the nerves "are not boring" as are the veins and arteries by microscopic analysis of their anatomy, so he specified that, instead, the subtle humor that carries the spirits is transferred in their surface morphology. Willis thus reiterates, "since animal spirits do not have a clear cavity in their brains to expand; nor do they need the like." Therefore, the ventricles "should be assigned to an office other than this." Willis also specifies in this case that animal spirits pass through the structures of the nerves as "so many lucid particles" or "so many diverse beams of light." As to the pineal gland, "We can hardly think it is the seat of the Soul," he argues, because it is present not only in humans but even in beasts of four feet and also in fish and oats. Willis suggested, most importantly, that people be endowed with several intangible and corporeally unified souls. "Human governance" was made up of two distinct human alms, which Willis defined in the two discourses concerning the soul of brutes as logical and responsive. Even the logical soul appears in the body in the previous text instead of being in the head: the bodily soul "joys at the body, at once, intimately bound" with the intangible and eternal rational soul that inhabits the body, living in its bosom [6].

Willis' careful change in stressing the cosmic consciousness of the spirit was a bold proposal as the prevailing Christian religion believed that the eternal soul was not tangible. Following the reconstruction, the Anglican Church has been granted much of its ancient rights and restored the strong links with the Crown: it seems that the Church's soul authority may not be wise, but that Willis was, in reality, a faithful individual. The continued comparison all throughout anatomy and recurring in two discourses, however, reflected only lay politics: the materialized soul gave the actual people in the corporate government those powers. It is difficult to tell whether his commitment to the Archbishop of Canterbury and the defense of "I'm reading only God's book" was to prevent two philosophies from objections to his political organization analogies or the discoveries of Willis concerning human anatomy and personality. Another concept that would ultimately rule medicine and science was Willis' corporeal souls: consciousness, mind, or personality is the product and dominated by the material nature of the brain and the interactions between some of the fine particulates that we call molecules. Willis' works are made so significant today by their position, content, and practices, when our medical technology community considers thoughts, remembers, and thinks primarily as bodily mechanisms rather than the metaphysical, which makes them too sensitive to release and understand Willis' time [7].

Willis apparently had more intention than the soul system in the anatomy to describe the frameworks. He claimed that the rational, eternal, or intangible soul is special to man and claimed that it is composed of the essential or "flamy" blood soul and a delicate or "light" mind soul as well as a third genital soul, which is common to the human being and the animal. From the two other bodies, hidden in the semen bodily components waiting "into a different Vital Fire."

There's a new guy. The responsive body soul is composed of a "very subtle small body" and resides in the brain, nerves, fiber, and blood and interacts within a circuit that is characterized both by movement of troops and trade in the brain and the body from and to different regions. The body is portrayed by Willis as a country populated by the animal spirits' blood vessels (atoms or particles). The brain is a city or

Cyborg Communication

a palace, medulla oblongata (MO) a street, the Cerebel is an Oblong Marrow. The brain is a city (cerebellum) The reserves are tense, the free and local towns. Willis defines the brain as a "'Castle' or 'Metropolis,' separated into many towers." The metropolis creates the forces of nature and transfers them to the "callous body" by a technological mechanism of blood distillation (corpus callosum, the thick band of nerve fibres that connects the cerebral hemispheres). Incidentally, however, the source of thoughts and acts is not governed exclusively by the ruler or by the rational soul. It is also regulated by the delicate flow of trade and communication between the animal spirits, the subtle small bodies which populate and revive the political human body. Essentially ancient creatures are present and vital in the brain and the nervous system. These vibrations are not sent to them by the brain or the sensible soul but rather are not conditional on the brain for circulation, contrary to Harvey's early picture of the heart as a ruler. Willis opposed, as some philosophers had supposed, the idea that the brain beats like the heart, "We reject completely that it has a constant Systole and Diastole," he wrote of this "vulgar opinion" [8].

In the battle of influence between both the emperor and his non-passive people, the central picture here, perhaps not unsurprisingly considering the chaos of Willis (who had enrolled in the emperor's forces in Oxford in 1645), is organized distribution, where each "atom" has a specific "office."

The words trade and distribution used here by Willis appear both to refer to the concept of "communication" entities and to a trade meeting. The "callous body" is a spacious area from which [ancient creatures] meet and live as in an emporium or a public mart in an open and accessible location, from which it functions as an opportunity. "They are created and positioned for each Faculty's needs."

From the brain and the skull, the MO extends forth into all of the nerve sections of the body, as seen in Figure 3.2. This pathway is sometimes referred to "as the Kingdoms Highways" or "as the high road to the kingdom." They're here "ordained

FIGURE 3.2 Transfer consciousness in cyborg.

as they were by sequence and commands" and then flow to the nervous, "while they are exercising the locomotive faculty, or within their fountain, while they perform the actions of meaning or even the frustrations of sensuous matters." In this medullary trunk, there have been several points of diversion and lesser directions where the creatures appointed for those particular offices go apart: such that all the personalities that are on such journeys do not meet each other and disrupt each other's workplaces.

3.3 EXTENSION OF SOUL COMMUNICATION

Willis had ignored the deep insight in the matter and structures of the soul in his description of the brain and had instead explained quite briefly that the nature of the sensitive soul is the "order and ordination" of animal spirits. He had said, "But the analysis of that soul and its force demands a peculiar tone which, afterwards, (God willing) is our intention, the faculty depends on the 'gesticulations' of these 'atoms or subtle particles.'"

Willis' two speeches on the soul of brutes were released almost a decade after the *Anatomy*. It was an important treatise, written in Oxford, London, Amsterdam, Lyon, and Cologna and translated into the English language in 1683, which went through eight editions between 1672 and 1683.

The two speeches were both a cautious and a visionary explanation of the logical spirit, which ultimately shared its strength and power with the reason, responsive or bodily soul. This doesn't mean Willis was in favor of more influence and voice in the political arena for everyone. In reality, he had been calling for the already defined hierarchical organization to preface his work. In the foreground is a warning for the Most Knowledgeable and Venerable. The comparison of political bodies is present in *Two Discourses*, but it is not so long-lasting as the earlier *Anatomy of the Brain* [9].

In this section, Willis struggled to explain carefully the materiality and strength of the lower spirit. By plainly talking to the body soul of animals (which is humanly the lower soul) he apparently tried not to attribute any connection of material or mechanical behavior to his divine monarch in the brain. However, the abilities of the two people that rule the body are actually necessary, a more complicated relationship that sometimes compromises the logical soul's authority and autonomy. Willis' preface, which contrasts the logical soul of men with that of the corporeal person and animal, carefully distinguishes mechanistic action, meaning and finite intelligence from the rational and eternal aspect of humanity from all living bodies.

In addition, "If any man asserts that the most subtle, and wholly etheral substance which is of immaterial importance in the vital economies or governments is a lenient disposition of inanimate structures, let him recollect to be indulgent with me, if by accident. I call it stuff, though the Prerogatives of Reason remain much lower than them." He hopes that his thesis would not offend an orthodox conception of the sacred soul of man. "We are hoping it'll be orthodox in the first place and that he's agreeable to a decent life, and a pious institution. Thence the Wars and Streams between our two Appetites, or between flesch and spirit, both morally and theologically instilling, are also physically known." He said, "He does not worry that his work would be 'censured for Pernicious or heroic."

Willis was not very careful in his repeated reassurances that the artificial body was governed by an immaterial spirit. The lower soul of man, Willis claims, is intangible and subject to manipulation, "co-extended" to the body, and to some degree independent of central nervous systems within the different sections of the body. This can be seen by the cutting into bits of worms, felt, or vipers because they keep moving and are sensitive when picked [10].

Both by the blood of the systemic circulation and the nerve jus of the brain and its annexes, the material spirit has provided the body with its own innate movement, resilience, and also reasoning. Willis thus reiterates, "since animal spirits do not have a clear cavity in their brains to expand; nor do they need the like." Therefore, the ventricles "should be assigned to an office other than this." Willis also specifies in this case that animal spirits pass through the structures of the nerves as "so many lucid particles" or "so many diverse beams of light." The editor Elie Luzak was convicted by the authorities of religion. Man's computer inspired great animosity and countless published criticisms and refutations (one of them by Luzak himself). It was banned from Leyden and burnt on a public square in The Hague; the sale of the book was prohibited in France.

The logical soul is only the delicate soul applied to the reflections of thoughts and reasonings, may "La Mettrie proclaim!"

This argument that in the human race the main ingredient is an inherent quality of the matter itself and not a separate soul was strongly a political claim: if the individual consciousness is excusably mechanical, residing within the material body instead of the divine steering officer, it would be possible to exclude all appeals to the body government on the basis of the religious hierarchy.

Nevertheless, Volis said that two distinct souls existed and affirmed that the body's two "imperial chairs," the former in the blood—providing the body with its nutrition (the essential soul)—were mechanical comprehension. The second one is responsible for unintentional roles of anxious juice (the animal soul). The logical soul lost control over the body's "people." Therefore, no wonder that the lectures of Willis, which were published more than fifty years before the scandalous work of La Mettrie, stressed over and through that the man-machine was the work of God and was moved by God: "I profess the great God as the only Worker," so he was not able to impress strength, Powers and faculties fitted for the offices of the first mover and fortunately to the place of almost everything. In reality, this is a highly nervous introduction to a tract that shows that certain aspects of information, intellect and character are the result of material processes of the body both in the human and the animal.

Although Willis traditionally reaffirms the mechanisms of animal administration and reaffirms the activities of the *Anatomy*, mechanical imagery takes on more importance in the *Two Discourses*.

The animal spirits, including pipes and other machines, are here shut up with both an impartial virtue of sensory ability and an effective one of "loco-motive" control and "Spasmodic." Willis points out that while human beings are an energy-driven machine similar to human industry, they are nevertheless superior machines due to the excellent workmanship they have created, which has given them "surplus" properties to the particles in their souls [11].

These characteristics, such as heat, air, and light, are mostly "energy" in mechanical things in humankind. He wrote that, like that, we might assume that from the very beginning the Great Worker, to the Chief Creator, made out of their Particles the most highly alive, and even subtlest Souls of living beings He was active; to which also he gave the most beautifully worked, above the workmanship and artificialism of any other machine, a more surnatural virtue and performance.

Matter, say Willis, is not "substantial passive," even though it is normal to assume the matter is not moved by something else. Atoms, he said, are alive and self-moving—the smallest components of matter and therefore of the body soul—a material soul may move and perceive.

The human being is significantly more than an energy engine: it represents images and impressions. "The Brute Soul is solid as a machine in meaning and movement," Willis writes, as the "rational soul" presides, as it were, looks at the images and impressions embodied by the delicate soul, and performs acts of reason, judgment, and will according to its conceptions and ideas. It is quite clear, though, that ancient creatures are the "Authors of the Function of the Animal Function," says Willis. They are not comparable with flowing spirits like turpentine or champagne but more like "beams of light" from the fire of the blood: "for the impressions of every tangible object and the air of every auditory thing are present as light; therefore, the animal spirits obtain the impressed images of them." While metaphor is extracted from the printer, Willis does not define any grossly physical stamping or imprint of the brain in this respect: it is a premise on the basis of the observable properties of optics, that light will transmit images or sensations from the sender to the target. He'll clarify, using the cryptic camera analogy,

> I say no one can ask. For visual perceptions to describe 'who has seen all the objects of the hemisphere, has enabled him to cast a hole into the dark part of the continent, then on the Paper, as if made by the Artist's Pencil: so, the Spirits cannot, except as Rays of Lights, frame the images or forms of things without being confused or obscured, so as to illustrate the perceptions of a visual vision [12].

The part of the soul that is the perceptual part of the mind, then, has "the fastest correspondence with everything, but also with the different bits" of the soul that chooses those appetites or deeds due to "impressions there got."

Willis' is a clearly anti-Cartesian image of man-machine, and indeed this bodily soul inhabits a mechanical body, depicts the world to the consciousness as a mirror and reflects the light that shares a camera that is obscure, extends in space and animates a "energetic" force such as ashes of light an Ethernet machine that functions independently of a human host. By studying Willis' development of the nervous system as a communications system that receives and communicates experiences or news, this proposal is more convincing. In two speeches Willis reinforces the picture of soldiers executing commands for the movements of the body. In *Anatomy*, however, animal spirits were represented as troops "starting to flow" through the nerve passages into the fibers interwoven with the muscles and then set into the muscles as in a watch-tower. Ten years later he described the spirits as spread throughout *Two Discourses*. The brain's nervous store, though, was positioned in the machine to transmit "news" into and out of the brain. The spirits are not going from one end of the course to the other, as is generally thought, in this later text; it is rather the material that flows through the circuit:

> "It is instantaneously transmitted to all parties inside the Head any pulse or stroke inflicted from without onto every member or sensitive body. If a perception or force tends out of the brain, throw the neurons into the moving parts, there is motion; yet it is guided externally into the working components." Brain, Sense. . . . the soul is extended, throw the whole, with a certain continuity, the objects, viz. spirits adjacent to each other are set as an army in Array, do not move from their stations . . . and if they are set in battel array or on the watch, they carry out the commands made in the brain, they are nearly immovable and effect motion, and de Willis explains that the sensitive spirit is in fact in the head, as is in the nervous system, "only in all his power and exercises sustainably, and so expanded, and in a body way." The rational soul is "purely spiritual," however, he recognizes, it is especially difficult to distinguish between the material body and the immaterial rational souls when we think that intelligence, the decision-making and the willingness to act in such acts of animals can be perceived. What mechanism does the soul of the rabbits come by their intelligence when it is 'much inferior and material?' For we know, Willis remarks, that beasts can even 'end up choosing acts that appear or a certain deliberation to flow from the Council.

He comes to the conclusion that animals don't have an immutable logical soul. Conclusions, acts of logical thinking, are not directed by a rational, eternal spirit ruler but by a tangible soul distributed through the body.

"We do not envy this 'rude and plain tone' with which the wind blooms into a pipe, but we are surprised by the sophistication and harmony of the musical organs whose influence exaggerates both 'the matter of the instruments and the hand of the musician who strikes them.'" A computer can be designed without guidance of a musician to play a complex harmony, but it can play only what is recommended. Again, this little governor (or musician directing) is the difference between man, beast, and machine. Though brute acts could match the harmonic tunes spontaneously provided by a water organ, Willis concludes that they "are almost always willing," i.e., preprogrammed and free-will. In reality, he says, that more imperfect animals are inscribed in their natures or souls. "They don't behave as much as they do."

The administration of people is another matter, where peace cannot be coded. Since "the Spiritual Soul is the absolute focus of the ultimate reality, whose act, perfection, completion and structure, is itself, the rational soul often affects its form, and human acts will easily govern the entire human being" sometimes they are willing to dissent; sometimes they are willing to disbelieve and travel rather than wage civil wars. The main cause of the civil war is concupiscence, the need for worldly gratification. The physical body spirit, "professional and non with the rest of the body" and often distributed further than the body, could spread "his diseases" to the body and corrupt the latter through "failings and defects." This effect on the intellectual soul force as a neurological disease extends across the body and is characterized by Willis. Then follows an appreciation of physicality, "Wars and struggles with all our Inclinations or between the flesh and the Spirit," he writes, "for, see and approve of the best things and obey the worst things." In particular, these shortcomings happen in us because "the corporeal soul, adhering to the Flesh, is inclining man to sensual pleasures, while the rational soul, being assisted by professional ethics or divine favours, is inviting him to the common courtesy and the functions of piety," he said. What Willis geopolitically suggests is that the king is fallible and conditioned by material processes, although religiously superior. Unlike Descartes, a dropsy guy who drinks and damages himself like he's like an underdog time, you won't find me drinking and hurting myself in Meditation sessions on the First Philosophy. In other words, the movements are induced by an inferior corporal process, even though there is no mind at hand—the contact with the material, the corporeal soul could literally pervert Willis' logical soul. "Knowing POWER," therefore, is the two things—of mind and body—of the brain (the handmaid of the rational soul).

There is still a possible war in the "Empire of the Mind," says Willis, in which reason could surrender to corporeal spirit through its "proper power." Is it unintentional that Willis remembers specifically that "it does not follow," that the "rational soul" is promulgated ex translations, that is to say, the rational soul is not producing a rational soul, "as often as often the sound is exactly like the father, as regards wit, personality, naivety and other faculties of animals"? Willis comments this right before announcing that he "would accept now the conflicts and distortions that often occur in relation to their power." Could he elaborate on Charles II's continuing power fighting with Parliament among increasing parliamentary factions? What else may cause political groups to emerge from speculation about Charles's illegitimate son and probable successor to the throne, other than the lawful succession of Charles's brothers, the catholic James II? Although Willis has previously claimed that the rational soul completely rules the body, in the *Second Discourse,* he admits that the corporeal soul does not readily comply to the rationality in everything, not so in things sought, as in things to be learned. Affinity to the flesh and its benefit and preservation is bound entirely. However, as the bottom bodily soul is accountable for the nutritional tissues, it will engage in leisure and become "sour to reason," and, furthermore, Willis avers: "the lower soul, being tired of the Other"s yoke, frees itself of its Bonds, affecting its bonds if an opportunity arises."

3.4 LITERACY COMMUNICATIONS IN MECHANICAL OPERATION

Where Willis is perfect for an entirely liberal 18th century society of disorderly economic growth, trade, benefit and rising national prosperity, a spiritual hierarchy of monarch, trade, and the army interacting by regulated and moderate transactions on an organized march, modern hierarchies will characterize government and business both in the government and in the body of politics. Increased representations of the delicate, anxious group of higher order people are more affected by a disorderly and unregulated printechnic of journals, pamphlets, booklets, and broadsheets in a trade than by the rule of the king and the aristocracy. Throughout the 18th century, moral authority claims rely on the distinction between the mechanical intelligence of the lower order and the nervous mind of the higher order, that means the gross, gross, bodily nature of refined responsiveness.

Furthermore, it was a worrying decline in the status and earnings of the learned member to imagine academic work as industrial, manual, and mechanical work. In the years between 1640 and 1660, England saw a collapse of the regime of surveillance by governments and the church authority, the establishment and consolidation of a Republican commonwealth by the Client state of Cromwell impressed content in the country than ever before. The pamphlets, journals, newsletters, and textbooks for political, religious, and business purposes contained a misunderstanding of both traditional and unorthodox thoughts and opinions. Parliament first had press access, resulting in what Sheila Lambert called "the first new mass propaganda machine."

Whilst the progressive works written by Quakers, Ranters, Levellers, and those with an unparalleled independence can have a dangerous consequence, you are the head of the Church and the Supreme of the law. Shall the body rule the head? According to Atkins, the fine (spiritual) publishing art was "depraved" by the imprinters forming a body—"Consortia," incorpora—with the traders, booksellers, and type founding companies: "they're forgetting the Head, and by degree, they're kicking against their life-giving power."

The hierarchy of Atkins is explicit: mere technology usurped what a greater metaphysical "art" should be. It should be remembered that Atkins' brochure was actually less a heartfelt applause for the King than an effort to assert his own place in the legal dispute over a number of lucrative patents with the London Stationers' Business. Purely on the basis of his own objectives, Atkins metaphors, however, are a distinct conflation in the early modern print community between material (commercial) motifs and material (mechanical) spirit.

The act was only permitted to expire in 1694, which was too expensive to carry out, since the Licensing Law in 1662 had already been created to restore State regulation of printing that had lapsed after the civil war and its replacement in 1685. The number and the need to register books with Stationers' Company was increased, due to the fact that it had been limited to twenty in London. Authorship is a trade in which less time was spent on studying and patronage of the subject matter was determined by production and demand—by booksellers and the public buyer. Before printing, the censorship of print content was no longer a barrier to correspondence. Printing was a free and unparalleled business expansion. In England in the 1620s, there had been about 6,000 names, in the 1710s 21,000, and in the 1790s it was over 56,000.

How did the rationalist trade flow combine with an intelligence and brilliant idealized trade? As print culture grew through increasing marketing and professionalization in the late 17th and 18th centuries, around the same time, the modern literary market was a force for democracy and rationality, a turbulent traffic of ungoverned thoughts, which both destabilized and undermined the existing letters, power, faith, and civility.

As may be anticipated, an increasingly growing number of heterodox publications were published in England in the 18th century. As Martin Battestin pointed out, the literary turmoil also provoked Laurence Starne at the end of the century to defend biblical Orthodoxy by writers of fiction from Jonathan Swift who, "acknowledging the risk that Hobbes and the deists raised," "strike," against these recent newspapers, like materialists and mechanic theories, as "dangerous to the moral and spiritual welfare not only of their subscribers, but to the very social structure themselves."

A threat to traditionalist opponents was the modern printing society because of the lack of centralized power. In his study *Licencing Enjoyment: The Advancement of New Reading in Britain*, 1684–1750, William Warners finds a remarkably similarity to that of the 18th-century print culture: "the print marketplace in which Pamela appears," he suggested "can reasonably be defined as an open system" (meaning open-source creation and dissemination of computer programs that make "text" available). Moreover, "I'm not saying" it's random, unpredictable, and not restrictive. Neither is it a personality entirety that maintains a certain fundamental character by the kind of homeostasis that, for instance, many biological processes are characteristic of. The print industry is a producing and consuming environment in which the meaning of its texts cannot be regulated or guaranteed by anybody. The body is open to earthquakes and disturbances. The competition is influenced by everyone who can print his writing whether centralized censorship or qualification is missing. There are no universally accepted guidelines here and noticeably few restrictions on what can be said and written. The fact that print culture tended to flourish and survive, amid various upheavals and fluxes, shows an inclination towards a kind of tight, yet continuously troubled, equilibrium. One might wonder if the print culture showed immune function:

> However, Warner's example both emphasizes the essential and enduring metaphorical connection between body control structures and the administration of human speech, and reminding us how important this change was during the eighteenth century in media culture.

Some authors draw a shaky link between unchecked man-machine technology and underdeveloped visual culture. The connection between mind physiology and democracy in human speech was made clear in time. Period. As we saw with the complaint by Richard Atkins, a "mechanical" world view was one of the most troubling facets of the unruly communications of consumerist culture, since the unity of the body politic, as well as the solid social order founded on church and monarchy, will not only be endangered by uneven telecommunications for benefit. "We believe, and even we are honest people, just as we are alive or brave; everything depends on how our computer is built." In the more provocative publishing environment of French writers at the mid-18th century, La Mettrie wrote: "words, grammar, rules, science, the arts" are all an education system stamping ideas in our minds. "Man has been trained as a dog," says La Mettrie, "in the same way that he was a dog poet." A man-machine without the unified divine sovereign Hobbes and Descartes actually destabilized the old Hobbesian hierarchy of the mind over the muscles of the empire. "This is therefore just a mess of words and all windy writing, which is inflating our haughty pedants' ball-like minds," La Mettrie writes, "Figures that make up all the remnants of the head from which we distinguish things and remember them."

A mechanical power with major socio-political consequences is a matter of relative equity between writers and porters. "According to La Mettrie's slightest theory of movement, animated bodies will have what they need to move, experience, perceive, confess, and act in one word within the intellectual and emotional spheres that rely on them." His demonstration is a list of tests that the body or portions of the body, isolated from the brain and central nervous system, may transfer or reanimate themselves or polyps, which are also spontaneously regenerated into tiny fragments.

"To beyond doubt, I have provided more evidence than enough to show that any small body of fibers or part of the ordered organisms move by its own theory whose behavior, including voluntary movements, does not rely on nerves."

In La Mettrie, intelligence and authority are no longer an object of the conventional influence if there is no divine master in the brain who controls the movement of our organisms.

Whether awareness could be the product of this physical conformation would be a physical mark on the mind, judgement, cause, recollection, all those things that humanize us. So, while the picture of man-machine meant a certain equality of ability that negated a previous hierarchy of aristocracy and birthright-dependent rank, it was also now a hierarchy that was based on education and institutional dominance. A sensitivity of the nervous system in England and on the continent will strengthen an older political hierarchy of the bodies and minds of the United Kingdom.

In *Tale of a Wrote Tub* on the fundamental development of men and women, in which Swift made both his criticism of the arrogant faith in new science in explaining human identification and the involvement of higher understanding and Christianity as a mere mechanical method, Swift recalls the aims of satire early into the century: "the various and gross exploitation of spirituality and teaching." He produced educational media with the premise that the average person may benefit from absorbing "a small, compact, universal system of all to be known, believed, or conceived, or performed in daily existence." You inhale the elixir via your nostrils, where it "is going to dissipate through the brain in 14 minutes and you are immediately going to experience in your mind," the condensed knowledge in its entirety. Take the set of educated texts, distill them, and make everything that is volatiles vanish. The "technician" was often linked to the materialist roles of the workplace in lesser order for Swift. Recall that although the (now extinct) critical had been a restaurateur of "Ancient Learning" and a virtuous hero, for example, his Grubstreet Hack comments. The "real contemporary" critique "is a kind of mechanic set up with stock and instruments for the trade as little as a tailor; and . . . the utensils and capacities are a lot of comparison." Swift's Hack argues that "the good attributes of a man would cost him all that is good, maybe a less buy will be conceived of but an undisputed deal." The contemporary critical is utterly opposite to true intellect: "before you can begin a true critic." In contrast to the learned intellect, the combination of mechanical laws that characterize the mind and the mechanical laborer represented the elite's lengthy apprehension over management political agency correspondence. It may have satirized the materialist tradition implicit in Hobbes' works, but Swift definitely combined with Hobbes in confirming a need for a pyramid of mind and muscles in the political world.

For example, the "great grand Princes" who are evoking Hobbes' monstrous mechanical man in an attempt to raise a powerful army and unstoppable navy, inspired by the influence of pent-up sexual spirits on the brain. The same ghosts, the narrator of Swift says, "who will conquer a kingdom in their higher growth, coming down on a fistula." The satires of Swift did not only exacerbate certain reductive structures that characterized and reduced spiritual, characteristic, and moral complexities to mechanisms. He also stressed his irony about "new" cultural shifts promoted by ill-trained follies that threatened his marks of power as an Anglican clergyman and as a highly accomplished reader and writer. He was sensational in caricaturizing of the influence of a mechanistic spirit on the body politics.

Swift made probably his most disgraceful comment on the mechanical structure of the spirit seen in Figure 3.3 on the mechanical operation of the Spirit, added to *A Tale of a Tub*: it is a sharp invective at once against mechanical soul representations and a satire that Swift views as a bogus faith, asserted by the "mechanic,"

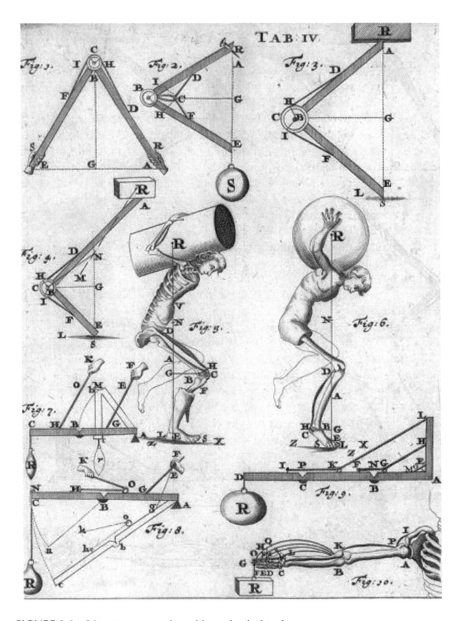

FIGURE 3.3 Literate conversation with mechanical tools.

unauthorized preachers of the "Enthusiasts," who demanded that the Holy Spirit be enlightened directly and immediately. The components at play in this work defend the status of the true souls, spirits, and voices of the United Kingdom against the laughing depth of its untrained "handbook" tradesmen. Through digesting "the theologic polysyllables and mystical holy written texts, added and digested using these methods and already associated mechanical operations," he writes, one may acquire a "skilled share of inner light"—that then becomes liquid by the rubbing of the nose, trying to peddle, spitting, or blowing or snuffing of the congregation.

Swift is a double sound with his most barbed connection with real divine enlightenment and artificial physicality, combining the human spirit (this is the religious being) with sperm (which creates the material being). This kind of worship offers three ways for the writer to ejaculate the soul or to carry it into the Sphere of Matter:

> That's why, according to the British workers, I mean, it is through this mechanical function of the Spirit that I am dealing with. I am going to provide the reader with the results of several wise comments, tracking the complete path of this trade as near as I can.

Swift's satire is geared toward his view of the deluded and self-obsessed blindness of fanatics, the concepts of which he reduces to the crassest materials and physiologies of imagination. But even the lower class, the tradesmen, the workman, and the machinists, who had little power to express their own kind of spirituality to Swift, are a haughty ridicule. The characterization of the human being as mechanical was often linked to natural law and could equalize all humanity's ability to think, argue, decide, and convey ideas. The emphasis is not just on how we get across and whom we owe, the intellect, and whom we ought. Awareness controls our behavior but also its relative importance.

Swift's works mock an overwhelmingly secular and entrepreneurial society that is concerned not with moral gain but with material and economic benefits. Modernity exemplifies Swift's acceptance of intellectual and moral facilities and crass hedonism. In his Peri Baths, Alexander Pope embraced the image of process and materialism in a contemporary thinker several years later: Or, Martinis Scribblers, in his treatise on the art of descending into poetry (1728). The ironic formula from Pope produces an epic poem.

3.5 MAN-MACHINE AND INTELLECTUAL ELECTRICITY

The idea that electronic text acts as a prosthesis or extends the mind has been used as a metaphor and as a basis for real forecasts of societal change and connected identities for quite some time. However, Chris-Hables Grey and Mark Driscoll argue that new cybertechnology will make the body an obsolete; destroy subjectivity; generate new worlds and universes; transform the economic and political. However, the transformative impact of "cybernous" or social technology on the human personality continues to be explored. The mitigation strategies given by Bolter and Richard Grusin found that "we use media to define personal and cultural identities. When such media (movie, broadcast, artificial intelligence) simultaneously become technological and social analogs to our identity, both we are the focus of today's media and object."

"This is not a modern phenomenon," they comment. The historical relations of mind, body, and technologies, as I suggested, ought to be examined more closely not only as the "dissolution of the cartesian ego," as so many scholars have as regards representation, detailed description, and imagination. The "real contemporary" critique "is a kind of mechanic set up with stock and instruments for the trade as little as a tailor; and . . . the utensils and capacities are a lot of comparison." Extremist opinions on political physical fitness or cancer. Our exchangeability of the analogy of the activities of the human spirit and the electrical/electrical gadget seems to imply that the soul, the personality or the human being, will flee entirely from the network of the body into the unsupervised computational pattern of connections.

As a "tower of Babel," our "modern electrical technology which stretches our senses and nerves into a global environment" could improve the language's "unity and differentiation." The machine promises, in short, a Southern Baptist state of Universal Vision and Union by science. Marshall McLuhan has enthusiastic religious polysyllables. The next obvious progression would seem . . . the languages for a universal celestial consciousness to be bypassed.

McLuhan's concepts of man-machine linkage with modern global awareness have re-emerged in the 1980s and 1990s, realizing the web as a networking device not just for strategic use by the U.S. government but also for civilians of different countries worldwide. In several cases it was a deeper view, expressed in particular by scientists and critics confronted by networking technologies or technologies usurping the written word's conventional authority.

Conversely, La Mettrie banned it from 1747, and then, as we have shown, in the ever more secular research in human physiology, this "steersman" or pilot in the brain.

La Mettrie had protested against Descartes' interpretation of the man-machine as a clock or the flutist car of Jacques de Vaucanson (1745). He said that man looks like a "Vessels without a pilot in the midst of a sea, which has capability to sail by its design, but is dictated by tidal currents" or a "true automaton," a system subject to the most constant need, pulled by impudent irrationality like a vessel by Water's "currents."

However, La Mettrie had fully and aggressively changed this view by the time he wrote *L'homme machine*. "Let us finish proudly," he writes, "the man is a machine and there's only one changed substance in the cosmos."

Diderot also suggested the effect that human identification would be exclusively substance at the end of the 18th century. His Rêve D'Alember metaphorically described the central nervous system as a delicate "network" consisting of long legs of the spider stretched around the body that sent signals to the spider in the brain. "The discrepancy between both the psychologist as well as the machine the method," he claims, is called clavichord: "the philosophical instrument is sensitive, being performer and instrument at one time."

The body controls itself: Diderot implies that our keys can hit themselves, or objects in the nature surrounding us might hit them, but a separate musician is not required. (As observed by Sergio Moravia, the work of Diderot in turn was inspired by Théophile Bordeu, who regarded the body not as a centralized dictatorship but as an "organ union."

This play, which was an offense against morals, was unpublished by Diderot. Not surprising are the two of Diderot's plays *Julie L'Espinasse* and *Jean le Rond*

D'Alembert. There was only one copy of the manuscript, and the characters ordered to destroy it. The interrelationship between both the human mind as a distributed (and material) network and the frameworks intended to control it or govern it was a significant and controversial problem: the censorship of the written works of La Mettrie, or of Diderot's imprisonment for his *Lettre sur les aveugles* (1749). During this period, different cultural interpretations and representations were derived from the mechanism of Man's paradigm in the modern medical sciences, used in tandem to strengthen a hierarchy of mind and spirit in ruling the nation's material muscles. The modern medical and philosophical methods for the body led to the new interpretation of the body politics and stemmed from the spiritual, political, and aesthetic roles.

Electricity was not animated by boring mechanisms, and neither were the ideas of an interface between human mechanics, communications and individual and social bodies, since the mechanism explaining susceptibility for Bordeu and Diderot was also the discovery and theorization of the "electric fluid" to have an explanation for the human being of the material type of government. For example, in his *Philosophic Essays*, in England Richard Lovett believed that "evolution's time and cost overruns and chief agent, or what . . . rules nature" was an electromagnetic fluid—special relativity subtle mechanism or ether. Lovett clarified that there was reality in Newton's *Principia* describing a "a very subtle spirit . . . by its strength and motion . . . the whole feeling is animated and animal bodies are moving in the direction of the Will." Lovett clarified that in fact matter is infused with movement. Along with other mechanical causes, motion was discovered to take place with the effect immediately of the first cause. Before this discovery, we appeared to believe that all substance was inert, whereas now we have found that we were seriously mistaken and that there are primary and secondary parts of matter.

It is not surprising inference that there is movement in the smallest parts of matter, what La Mettrie has termed "normal oscillation" in all of the elements of corporal fibers, today as atomic structures are learned in secondary school: Cyborg is an integrated man and machine.

The physician Erasmus Darwin proposed in his *Zoonomy, or the Organic Life Laws* (first published in 1794) that spirit was a substance of the same class as electricity. He wrote, "I beg to be understood," in preventive appeasement,

> that I do no longer want to contest words and that I am ready to enable gravity forces to construct matter of a finer sort, a peculiar attraction and energy, magnetism and, indeed, the spirit of animation; and to accept, together with Saint Paul and the Malebranch, that the supreme cause is immaterial only to all movements, which is God.

Darwin had a materialist version of the body but not purely mechanical. Previously "idly inventive" thinkers, he demonstrated in the opening section of his preface, had made the error of "trying by mechanisms and chemistry to justify life's laws; they presumed that the body was a hydraulic system, the fluids were moving through a sequence of chemical transformations, ignoring that imagination was its central feature." However, although he stressed the errors of mechanistic philosophies, Darwin's ongoing matter nonetheless played a decisive role in the definition of the individual. Spirit thus had the ability of moving, whereas matter was a kind of transmitter and receiver: "there should be two essentials, one of which can be called spirit and the other materials, supposedly both of existence. The former has the power to start or to generate motion, and the latter has the power to obtain and to pass it on." Darwin has categorized movement into two classes, supplementary or mechanical, and primary movement in three classifications: gravity, chemistry, and life. The fourth division was "the alleged etheric fluids of magnetization, energy, heat which light" and was not widely recognized to be classified sufficiently. The animal spirits are no longer subtle corporations in the political architecture but Darwin's "animation spirit," a kind of mechanics that he described as

> the medulling portion of the brain, the longitudinal material, the nerves, sensory and muscle bodies, but also at the present time the living concept, or movement spirit which lives throughout the whole body and which is not known to the sensations but in its consequences.

The brain secreted a substance for motion and touch, which may be "subtler than the electric aura."

The electric fluid did not require tubes (it was seen that nerves were strong even before Darwin) and because electric shocks could activate a disabled extremity, he reasoned that brain was probably like the electrical fluids. The head and back, the primary areas of the senses, are a communication between both sensory and muscular organs. In this materialistic statement the governor or singer was no longer present, but the mechanical words of contact were interpersonally used: "external sensory organs are the coverings of immediate sensory organs, mechanically tailored to receipt or transmit peculiar or qualitative bodies."

Then Dawkins could only eradicate the soul from his medical discourse on this body: "through the terms spirit of animation or sensory force, I only mean the animal life that humanity possesses, in some degree, except with vegetables, and leave the contemplation for those who treat revelation to the everlasting part of us, who are the focus of faith." This was considered an immaterial agent by philosophers "External sensory organs are the coverings of immediate sensory organs, mechanically tailored to receipt or transmit peculiar or qualitative bodies."

Then Dawkins could only eradicate the soul from his medical discourse on this body: "through the terms spirit of animation or sensory force, I only mean the animal life that humanity possesses, in some degree, except with vegetables, and leave the contemplation for those who treat revelation to the everlasting part of us, who are the focus of faith." What was considered an immaterial agent by philosophers The subtle essence, the ether and not the mechanism of divine rule over the limited single body as the mind, was now expanded through and out of a material nature such as magnetism, fire, light, and electricity through the intimate nervous system of the animate bodies. The body governor becomes a global soul that comprises the world and that, by nature's laws, leads motion, sensation, and will. The body science left behind the problem of whether the governor owed any loyalty to any eternal and intangible authority quite uncomfortably.

The description of *Intellectual Electricity, Novum Organum of Vision, and Grand Mystic Secret* says:

> this state of affairs has the consequences of an initial change in human nature itself: a revolution that, and the more the spirit pulsates, its android devices, and flames, the

longer it is twisted and crossed; a demonstration of its current existence of man's sense or faculty; The relationship here between physical and metaphysical world is clear; the medium of thinking is made visible; the intuition seems to be moving forward and politics is assuming a kind of Magic Intimation.

Quoting Newton, David Hartley and Erasmus Darwin, among other things as an indication of their philosophy, Belcher suggested, while well known, that a maze in which the deepest thinkers lost the following question constitutes the complexity of the human machine:

Can electricity (and therefore lightning) not communicate thoughts in certain ways? Figure 3.4. Seen.

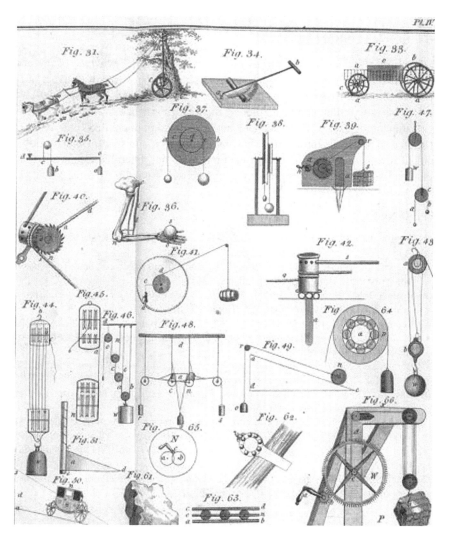

FIGURE 3.4 Man machine interaction.

1. By electrical and nervous fluid identity. 1.
2. Whether they are similar or not, their behavior on nervous fluids. . . .
5. By its subtle sensory contact.
6. Through their exclusive soul access.
7. By being the spirit vehicle.

Whether fanciful Belcher's understanding of corps-to-body interactions appear, Willis' and Newton's early speculations of the existence of ether in the nervous system would not be illlogical if an intellectual electricity permeated the body and continued with the natural environment. Besides that, the cybernetic assumptions appear to emerge from these tropes of the cohesion of social communication with material expansion.

Too much influence of the mind of the human being, particularly consciousness as the mechanical system steering, governor or pilot, has been linked to it.

3.6 CONCLUSION

Allison Muri provides in *The Emancipation Cyborg* cultural evidence of the presence of mechanically driven or cyber human beings in the works of thinkers of the 17th and 18th centuries, from historical, metaphysical, science, and medical texts. Muri shows how Enlightenment explores "man-macht" theories, which demonstrate the relation among scientific philosophy and social and political thinking, inextricably linked to ideas of reproduction, law, human liberty, and spirit. Democratic and political revolutions of the late 20th century, including liberalism, feminism, and even populism, thus are not unusual in using cyborgs as a politized topic.

REFERENCES

[1] Bataille, G. (1985). *Vision of excess: Selected writings 1927–2939*. (A. Stoekl, Trans.). Minneapolis: University of Minnesota.
[2] Bataille, G. (1988). *Inner experience*. (L. A. Boldt, Trans.). Albany: State University of New York Press. (Original work published 1954).
[3] Beniger, J. R. (1986). *The control revolution: Technological and economic origins of the information society*. Cambridge, MA: Harvard University Press.
[4] Bennett, J. (1993). Primate vision and alter-tales. In J. Bennett & W. Chaloupka (Eds.), *In the nature of things: Language, politics, and the environment* (pp. 250–266). Minneapolis: University of Minnesota Press.
[5] Bernardi, D. L. (1998). *Star Trek and history: Race-ing toward a white future*. New Brunswick, NJ: Rutgers University Press.
[6] Biro, M. (1994). The new man as cyborg: Figures of technology in Weimar visual culture. *New German Critique*, 62, 71–110.
[7] Boyd, K. G. (1996). Cyborgs in utopia: The problem of radical difference in Star Trek: The next generation. In T. Harrison, S. Projansky, K. A. Ono, & E. R. Helford (Eds.), *Enterprising zones: Critical positions on Star Trek* (pp. 95–114). Boulder, CO: Westview.
[8] Braidotti, R. (1994). *Nomadic subjects: Embodiment and sexual difference in contemporary feminist theory*. New York: Columbia University Press.

[9] Brasher, B. E. (1996). Thoughts on the status of the cyborg: On technological socialization and its link to the religious function of popular culture. *Journal of the American Academy of Religion*, 64(4), 809–830.
[10] Brown, R. (1996). Marilyn Monroe reading Ulysses: Goddess or post-cultural cyborgs? In R. B. Kershner (Ed.), *Joyce and popular culture* (pp. 170–179). Gainesville: University of Florida Press.
[11] Bukatman, S. (1997). *Blade runner*. London: BFI.
[12] Burke, K. (1966). *Language as symbolic action: Essays on life, literature, and method*. Berkeley: University of California Press.

4 Evolution of Woman Cyborg

4.1 THE INTELLECTUAL FEMALE CYBORG

In 1961, Anne McCaffrey's collection of short stories "The Ship That Sang" was one of the first human experiences of the 20th century as a logical intellect Cartesian driving a mechanical body. Helva XH-834 is the female brain of a titanium Intergalactic Scout ship implanted in the middle shaft. Their "extended" neuronal and audio, visual and sensory abilities are bound to and grafted into the hull of the ship to be the ship, to control the brain and to turn the ship as its natural body will be to complicate the Cartesian logical machine's stereotypical picture clearly as mind-body dualism; one of Helva's early tasks is to provide the planet Nekkar with 300,000 fertilized Ovans. Here the self-directed human-machine reflects women's preference of reproduction and empowerment. At the culmination, Helva reveals to her human scout, Kira, that she is not interested in procreating her own offspring from the genetic material contained in her ancestors, as Kira intends to do herself.

> No,' said Helva sharply [1]. 'After all,' chuckled the ship, 'there are not so many female people,' and Helva used the term boldly and knew that she had gone from girlhood to a woman's estate as surely as any of the mobile sisters 'who birth 110,000 babies at a time.'

This means that not only the embryos are released but even the delicate load is released. The modern freedom from "normal" reproduction finally appears to represent the female cyborg in either the feminist therapy in reproductive discourses, the feministic cyber philosophy like Haraway's "I would choose cyborg rather than goddess," or the fetishized images of the techno-female in film. As a man, a female cyborg invites a Cartesian understanding of mind-body dualism in contemporary literature and film. Contrary to a male cyborg, it has been constantly characterized and problematized by its reproductive and sexual capabilities. Helva is a remarkable portrayal in film and literature of women's human robots without gendered human body or avatar. The more common is the female-gendered cyborg in Mamoru Oshia's anime, Shell Ghost (1991). It propagates, if by no other means, only by "normal" means, narrow-waisted, broad-broken, attractive, dangerous, and logical (that is, pure consciousness), in this instance transferring from organic brain and brain stem to mechanically built. The fictional cartoon cyborg normally represents one of the magnetic sides of a continuum range from heroism to evil, but in any case, it interacts with a technology that is similarly dangerous. The female cyborg was described as a misogynist portrayal by Claudia Springer as the "fetishised phallic wife" [2].

Evolution of Woman Cyborg 85

It is a warning against both the amoral search for technological advancement that cyborg eve 8 of Duncan Gibbins' film *Eve of Destruction* (1991), such as a Maria duplication in Fritz Lang's movie, *Metropolis* (1927). As Springer states, Eve and Maria each represent an overtly erotic, seductive, and dangerous combination of technology and women's bodies. The Department of Defence has placed in her womb the atom bomb, Eve 8 (suggestion to the ancient history of the decline of man by consuming the fruit of the tree of knowledge) A techno-fall problem, a bad thing [3].

The cyborg female is constantly placed under a challenge to Christian ideals in particular. In Metropole, the false Maria, the Whore of Babylon, instigates pleasure and longing for the town's children and encourages the Metropolis workforce to a violent uprising that nearly causes their children to drown. The fetish version of the feminine machine visible in the Machine-Maria emphasizes the feminine robot or cyborg in comparison, as a symbol of malicious techno-sexual impulses, to the actual Maria, who takes charge of the children of Metropolis, who directs the staff in the prayer, and whose name invokes the Mother of Christ, the Virgin Mary. The two representations of Springer are divided by half a century, but perhaps their simultaneity of techno-lust and technological trepidation represents the most typical characteristics of fictional female cyborgs. Speaking technically, female cyborgs don't need mammaries, but they have a huge and lovely pair for apparent motives in the fantasy universe of cinema.

The razor-girl Molly reaffirms the motive of the attractive and dangerous techno-female in William Gibson's work. Indeed, *Necromancer* starts with this woman with retrievable razonrs in her finger nailbeds running through her existence. Decidedly nonsubstantial, Molly (remembering "moll" or prostitution) forms close relationships with Case but leaves Case at the end of a book; her only uncompromising response is: "hey, it's all right, but it takes my edge off my game, already paid the bill. No,' said Helva sharply [1]. 'After all,' chuckled the ship, 'there are not so many female people,' and Helva used the term boldly and knew that she had gone from girlhood to a woman's estate as surely as any of the mobile sisters "who birth 110,000 babies at a time." The Terminatrix is another variant of the fetishized female cyborg, which is obviously evoking the dominatrix, in Jonathan Mostav's film, *Terminator 3: Rise of Machines* (2003). The Terminatrix sheathed into skin-tight red latex was again clearly determined to eradicate John Connor at the brink of Judgment Day, who was spared only with the newly updated male T850 unit. She is romanticized and specifically unfertile again. If it is ultimately on the "right" or the "evil" side, the fictional female cyborg is a fantastic, possibly uncontrollable threat to the male hero [4].

Cyborg 2 (1993), head by Michael Schroeder, stars Casella "Cash" Reese as a cyborg, a "speak condition, top of the line" cyborg that is not only a master of martial arts but also a complete psycho-emotional program with 100,000 mg of unreleased bio-explosive chemical implanted into the organism.

The film, which was set in 2074, opens with a printed story that says: the very first episode showed some men who witness a live example of a female cyborg test unit and security contractors at a technology stand to watch this same experience through the window. The first scene shows a wall scale video in the boardroom.

A well-doubled, blonde cyborg kitten straddles a pornographic peepshow at first and happily fucks an equally splendid male cyborg under bright blue strobe lights. The objective of the presentation is to demonstrate the capacity of the "Osaka" (Cash Reese). The demo was fed to the circulatory system by a bio-explosive and was connected directly to a microbeam flamethrower. When all cyborgs are happy, the techniques continue to clock and the woman throws her head back and bursts at the time of what seems to be an orgasm. Both cyborgs have been spread around the walls and the curtains, in a feurishing (complete goodness of humans in developmental lifespan) chaos of the spread of body parts and fluids. Accordingly, the cyborg operation is "assertively successful . . . the latest in counterespionage technologies in a humanoid robotics application," states the delegate of research and development of Pin Wheel Robotics Inc. The remainder of the film involves the rescue of Cash from this awful destiny. It is anticipated that we sympathized with Jolie, a self-aware cyborg figure, but film representations of the seductive hazards of technical intelligence still play Eve's ancient history—the combined charm of the attractive woman/ prostitute and prohibited or prostituted awareness—as humanity's future destructor. And, as in all this modernism myth, mankind is finally rescued only with a hero's courage (here, with the aid of a rogue warrior called Mercy, "Colson" Colt Ricks) [5].

Likewise, the "impressive combat capabilities" of Gibson Rickenbacker in Albert Pyun's previous response film *Cyborg* (1989). The woman defends and releases the cyborg Pearl Prophet, who bears the details necessary to save mankind from a fatal plague (Jean Claud Van Damme). "It is weird," Pearl says, "but I think he's the only antidote to the universe when she actually appears." In *The Matrix* (1999), Case of Necrömer, even the Terminator of *Termination 3*, all bear this burden as savior personalities who are eventually the fighters, the heroes whose abilities go beyond even the most dangerous and strong sexy female cyborg protagonists, of courage, of intellect, toughness, and braveness.

Although in *Innocence: Ghost in the Shell 2* (2004), Mamoru Oshii subverts and challenges these representations in a certain degree, it may not be sufficient to emphasize that men and authors have continued to uphold the assumptions. By comparison, the campy film *Teknolust* (2002), composed, directed, and made by Lynn Hershman-Leeson, is deemed to be a fearful "one human, one part of the science" spectrum of the cyborg. *Teknolust* is the story she developed with her own DNA of Rosetta Stone, a biogeneticist, and "Mother and Her sister" to create three Self-Replicating Automatons. The SRAs are designed to remain invincible, but they need Y infusions to survive: the selection of sperm from the beautiful Ruby consists of the sensuality of suspected men who are consequently infected with a harmful self-replicating virus that renders them sterile, powerless, and branded with a rabid look. The cyborgs, though, are just human and this story ends happy with all the men and six of the major characters in love and in bed, rather than annihilation and ruin. Hershman's mocking Leeson's treatments is a rare criticism of *Teknolust* in popular fiction and film, reflecting a mistrust of the non-reproductive sexually engaging female and pursuing the organic development of artificial material.

Cyborgs are not knocked up in the movies. The womb is constantly and discouragingly in vitro as a reproductive organ. the womb machine in literature and movie

Evolution of Woman Cyborg   87

is both in significance and in horror, autonomous and extraneous to the "natural' body." *Alien* (1989), for example, Ridley Scott's film, depicts the human spacecraft as "Mother," with completely autonomous descendants as machines inside herself. Womb is a monstrous image of the mechanical or biological reproduction while the alien spacecraft is full of genital holes and glowing passages.

Mary Shelley's film *Frankenstein* (1994) by Kenneth Branagh portrays the technoproduction by overworked images of monstrous industrial equipment, hooks and sprockets, the seething, testicular womb hanging from the ceiling to release his organic sperm of galvanic electric aggregates into a series of oversized pipes and channels; a copper "womb" reminiscent of a coffin or "Robert Fulton" "Collected from pregnant women and absorbed by a fully formed pilot/spring holding its breath till the right time to accelerate." In *The Matrix* the sinister "plant" mechanically reproduces and recycles humans in individual womb-pads such as transparent amniotic bags, incongruously attached to a mechanized metal framework, whereas the cyborg Trinity, dangerous, lethal, and sexually desirable, is a grotesque and repulsive system that creates human bodies as "batteries" for the machine, that recycles dead material within a closed system. The "matrix" here is an uncanny womb that circumvents all "conceptual" logic and procreation before Neo is resurrected to spit out the machine (see Figure 4.1).

The behemoths seen in cyborg movies that look at issues of selective breeding may be less apparent, but they too respond to the evolution of the "romantized natural motherhood" coined by Laura Briggs and Jodi Keebler-Kaye. Briggs and Kelber-Kaye showed that the film *Gattaca* (1997) by Andrew Niccol "locates in a science fictional background that goes back to foreign films" and harmonizes with homosexuality and communism the "unnaturalism" of genetically engineered embryos.

Likewise, the combination of technology and the feminine marked both by desire and fear, as Mary Ann Doane showed in "Techno philia: Technology, recognition and the feminine." "So much a longing of and fear for the maternal role reveals the proliferation of such inconsistencies in the genres of science ficion and horror, she says, connecting and getting divorced it from the concept of the machine woman."

The fusion of no reproductive sexual gratification and a cold character in the sexually attractive nature of Kitsune, as I have observed elsewhere about the cyberpunk author Bruce Sterling's *Schismatic*, is an indignation for hypothetical gender identities: Kitsune explained to his lover, "They sent me the surgeons. They picked up my womb and placed it in the brain. Pleasure Core Grafts, sweetheart. The butt, back and throats I'm designed, 'and stronger than the God.'" Kitsune's intellect is merely a "a fusion of cold-pragmatic logic and uncontrollable gratification" and context. In the end, she uses female-specific hormones and erototechnology to transport herself to a building with flexible skin and broken doors—beds of flesh for the enjoyment of male tourists. Kitsune is almost not sentimental but demonstrates strong sexual satisfaction, cold reasoning and reproduction control once again illustrates the fear and attraction that cyborg stories record in the woman's biological matrix for the sake of female spirit, sexual autonomy, and enjoyment without "natural" sexuality [6].

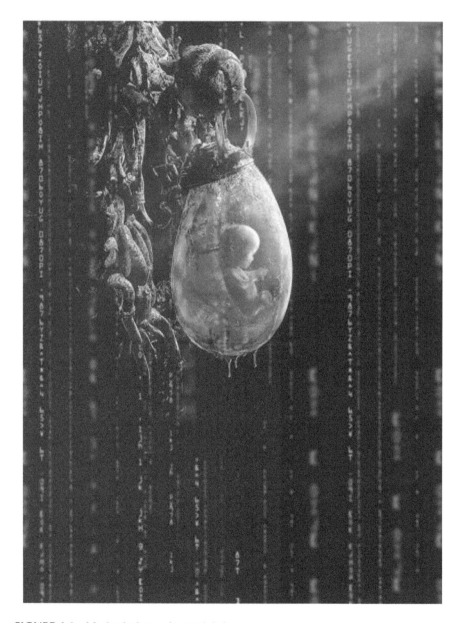

FIGURE 4.1 Mechanical organic womb baby.

4.2 CYBORG REPRODUCTION TECHNOLOGIES IN MODERN ERA

The noticeable contrast to the pornographic sex machine or horrific reproductive engine is the depiction of the living woman inside medical technology in contemporary film and literature. The fetus is often described as a self-employed pilot, astronaut, or passenger in the female body in biomedical and traditional pregnancy speeches. Some authors like Donna Hardaway praised the cyborg as a hope for freedom of reproduction. Mostly, however, passive female procreation has digitally manipulated roles compared with the superior ingenuity of techno technology. In the context of "Modernity, postmodernity and reproductive processes," Adele Clarke, for instance, delineated cyborg pregnancy as a phenomenon controlled by surveillance technology in approximately 1890. Clarke suggests that contemporary reproductive methods "were and remain anxious to achieve and/or strengthen regulation over these bodies and systems."

The "moment of women's disembodiment," as she continues, is when pregnant women "are eased to make room for the one true male, the fetus" Postmodernity the Postmodernity (https://www.amazon.in/Condition-Postmodernity-Enquiry-Origins-Cultural/dp/0631162941) (citing Rosalind Pollack Petchesky [1987, 1985], Barbara Duden, and Monica J. Casper [1994a]). Like wombs, Janice Raymond observed that reproduction devices make women observers rather than players in the entire process of breeding; they restrict women to fetal status; physically, scientifically, they sweep the fetus away from the pregnant woman. Computer technology reproduction appears literally and symbolically to have the fetus isolated from the mother, but not semen, "physician, and society."

Repetitively the female cyborgs were portrayed as vanishing or unseen objects as well as passive technofetal containers, just like Monica J. Casper claims that all fetuses and mothers become cyborgs by means of the various available technology in science and medicine. For example: "technologization of fetuses," which makes them more "normal" as the fetus is, Casper argues, as the fetus is an actual patient, while the mother is considered to be a "complex maternal health care service" rather than a particular person or performer. Furthermore, she proposes that they appear as "corps that are made visible by imaging technology . . . as dead techno capsules for uterus."

In feminist lectures, the subsequent propensity to assign agency or personhood to a fetus is one of the most important trends in the field of simulation technology such as ultrasound. As Rosalind Pollack Petchesky claimed in 1987, obstetrical imagery makes the fetus a "mini-spacial hero" and not just "an infant" but rather "a baby guy." Here, it reflected the 1986 discovery by Barbara Katz Rothman that "fetus in utero is a metaphor for 'man' in existence, floating free, only linked with the spacecraft by the amniotic sac. But in this metaphor, where is the mother? It's gotten empty room." In the 1980s and the 1990s, the picture of the women's anatomy as a passive system containing the fetus as an autonomous passenger or pilot was studied again by feminist evaluating the culture of reproductive technology. Susan Merrill Squire confirmed these interpretations in *Baby in Bottles* (first published in 1994), presenting a series of images representing conception and gestation without the womb, emphasizing once again that a gestating mother is almost totally eradicated from the childbirth narrative: in these images of artificial wombs and reproductions, the mother's image is not very much present. In a later chapter, cyborgs have a tradition that goes back to the early years of this century: "The Cartesian fantasy of liberating man from the mortal corporeal starts with the concept of liberation of fetus (as proto-man) from the female body. Briefly, the cyborg comes through exogenesis" [7].

Squire suggests that, given that the fetus itself is presented in the "complete man" (cyborg), Hardaway's paradigm of unmasking the philosophies of organic human reproduction is untenable and should be reassessed. Her arguments are that

This fantasy/image of cyborgs is a challenge as a solution to the feminist dilemma of birth, including its emancipatory overlay. We must look at what it implies to think of the pregnant woman as a cyborg if cyborg itself is linked—by the shared origin in the exogenesis imagination to several troublingly repressive contemporary pregnancy: from the photographs of the fetus/patient who is at risk in the hostile, hazardous environment of the preñed woman to the publicity photographs that are requested by her.

She also writes that "the fundamental to the imagination of a cyborg, with the negation of its connection between mind and body, is a previous negation of the association between some of the fetus and a gestation period woman."

Most importantly, Squire (except with the word cyborg) affirmed that applying for citizenship of reproduction technology leads to the treatment of the child as "outside, a lawful medical, social and moral interference," while the mother is "less a civil, legal or health care subject than police subject to all three organizations." In comparison to the inherent intellect the representations of the machine woman/cyborg or of the techno-reproductive woman bearing an automobile fetus repeatedly indicate that the female body is irrelevant [8].

In "The Virtual Speculum of the New World Order" Donna Hardaway argued that, while technology-visualizing created the fetus for a public purpose, female cyborgs should not have to play a significantly passive part in the monitoring phase of the fetus. Hardaway reflects on the technological "technical dramas of origin," in which techno-science reproduction images are "connected to the cognitive organs and Writing," adding, though, that the picture of a cyberplace extra uterine fetus is a symbol of an organization formed in different cultures of practice: a woman who views a technologized fetal image may have several alternatives. In recent years, after looking at a photograph of the embryo, Rayna Rapp has studied the trend of expectant mothers to identify the fetus as the stereotypical floating space traveler. "Fetuses, who were seen as individuals, have been a major part of our social worlds," she writes in disputed feminist and antifeminist policies. In any case, considering the pilot's historic combination as a spirit or as a logical intelligence in a man-macht, and the insane woman's womb as analog of the brain of the rational man, the prominence of the fetus' image within the woman-cyborg body is striking.

4.3 FEMALE CYBORG ORIGIN STORIES

As in many other studies on cyborg heritage, the female cyborgs are consistently placed in the evolving philosophical environment of the Scientific Revolution and the Emancipation. For example, in the article of Petchesky there is a short summary of the early medieval period of how the floating free self-going fetus "extends the Hobbesian view of born man as separated, lonely being."

Hardaway originally brought forward a history of post-modern technology of reproductive surveillance in Renaissance perspectives and improved artistic imagination. She explains that she shares "conventions for redemption and epistemological assumptions" through new graphic methods in the 16th century and computer realism in the late 20th century. Thus Hardaway regards the *Draughtsman Drawing a Nude* (1538) of Albrecht Durer as a metonym for "all the range of visual strategies of Renaissance," the tale of a device for transforming "disruptive bodies into regulated art and science" and points to the distinct screen between women and artists as a not-so-subtle division between art and entertainment. The combination of the physically attractive picture of Hardaway and all its derogatory ramifications of pornography is, as always, a challenge—as is her anachronistic combination of Durer's gynecological examinations with a perspective machine, which she accepts was not standard practice until the 19th century. Hardaway's point, however, that "icon specific examples of European architecture, humanism and technology from early modern Europe" have inspired techno-scientific depictions of pregnant and fetal organisms calls for further investigation. However, interestingly, this brief history of fetuses and mothers in "new world order" is a lack of the representation of the female reproductive body in advanced conventional technology and anatomy. The history of female cyborgs was also confined by Robbie Davis-Floyd and Joseph Demit. In *Cyborg Babies: from Techno-Sex to Techno Tots*, they concluded that the analogy of the woman's body as a dysfunctional mechanism "shaped the intellectual basis of contemporary obstacles" and is based on the female cyborgs roots in the Western philosophy of the 17th century in their 1998 book. Floyd and Davis proposed that the cyborgifications of women and fetuses should commence with the Euclidean paradigm of the body, which reads, "In his philosophical divorce of body from spirit and the ensuing detachment of the body from religion, the functional usefulness of this metaphor of body-as-machine is to be explored for scientific research."

In Western Europe in the 17th century, they said that the "catholicists" held women as inferior to men, with little or no piety, which led to the idea that "the male body is a humanoid "computer's template." They claim, in part, that the female computer was conceived of as "abnormal, intrinsically flawed and dangerously influenced by birth, that inherently faulty and that man had to constantly be manipulated." They argue that alongside "the strong societal embrace" of the male machine, it is a "dead-facing of the midwife and the emergence of the masculine, technologically manufactured birth," leading to "an ensuing acceptance," a metaphor that has finally formed a metaphysical basis for contemporary obstetrics, of the metaphor of the female body as a dysfunctional machine. Summaries, nothing larger than one paragraph or two, are much more cautious: all medical treatises and cultural theories must be examined more thoroughly in the complicated past of the early modern human machine.

In the early days of modernity, the female machine was actually quite an elusive personage, and certainly not inevitably inferior (after all, the male machine was seen as seriously malfunctioning or failing and disintegrating even in medical areas); however, it is generally a body where the mechanical appliances of sexual attraction and reproduction have been used to rewrite the old genus. The feminine computer was the default man-machine with customized parts, which in turn made it more subtle, tense, emotional, and compassionate than the male in some depictions. In spite of "the hierarchic model of the women's body that was constructed as a lesser, turned inward variant of the male's hierarchical model, the scientific progress of the 18th century provided a background in which cultural interaction with radical distinctions became indispensable."

In the mechanical terms of reproduction and nervous parts, this disparity was seen in large part. If in terms of technical thinking or the soul, the metaphysical issue with the medical conjectures of the man-machine was the creation, then the same problem was in order to fully comprehend the feminine machine, the reproductive system that regulated all emotional and sexual functions mechanically. The female's nerve sensitivity and "hysteria" during the 18th century was a technological mechanism between the mind and womb where sensitivity or sensuality continually fought against reason. To recap, the female cyborg in her behavioral and cognitive abilities and her openly seductive yet non-reproductive sexuality, in the post-modern myths created by fiction and film, are risky while the artificial womb or matrix are frightful, hideous, grotesque. The reproductive woman is depicted as passive, artificial, and meaningless by emerging developments in the medical profession of genetic engineering studied by feminist researchers, and focuses our gaze on the fetal patient as an independent patient pilot. In these reviews, the stories frequently focus on revolutionary philosophy and medical methods of the 17th and 18th centuries. However, the succinct original history of the female cyborg, as mentioned in the past chapters, is also unsatisfactory. Durer's drawing machine is often used for men's and women's bodies, and he also used loads for this purpose (sometimes known as Alberta, Leonardo da Vinci, and Brabantio), though Durer's popular measuring tree woodcut *Under we sung der Messung* (1527), definitely refers to a pretty clear heterosexual male imagination; it seems insufficient. However, interestingly, this brief history of fetuses and mothers in the "new world order" is a lack of the representation of the female reproductive body in advanced conventional technology and anatomy. The history of female cyborgs was also confined by Robbie Davis-Floyd and Joseph Demit [9].

In *Cyborg Babies: from Techno-Sex to Techno Tots*, they concluded that the analogy of the woman's body as a dysfunctional mechanism "shaped the intellectual basis of contemporary obstacles" and is based on the female cyborg's roots in the Western philosophy of the 17th century in their 1998 book. The response is yes—but it is a surprisingly educated yes as the feminine machine is almost totally absent from the literature of anatomy, philosophy, and fiction. Counterintuitively, the personality of women was the mechanical result of the deficient body and nerves. The traits of domestics, appetite, and fertility are usually addressed in the distinctly "female." In Edinburgh surgeon John Aitken (d. 1790) said that the close source of acute hysteria in women "seems a disease of the nerves that gives morbid sensibility," which most often arises from the "extreme desire or angry feelings of the soul, as consequences of disappointing affection." By comparison, the mechanics of the theory of human physiology as regards matter of movement, according to James Grantham Turner, posed the paradox of a man of sense who both praised and dismissed a mechanical body independent of conventional morality. "In inducing such ambivalence around the question of methods, mechanisms and imitations, libertinism indeed questions its premise—the total sovereignty of the single person," Turner says, "and yet the human is a consequence of blind drives and unpredictable circumstances in the materialistic theory of the libertines." For the man machine, mechanical behavior was not ideal. Moreover, "libertine sexuality cannot clearly be understood as a concession to accidental physicality; it cannot be separated from the brain success over the opposite sex, from the superiority exerted through "tactical reason."

4.4 INNOVATION OF WOMAN-MACHINE

Walter Charleton's *Enquiries into Human Psychology* (1680) starts with the preface of London's physicist and naturalist Walter Charleton describing his subject as "the most abstruse of all kinds of natural economies on a human being." In this regard, Charleton endorses the picture of the man-machine but also strongly resembles the human spirit, by clarifying automatically afterward that these disparate processes are "adjusted in a common end, that is, the creation of a living epigastrium or workhouse where the rational and everlasting soul will do its spiritual work." Just like Thomas Willis, in man-machine Charleston portrays the spirit as feminine, but like in Willis, representatives of the human being are categorically masculine for his talk about a human being. Nearly a century back, in *Sentimental Journey* (1768), Laurence Sterne's Yorick exclaimed that "what a weird machine is a man, framed by Nature's hand with such a beautiful framework, that every aspect hinders its complete movement." The playful invocations of Charleston and Sterne of mechanistic theory in these texts draw attention to the following fundamental hypotheses of the 17th and 18th centuries concerning human machines. First, the relation between logical and eternal spirit and merely the artificial movements of the human body were eventually problematized, as we have already seen. In the later part of the eighteenth siècle, in works such as a revised edition of William Emerson's manual, the man-machine (that is, the living body or the laws of motion) represented a calculus of labor and often appeared with many other moving concoctions (published in 1758). Emerson quotes Borelli, who calculates the strength of the muscles to bend the arm and the elbow: a healthy young man is able to support a 28-pound weight in the arm end [10].

And it finds that the duration of CB to CK is more than 20 to 1. That's why he gives these muscles so much strength that they are at least 560 lbs long. He also concluded that "It is obvious that robots are just "animal bodies," as shown in Figure 4.2. The following example illustrates the dynamics of "man's movement, walking, riding, etc." "Will run 400 meters in a minute, a fine footman," he said.

Similarly, in his *Survey of Experimental Philosophy* (1776), Oliver Goldsmith writes that

> The strength of the organs of the shoulders to stretch the hand and arm is awesome . . . these natural workmen's machines are infinitely stronger than any artificial engine which one can build, whilst the hand carries a weight approximately 20 livres, thereby forming the Grand Workman.

In a number of addresses, both specialist and commonplace before the end of the 18th century, the male human-machine, if entirely omnipresent and definitely permissible as a model of the parts of the organization, remains "animal machine," merely "machine," and "mechanisms" in the accounts of many bodily structures. The inevitable concern arising from the man-persistent machine's vocabulary is: was there a machine for women? Predictably, the laws of motion controlling a working woman-machine do not seem to be equal in language, and neither can the soul's universal applicability in the logical woman-machine be debated noticeably, since women can fit in a non-gendered fashion into the category of human machinery, as

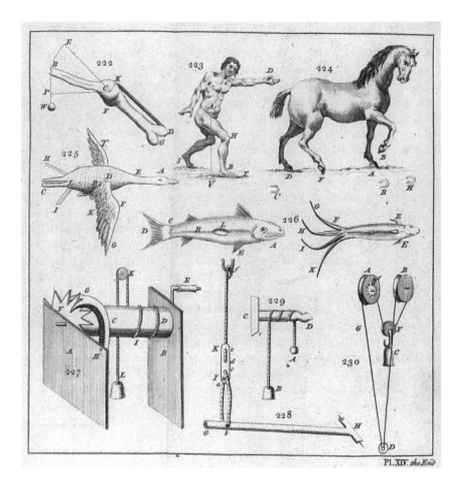

FIGURE 4.2 Animal bodies machine.

Richardson's Clarissa told Anna: "O my love! What a bad passive machine is your body when your mind is distorted!" A woman machine appears to be a word not found in the English time literature. How can we explain this counterintuitive disappearance of the machine, as equal sex is described uniformly in nerve and reproductive material terms? In his notorious text, though, La Metria had very little to tell for women, as we might refer to the infamous author of *L'homme machineto* as to whether the female machine was in itself an individual. "Fair sex" is the traditional dish of the women's nervous system that he wrote with no real uniqueness. He claims thus, "the tenderness, love and animated emotions dependent not on reason but on passion, The mechanisms of their lower content have led to biases and superstition, the depth of which women's imprints can scarcely be erased." "The spirit, as its characteristics are, is more vivid in men," he wrote, "whose brains and nerves have the firmness of all the solids [11].

"The education lacked by women, strengthens [men's] souls more, how could they not be more thankful, more charitable, more friendly, more resolute and adversarial?"

Women should not envy the nervous system and human education "twice as much as they are," because there is no one else, he implies, who is affected even more by the allure of physical attractiveness. "And just the someone else to administer to their delight all the better."

For the sensible male, design is not the disappointing or transient material vanity of women. For men, art as a commodity of learning allows creativity or cognitive brilliance to exalt humanity above all other creatures. The man-machine indicates a complicated conceptional hierarchy: male's intellectually honesty and intangibility showed the proposed system of a "natural" emergence to the physical conceptions of women with their lower nerves and unreasonable aspirations. He concluded that the pregnancy and the illness of a young child, known as chlorosis, would also wake "despicable senses" and cause the spirit "to perform the much more absolutely horrendous of the complots "cause the brain, the womb of the soul, is subverted . . . along with the body." Of course, for La Mettrie the human-machine is a superior device that has benefitted from a more vivid mind supported by superior material and education. The key point is that for La Mettrie the man-machine is alive, committed, imaginative and imaginative—yet necessary.

The Mettria was not exceptional since women were more vulnerable to their fragile anatomy's natural disability: it is well understood that women are represented as an individual dominated by body functions in the literature of sensitivity. In fact, the feminine conscious experience, quite so many medieval theologians of history, analytic philosophy and gender studies have demonstrated, has been more often subject to the lack of confidence of the body's natural anxious material strategies than males, women were constantly supported throughout their periods as adversely affected by their especially inferior embodiments, their lower nervous system. Nonetheless, except in the most obviously mechanical literary, medical or metaphysical works, the express language of the computer or women's machine is almost entirely absent: instead, "woman" is more often related to the nature of a mechanical world view where man's autonomous reasoning contrasts with the automatic reflection of a mechanical woman. The woman-machine is usually quite implicit and is distinguished in the present case by natural passive processes that distinguish her from the active and sensible man-machine. In addition, the artificial woman would appear incidental to the argument about mind and matter between men and machines; when the mechanism is alive and moving, her artifice or ego are characteristic of it. Though the word mechanical is seldom directly used, the mechanical woman is not shockingly implied by many famous depictions.

4.5 FEMALE VANITY

Unlike a preferred "normal" state, the modesty of feminine artifice is one of the female versions of a false construct, in which she ambiguously, often grossly, is changed by protest action. The 18th-century artificial woman stands for female beauty, lavishness, and arrogance and at the same time evokes sexual and disturbing responses. The representation of femininity, for example, in a print entitled "Work or Female Fortification" (1786), shows a crueler jest toward women in the needless "extensions" on the body of feminine dress. Artificial Bums: The Biography of John Denton, Eccentric London librarian and hack writer: in a comment that essentially registers disgust for fashion, commodities, and marriages, a touch at the *Lady's Tails* (1707), which plays a joke of the monstrosity of the prosthetic extensions of human genitalia. "Lady, this year I did (being at the Tunbridge-Wells)," he'll start, "Endulge me the freedom to see your monstrous tails." The lamp is influenced by "Lady's Thails . . . being cultivated so high and high that a walking A-Breast of three ladies will block the walks in the wider place." The attraction conveyed in the present case is the "compound" organism, whose attraction comes by expansion and alteration of the natural body, and whose threat is emphasized by combative violence ("battlement pistols"; Figure 4.3). An antithesis to cleansing and sometimes recording fear or elation for gender distinctions breaking up, cyborgs have specifically traditionally marked "clean" flesh and blood—"pure" existence—a more relaxed version of "normal" feminine household responsibilities.

It may be a jest, but the terror of the "grotesque" misrepresentation of trendy women expresses misogynistic fear at the transposition of the tidy ideals of feminine/mental and sometimes even spiritual/material binaries: the female's sexual beauty could, horrifyingly, seem "normal" to the man who was awakened, but he found that it was a tricky creation to his grief. Thus, she has been adulterated not only by the debris that makes up her virtual body but also by the vile deception that uses cultural and materialistic explanations to trick men into sexual attraction: con't be

FIGURE 4.3 Aggressive cyborg.

loved, Dunton slowly advises his male readers, giving a formula that begins with boiling greed.

He recognizes that there's someone like an honest lady, but "Wrongdoing is in the matrimonial competition, there are a thousand blanks for One prize."

Dunton ends with a "Bumographical characterization of a Good Wife" marked by business, frugality, and chastity: she recalled that although her Taildoes conquered its vulnerable side, "God and law still designated him their heads." She also thinks she is called a working woman and is trying to get the title right and refrains from "Abroad Gadding" and is not "Sluttish" in her family home; It has one eye that nothing has been misplaced and remembers that an ill managed Kitchen has destroyed many a Noble Hall; it's spending more time in the priesthood and devotional exercises than between glass and the dressing box; it's adapting its dressing and apparel (but particularly its Rump and Top-Knots) to the husband's high quality. She "is not dumping her bag for fashionable vanities" forever. The good wife distinguishes everything from the artificial being. In the satire of this the discomfort of a self-determined woman whose presumed ambition to find rich husbands in such elegant spas as Turnbridge Wells with their vanity and ambition makes finding a family provider who will allow her husband to think all the more difficult and not a little stupid because of the promotional video. In his disgusting "Progress of beauty" (1728), Jonathan Swift's devastating cartoon of feminine pride also clearly states the moral stuff about spiritual binaries and mechanism, in this case the aged slut. A vile example of both the flesh's death and the hideous sin of the women's manufactured pretension for economic benefit is the transfer of Celia's "clody, wrinkled face" to its "artificial face."

4.6 DOMESTIC WOMAN CYBORG

One could suppose that, where the results are examined, the man-machine is defined by his labor, and women can be defined as a mechanism in the wider machine of a national economy but these pictures are still fairly rare, perhaps partly because of such a reluctance to turn from what Ludmilla Jordanova has called "the common nature of the community."

When homeowners are mechanically represented, a familiar normative structure for the relevant events within the field of domesticity seems to be determined systematically. In 1711 Joseph Addison contrasting the perfect female in a normal household economy without using the word "glowing Gew-Gaws." . . . Too many attractions to females with poor minds or knowledge, in a daily rural community. In 1711 he went without any of the term machinery. In his hours of commitment and repast, employment and diversion, Addison portrays the good families of the Aurelias as the perfect microcosm of the broader European union, "as regularly an economy, it seems like something in its very own right."

They are the "envy of everything that knows them, or rather the delight of everything." Fulvia, on the other hand, is one of those light wives. "Disposition[which] sees her husband as a steward, and sees discretion and decent holiday home as small domestic virtues, a woman of performance unbelieving." Though Aurelia's family seems to be working as clockwork, Fulvia lives "in a permanent movement of mind and sleeplessness" mechanically. it is a mechanical being.

Pamela: or, Virtue Rewarded (1740), portrays the female as a home automaton device in a more deeply articulated form. The budget here is like a mechanical body of two ministers. The women's system is an administrator, and essentially it is subjugated to the man or the sovereign: that is, the most significant role of a woman in the well-working domestic machinery is directly reproductive. Finally, the comic illustration provided by the British Museum catalog to George Bickham indicates a woman's machine designed from the point of view of servants' domestic equipment.

The rest of the story here involves mostly the working girl in reproduction and sex: the following text contains a letter of recommendation for Moll Handy asking if Lady Crosspatch would hire a maid who, by the time she got Crack'd and changed into Pot Belly'd, was already a common flaw in our sexual encounters. I hope it's going to be overlooked. P.S. For the little wages she is going to come. These photographs present a feminine version as a mechanism in the household industry: although the technological arts of good housekeeping are linked to the female, they designate their major function not with regard to the working mechanism but with regard to its "normal" sexual and motherly requirement.

4.7 CYBORG MECHANICAL OPERATION

There's a "mechanical jade," as the wise frank tells the wonderful Katy in the incidentally written Venus school, with candles of around four a pound (1680).

The portrayal of women in the domestic realm is somewhat unlike the representation of an individual sexual woman who defies the natural-born citizen house and family definitions. A limited percentage of "mechanical" women are women who handle their own cunts manually. "I squeezed and tightened the lips of the virgin slit and mechanically followed the precedent of Phóebe," the legendary Fanny Hill described. It was manually operated . . . "brought the vital bliss, the river of melting at last."

Will we end up in mechanical sex with a definite woman-machine? One curious case is a book by a presumed Bath doctor, Mr. Lobcock (i.e., large relaxed cock or dull fellow), in the territory of the Republic of Venus, in safety.

In the Practical Prevention Proposal and Final Eradication of a particular Lobcock Disease it proposes prudent legislation and a sound prostitution administration because, after all, it is "interwoven with the very existence and web of mankind." He says that we can't save girls from being scumbags until they're twenty, but that we can't discourage them from whoring themselves. Notwithstanding these production, the author's assertion relies on a changing basis for prescription of effective human behavior codes—existence.

> Without futile efforts to chain the waves, or to overcome the decrees of Providence and to grief before we have had the courts, the people who are wealthy on this planet will spend their time on satisfaction, whether it's buying a maids' head at the age of 13 or pots of green peas at Christmas. pain, since we cannot do miracles; let us take care of our true task to prescribe and observe a just and philosophical measure to others. Nor should the agony of suspended animation or the damnation, even of the common stock which Nature appears to have stockpiled, be imposed on it to fill a normal hunger while it craves;—for a venial offense which, if properly done, is as harmless as consuming flesh in a lent.

This, however, should not be thought to be administered. Indeed, "Lobcock" claims that the simple purpose and meaning prove that the sexual freedom of romantic men depends on the virtues of women, family harmony, and protection. Sexual lust as the oceans become deaf and enraged: you can adjust it, but you can't hinder it.

It must be prostitutes who finish with an eloquent flora, maybe "chastity." Yet "solitary drains"—a very self-serving statement for the mistrustful reader—makes good human beings mechanical and sick; in the male example, an automatic one. This paragraph is a change in Richardson's Clarissa in relation to the libertine figure mentioned earlier. Whilst the technological requirements of corporal impulses are a human-machine feature, man must never allow himself to become simply automatic with consciousness. Moreover, "The Onanist pallid and sallow is seen to be retroactive and to be basically timid." Automobiles are not male machines: when he meets a true woman expecting love, he becomes "unable to enjoy the true and legitimate delights of nature," his penis reduced to "a simple lilly, comfortable and worthless for any intention of enjoyment and peace, like a wet wool roving." The woman in this "illegitimate gratification" is likewise deprived of "normal" satisfaction but does not become required calculations: "by a slow excess and debasement of artistic juices, for the sake of the generation, it invalidates them and makes the warehouse of nature too loose and flaccid to maintain her jewels."

And, in the same manner as male sex in the same way, women do not have the normal way to love and cannot have any fun other than their own violence.

These passages illustrate the conclusions of the corporal roles of creative womb stores as the contrapart of creative womb stores. Quite evidently, the disease allegedly caused by mechanics is not only a deficiency of "normal" sexual gratification but also a drain of mental capacity; the disease caused by women is a drain of reproductive capacity. This out of the way example therefore summarizes several traditional features that delineate the roles and governance of men and women's bodies. The man-machine is alive, virile and sensible, in contrast to the effeminate automatic; again the female of the 18th century is connected to sexual gratuity and legal breeding. Not unexpectedly, her effortlessness remains untouched.

Catherine Macaulay's mechanist claim is in favor of marriage at the age of forty-seven to twenty-one years; William Graham is at the opposite end of gender conceptions of the sexual technological female. She declared her marriage to Graham less than two years after she moved to Alfred House, residence of Rev. Thomas Wilson, the rich 73-year-old (younger brother of the infamous quack James Graham with whom it was rumored she had had an affair). The overt relation to the sex-machine acts as a tool for adapting and reverting preferences for the female role.

Cleland's *Autobiographies of a lady of pleasure* (1749), a book in which we can find one of the few clear images of the woman-machine during this time period, is among the most renowned sex machines at this time.

Here we will find Louisa's fabulously loving, good-natured Dick, made a "meer" rig. Fanny herself mechanically gets to the orgasm more than once, however, as we shall see, the reactive or non-rational "mechanical" responses of females vary significantly from the responsive, powerful, and man-machine logical; rational. Though Cleland casts her as a mentally active, intelligent being, Cleland describes in her insalubrious profession a clitoral mechanical imperative over a rational will: "there is in Fanny a regulated seat," or a Queen seat, which governs itself by its own maxims

FIGURE 4.4 Cyborg mechanical operation think.

of state with which there is no one more powerful. Fanny is ruled by a leader sitting in his sensual center, not in the brain of his ruling tyrant Descartes and Volis. Man-machine technician—except in the clitoris that guides her desire and passion sin" (Figure 4.4).

Animal spirits racing mechanically to their own "cantor of attraction" are defined elsewhere by Fanny. These are the two cases in which Fanny is defined as a mechanism. Specifically, her body is not expressly defined as a computer. The sensitivity of Fanny, of course, ultimately confirms conventional buildings of marriage, passion, and morality: "I can't resist pitying." She says that her marriage is with her real love Charles, "those who are insensitive of such delicate charms of virtue, in a gross sensuality, then all those who have the joy of no greater comrade, no larger enemy."

4.8 REPRODUCTIVE MECHANICAL ACCUMULATION PROCESS

In addition to the emergence of mechanical thinking throughout the early Modern Era, today's invocations of the fetus as a passenger in a motor during gestation seems to reflect a rational hypothesizing that the 18th-century woman was a motor originated from "cyborg" devices (see Figure 4.5). Indeed, we see that the woman of the 18th century was never specifically marked as a computer, although its mechanical womb was a separate passenger's process. The womb is usually associated with the mechanical accumulation process of nature and also with the comparatively advanced health care and influential treatises. During that time, the uterus could be identified concurrently as the root or ground from which the fetus emerges like a

Evolution of Woman Cyborg

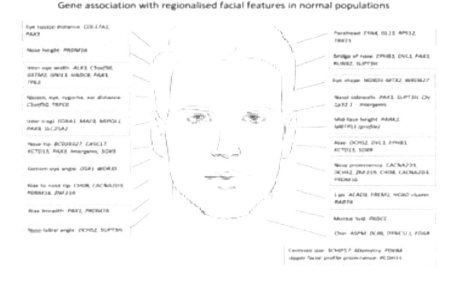

FIGURE 4.5 Reproductive machine knowledge of geometrical certainty.

plant and a fetus-containing bottle or vessel. In either case, male grain fertilization proved to be an innovative procedure using cleaning or distillation equipment (the tools implying thereby an act of deliberate creation).

The image of the embryo as a small, sometimes independent man and often a whole "engine" or "machine" separate from the mother contrasts with the image of the mother themselves whose completeness or rationality is overshadowed by representations of isolated, mechanical structures. For instance, ovaries, tubes, and tissues that support fluid and nutrient fluids or womb geometry. Towards the end of the 18th century, in the particular circumstance of ministry manuals, a reproductive woman may have been an "animal-based" machine, but, in those contexts, the discussions about the rationality, wisdom, and understanding dominating former dialogs about the man-machine are not surprising. Generally speaking, the womb needs to undergo a revision from the late 17th to the end of the 18th century The plant-like mechanical substratum for fetuses is a mechanical device that is mathematically defined and controlled by force, angles, curve, and torque properties.

While both the male and female seeds play a contestable and indeterminate role in the breeding process, the male partner's prevailing role as an involved, vital, and inventive power progressively opens up a new hierarchy in which the medical professionals' field is active participation and expertise, including the idea itself. It would be a mistake to say that the female reproductive system is more artificial than ever and is not openly and steadily imagined as a machine for women. The system meanings become even more mathematical, thereby establishing new fields for medical technology, in which the female is, more than an active and self-motivated machine, mechanical or automatic.

The woman's reproductive body is technologically fecund like nature at the end of the 16th century, its automatic womb being unconscious motion of matter limited by the omnipotent engineer's plan and intent. Thus, Robert Boyle likened the environment itself to a pregnant car or boat managed or held by natural laws of movement here,

> The mechanisms of these Engines are as little as they would be without the professional and none of outsiders; such as the sun, the sun, the air, etc. would be able to carry out their own functions as the great Mills that usually would be with us are to Grind Corn without the assistance of any external agent. A ship is a pregnant woman, a pregnant woman is an automaton that leads an outside active and vital force.

The brain of the living beings is also active relative to ether or light in its operation, as we have shown; however, as the engine is sensible, the powers are the agent of the ruling king. For passive or automatic womb existence, there's no other ruling pilot—just God's mechanical powers.

There are variations, as in all other generalized statements. A woman engine becomes a brunette joke with her poem "The Water-Engine" (The Water Engine), which appears as the anonymous miscellany of "A Pleasure for a minute" story, this depiction of women's generational organs as an "active and movements" is rare and maybe much less popular outside of comedic or satirical writings.

Since no other context for the term "female machine" existed outside of mining manuals at the time, she notes that much effort has been put into explaining how Nature performs this "service inside a female's machine" (menstruation) (George Walle's *The Art to Prevent Diseases and Restoring Health*). Despite women's inclusion in the human machine category, medical literature often ignores them.

4.9 CONCLUSION

The woman-machine is usually quite implicit and is distinguished in the present case by natural passive processes that distinguish her from the active and sensible man-machine. In addition, the artificial woman would appear incidental to the argument about mind and matter between men and machines; when the mechanism is alive and moving, her artifice or ego are characteristic of it. Though the word mechanical is seldom directly used, the mechanical woman is not shockingly implied by many famous depictions. Importantly, I don't mean that machines and Western worlds are antifeminist texts or do not provide convincing accounts of the dominance of female cyborgs. Rather, this critique aims to focus on how the female cyborg image remains an ambivalent place in terms of the techno future towards which it is called upon to make provision for publicity purposes.

REFERENCES

[1] Carey, J. W. (1990). The language of technology: Talk, text, and template as metaphors for communication. In M. J. Medhurt, A. Gonzalez, & T. R. Peterson (Eds.), *Communication and the culture of technology* (pp. 19–41). Pullman: Washington State University Press.

[2] Casimir, V. (1997). Data and dick's deckard: Cyborg as problematic signifier. *Extrapolations*, 38(4), 278–291.
[3] Chang, B. (1996). *Deconstructing communication: Representation, subject, and economies of exchange*. Minneapolis: University of Minnesota Press.
[4] Clayton, J. (1996). Concealed circuits: Frankenstein's monster, the medusa, and the cyborg. *Raritan*, 15, 53–69.
[5] Clynes, M. E., & Kline, N. S. (1960, 1995). Cyborgs and space. Astronautics. In C. H. Gray (Ed.), *The cyborg handbook* (pp. 29–34). New York: Routledge.
[6] Davidson, C. (1996). Riviera's golem, Haraway's cyborg: Reading neuromancer as Baudrillard's simulation of crisis. *Science-Fiction Studies*, 23, 188–198.
[7] Davies, T. (1997). *Humanism*. New York: Routledge.
[8] Davis, E. (1998). *Techgnosis: Myth, magic, and mysticism in the age of information*. New York: Harmony.
[9] Delia, J. G. (1977). Constructivism and the study of human communication. *Quarterly Journal of Speech*, 63, 66–83.
[10] Derrida, J. (1973). Différance. In D. B. Allison (Ed. and Trans.), *Speech and phenomenon* (pp. 129–160). Evanston: Northwestern University Press. (Original work published 1968).
[11] Derrida, J. (1974). *Of grammatology*. (G. C. Spivak, Trans.). Baltimore: John Hopkins University Press. (Original work published 1967).

5 The Cyborg Interdiscipline

5.1 INTRODUCTION

Let me begin with a provocative and audacious statement—our future must be mixed with artificially smart machinery! The topic of this book is how I came to that conclusion. I am not suggesting that we humans will look and be behaving like machines on an assembly line in the coming decades, but rather that we will be fitted with too many inventions, including electronic devices implanting in the brain, which have created us, rather than under the laws of the biological, a technologically dependent being, developing in compliance with technology laws. As we evolve into "machine like" or "cyborg as" technological advancements, artificial insight, neuroscience, and engineering technology encourage scientists to build intelligent technology that is humanly complex, practical, and increasingly intelligent, like us. I imagine a possible convergence of person and computer to be the natural result of technical advances in robot science, artificial intelligence, prosthetics, and brain implants. It is not a deliberate action taken by mankind but a slow and natural process [1].

And it is not so incremental that it will take decades but perhaps something that will begin this or an early future generation.

In a denominational context I may have played a minor role in this result (our coming merger with machinery), as I led a research facility as a faculty in engineering that aimed to develop functional, fully integral computing and sensors technologies in the living organism. In the mid-90s I started to officialize and I published in 2001 about the potential guidance for robotics and under-the-skin information technology in which my teammates and I advocated the deployment of sensors and cyborg implants for the purpose of restoring, repairing, removing, and improving broken anatomical and physiological structures. At the time, my colleagues and I were still dreaming about the potential paths for smart technology, making assumptions about today's innovations. But in retrospect, it does not seem that we have gone far enough and we have been too careful about how close we are to Singularity and subsequently to the posthuman period.

Many of my analyses on the concept and application of "wearable" devices was reported in various books, the *Immersive and Advanced Environmental and Functional Computers Basics and Augmented Reality,* which I co-published. The landscape of human improvement technologies and artificial intelligence have improved significantly since the release of the first edition of such books over a decade ago. I wrote an entire book to provide an up-to-date overview of recent breakthroughs in neuroscience, prosthesis, and brain enhancements and to address existing attempts to build artificially smart robots that benefit from human predictions and resolve issues.

Another aim of establishing credibility was to stimulate national conversations about legislation and policy to be enforced as human beings become better machines and as advanced artificial equipment achieves degrees of intellect from or above human beings. Due to the scope of this work, a broad variety of subjects such as genetics, science, ethics, and law will be explored [2].

The pace of technological progress in technology, medicine, and computer science is accelerating, as Google's Ray Kurzweil often reports. Science fiction was popular science in some places just ten to twenty years ago. The 21st siècle will also be a time of great transition, unprecedented growth, and unique problems for human beings, if advances in many main technologies continue to accelerate. The advancement in science and engineering, as projected by the informatic physicists, architects, and thinkers, would have contributed to such massive shifts in our bodies' composition at the end of the 21st century such as the very essence of what it means to be human. In this respect, science fiction author William Gibson, who invented the term "digital world" as in the short article "Burning Chrome," suggests a "cyborg" for mankind that incorporates silicone chips implanting into the DNA-modified human brain. Fast forward to the University of Southern California laboratory of Professor Theodore Berger, in which our cyborg future now is planned in the form of impressive neuroprosthetic instruments.

5.2   ENHANCING HUMANS

There are two key ways to categorize human artificial developments in Sidney Perkowitz' published essay *From Bionic Humans to Androids*: first, as practical prosthetizing devices and implants like artificial legs; replacement knees and hips; and vascular stents (which support blood flow across obstructed arteries) and second, as aesthetic or vanity implants. It was noteworthy that efforts of some pioneers in designing human robots may be viewed as changes in the aesthetic or vanity of the system, since those improvements are not functioning properly. This book concerns all types of improvements. Improvement technology may also take shape in several ways, assisted by a number of technologies that enhance people's appearance, capabilities, qualities, or functions. The improvements of the human body vary from medications and surgeries and silicone implants and beauty-enhancers to (perceived) bionic arms and improved human intellect [3].

Although "cyborg" enhancement technology distinguishes between enhancing human function over the "ordinary" or "acceptable" attribute, several enhancement innovations are used for clinical or regenerative purposes. For instance, a plastic procedure is used in the case of burn patients or prostheses for the loss of limbs.

In addition to the attempts to develop the human body through a number of innovations, robotics and artificial intelligence are making more substantial strides, even laying the stage for a convergence between human and computer. As algorithms and sensors evolve significantly, computers become much more automated, technology becomes more and more "intelligent," and robots start to look and behave like human beings rather than engines. They are being created. In reality, one field of research in robotics aims at designing authentic looking robots that represent an appearance of the human being (i.e., androids); others' field is to create facial characteristics that allow the robot to appear to convey emotions. The idea of an android in the spirit of Star Trek's "Data" will become true until "humanoid" robots are artificially fitted with intelligence and thus achieve more self-sufficiency from their human rulers. There are two intriguing possibilities that you can consider at this point: first, that the planet will be inhabited by more creatures of various kinds than we are seeing today, unenhanced and upgraded humans, cyborgs, machines, and androids among them, all of whom work in different but maybe identical ways in daily life. Second, one smart race, focused on the fusion of humans and machines, may evolve from advancements in technology. In my belief there will be many transitional types of man-machine hybrids, several of which we will call cyborgs before humans can eventually combine them with computers. Again, I mean when I talk about "getting together with robots" equipping human beings with technologies (usually IT) in order to increase their human body and mind, ultimately to be more "cyborglike" than they have in their present capacities. I use "cyborg technologies," "the future of cyborgs," "cyborg years," or "the coming era of cyborgs," to refer to the future in which we are being fitted with technology that facilitates the repair, replacement of our senses, and cognitive function. Throughout the book I reference the technologies that improve human body and mind. The legal questions that are relevant to consider for our unifying framework are therefore a "emerging cyborg law." In addition, it is a fascinating philosophical issue and the topic for discussions by many writers of whether a complete machinery comprises a human consciousness that is uploaded to an architecture of a computer [4].

There are many intriguing concerns regarding the idea of a futuristic planet of humanity, cyborgs, autonomous robots, and androids. However, one question is what this transformation entails for the fundamental or civil rights of the smart persons of the coming years? Would cyborgs still have "human" status as human beings? Can androids demand "human" rights whether they are the same way people or cyborgs look and work in society? And can people retain control over robots as they are becoming increasingly autonomous; in other words, are robots compliant with Asimov's three robotic rules, or are they going to resist, rebel and continue to control people like HAL in 2001? The physicist Stephen Hawking and Telsa Motors' CEO Elon Musk have warned the business of this very result. Some claim that since cyborgs are radical moves from the human race, the future generation is more likely to deem them human. As soon as numerous enhancement methods are embraced by a critical number, improved men will be literally the latest presence of the human race after the original pioneers. As a consequence, cyborgs are assumed to be inheritors of today's civil rights. Now presume that robots and artificially intelligent machines execute tasks that are identical to cyborgs, maybe even androids that look and work as cyborgs, but they don't have the exact right catalog. Substantial debates in society and law schools would be needed on this subject.

A further concern that digitally improved people can pose is if digital and social divisions between empowered and unenhanced people are being created. Human rights will play a significant role in this discussion because they set out the universal guidelines for people's care. The right to non-discrimination would offer important guidelines at first sight: non-enhanced citizens cannot be viewed universally. What is "nonetheless" if improved people in the future are substantially different from non-enhanced people? For instance, if an employer can choose between an IQ 120 non-developed and a cyborg with IQ 260 or higher, would the employer discriminate if the cyborg is selected? This is only one example of human rights issues that demand a public discussion. This is a very unique human rights topic.

5.3 HUMANS, BIONICS, AND CYBORGS

When we are filled with prosthesis and brain replacements, we go past our developmental history and are programmed into our genes. Since I believe that scientific advancement drives humans into a world with "cyborgs" and an ultimate convergence of computers, I must describe some simple words. Let's begin with a "cyborg," one of the principal characters in this book. A cyborg is typically a human-machinery hybrid that is assisted or operated by mechanical, electronic, or computer-based mechanisms by a certain physiological and intellectual mechanism.

"Cyborg," literally a term that derives from cybernetics and species, was coined in 1960 by Manfred Clynes, explaining the need for humans to improve biological functions artificially in order to live inside a hostile space climate.

"Transhuman" is a term that refers to a transformation of the human into the posthuman in order to add some other basic terminology. A Posthuman is a hypothetical future for transhumanist thinkers whose fundamental capacities are "so dramatically larger than those of present human beings that our present standards no longer clearly show a person." The distinction between posthumous and other complex theories is that posthumous individuals were once an individual, whether in their lifetime or in the lifetime of any or more of their immediate descendants. A requirement for a posthuman is, then, a transhuman being able to transcend the shortcomings of the human being while still identifiable as a being. In this way, the transition from human to posthumous can be seen as a continuum, instead of an omnipresent occurrence.

Communication and control mechanisms containing live species and computers are the domains of cybers. The artificial modules used by cyborgs, rather than substituting for an organ or limb's core functionality, incorporate, improve, or replace biological systems' computing capabilities. A person with a heart pacemaker may be called cyborg in a common example of a cyborg, as he/she is unable to live without the mechanical computing component. In a more dramatic example of a cyborg, others see clothes as cybernetic changes in the skin and they help us to live by using fabrics that are not naturally found in these conditions in radically different environments. I would say that the "clothing advanced human being" was a cyborg if the garments had computer skills that supported the wearer in my conceptualization of a cyborg. In nearly all of these situations, though, the "cyborgs" I speak about in this book are the product of technology being improved or merged into the body.

The words "bionic person" and "cyborg" are commonly used in popular culture as interchangeable terms with a human being augmented by technology. I discern thus that a bionic human being is a person augmented by technological or biological means but one step forwards—a cyborg has enhanced or technology-enabled computational processes with the purpose of enhancing actual cognitive and sensory capabilities. Interestingly, though many humans are obviously bionically improved, cyborgs now often exist among us. If the number of cyborgs or bionic humans is to be calculated by them, the number is based upon the description utilized. For example, if you use the word "bionic man" to describe an artificially modified person in a manner, then the digestion of medication will produce a bionic individual and tens of millions of those people will be actually alive today. But if a number of human components had to be exchanged for artificial implants and prosthesis to be a "bionic human," then there would be no tens of millions of these individuals. If, though, there are millions of people, one commentator suggests that many people today could be identified as "bionic," because 8–10% of the population of the United States, that is to say about twenty-five million, actually have some form of artificial share—a figure that is projected to rise with the aging population. In reality, only in the light of hearing, millions of cochlear implantations, including some of them in deaf people, are currently operational [5].

If one assumed that a "cyborg" was a technologically modified or substituted brain structure, so the numbers of those individuals would possibly rise significantly into the millions over the next ten years. As an explanation of the latest brain implant technologies, doctors introduced electrodes in the brains of patients from the late 1990s in the expectation that a computer-brain interface would be developed that would enable "locked in" patients to manipulate a robotic arm or to transfer their bodies on a monitor. In addition, technology is being developed that can digitally preserve memories in the brain. This is what Theodore Berger, his team from the University of Southern California, and Dr. Sam A. Deadwyler and Dr. Robert Hampson from the Waka Forest Baptist Medicine Center have already identified and tested for neuroprosthesis (artificial hippocampus). This is a successful reason.

Whether a person who is technologically equipped is considered bionic or cyborg is not a significant differentiation for many debates about augmented human beings—most people use the word interchangeably for someone who is technologically equipped. However, by regulation, the extent to which technology helps a person will matter. In disabilities laws, for example, a handicapped person will have to be accommodated by an employer; but, in a legal examination of the handicap and in the rights of the handicapped person, the kind of disability is relevant as well as the technology it uses. Consider sportsmen who have lost their legs but still start competing without prosthesis against sportsmen. Concurrent rivals also share the fear that the "cyborg" has an unfair gain over them because of the mildness of their carbon-fiber prosthetics. Although this case appears to be quite an outlier, with the advancement in prosthesis methods, the ability of prosthesis limbs to equal or even exceed the capability of natural extremities is fantastic. In comparison, prosthetic limbs can be heavier than average and allow the patient to withstand heavy loads. Alternatively, they can be more mobile or more accurate for some tasks—how many people have a 360° spinning wrist? Even when it may seem an idiotic instance, the prospects occurred in relation to non-enhanced persons for those who were deemed "disabled" once to become "over-capable." Will this provide those who do not boost their work with a comparative edge over others? Like any changed humans, can an engineered dominance over the normal build animosity between "enhanced" and "non-enhanced?" Would new types of discrimination laws then be needed? Within the United Kingdom Equality Act a person is "affected" if they are deemed to be damaged, which would greatly impact their capacity to perform ordinary everyday operations on a long-term basis; should we really deem them disabled if anyone is able to transcend the potential of completely qualified persons to do such work by using advancement technology?

Amusingly, a skill might have known a human already as a cyborg. Neil Harbisson, the poet, is entirely blind and has a vision disability known as acromatopsia, so that he sees the world in dark colors. Neil wears a visual amplifier as a head-mounted transmitter at the back of his brain in order to detect colors. As sensory substitute, "Eyeborg" transforms colors into sounds, enabling Neil to "hear" the color reflecting the electromagnetic radiation after a protracted struggle, with the Monarchy. A photograph of Neil's passport now includes a photograph of him with his cyborg device, acknowledging that his improvement is a continuous part of his personality to the authorities. For a passport photo that shows Eyeborg in Harbisson's face, people would find it impossible to claim that his Eyeborg is an optional accessory such as a camera or a hat, because anyone attempting to pick up his raise may undertake an assault with a deadly weapon that is equal to injuring himself. Incidentally, a "battery" can be carried out by statute even though the aggressor does not actually strike the defendant (e.g., cyborg) but touches something similar to his individual (like a cybernetic enhancement attached to the body). For example, courts found a person's battery may be touching the cane, even if the prosecutor does not touch the individual herself. In this case, the cane is like an augmentation of the body of the person so that it touches the body of a person. In certain cases, the person wearing them may even be appropriate to reveal battery clothes, caps and bags as part of a person. But the law in this area is evolving as a consequence of cyborg technologies, as we will see in a future chapter.

A significant argument at the beginning of this book is that as humans are being fitted to prosthesis and implants and thereby becoming more cyborg-like, robots will continue to be sophisticated and easily test human creativity throughout this century (actually our bias towards linear thinking). In reality, robotics with artificial intelligence and a variety of sensors, actuators, and algorithms will have contributed to the development by the middle and almost definitely by the end of the 21st century of machinery, which can exceed human intelligence and engine capacities. New models of individuals will emerge from various strategies to change human physiology, architecture, and cognitive systems as the technology progresses. All this may generate a continuum between intelligent human beings and machines, which will advance from human beings and bionic humans, cyborgs, androids, robots, sock puppet software, and machines [6].

5.3.1 Humans, Bionics, and Cyborgs

Technologies such as artificial intelligence can also result in disembodied machine entities that wander the internet, likely downloading their intelligence to distant robots or to android devices to achieve mobility at specific destinations around the country. One analyst used the word "digital individuals" to refer to individuals involving artificial and partially artificial entities, from mechatronic (mechanical plus electronic) robots to humans with BIIs. Martine Rothblatt claims that the brain can be replicated by software and computer science in her book, *Practically Human: The Theory and the Peril of Digital Immortality.* The perception that multiple forms of artificially intelligent organisms will coexist in the future is conveyed from this debate.

5.3.2 Brain-Computer Interfaces

In order to treat and substitute human anatomy and physiology and to redress, the technology is used to develop and improve human cognitive and visual capacities on the basis of medical necessities. For starters, interfaces between brain and machine assist individuals with weakening neurological conditions in order to "lock in" their own body. An interaction between a brain and a computer, comprising of electrode registration that is mounted on or incorporated into a person's brain, helps others who are locked into the capacity to interact with the environment through thinking alone.

In other fields of brain-computer interface architecture, further development is being made. For example, according to a report published in the journal *NeuroImage*, researchers used brain scanners to identify and recreate the faces people think about. Yale scientists hooked students up to the fMRI machine learning method that tracks blood supply and shows them photographs of faces, assessing the behavior of various areas of the brain. Then Prof. Marvin Chun and his team were able, using only the brain scans, to produce pictures of the people's faces. In the future, one might picture a witness to a crime recreating the profile of a defendant on the grounds of "extracting" the picture from his memory. Yale researchers noted that the big technical disadvantage that remains today is that such technology can only read functional regions of the brain; it cannot read passive memories—this is the way you have to visualize the mind to read it. Yale researchers found, strangely, that scientists are going past "reading" theories at California Berkeley University to foresee what everyone would decide next. And clinical psychologist Marcel Just of the Centre for CogNitive Brain Imaging in Carnegie Mellon University, Pittsburgh, has a vision that will make Google Glass and the like look very last century: installing a mouse rather than using your retina. Only imagine a computer that dials your feelings by reading them. But what if all our proposals were public? Dr. Just plans a dreadful version of the future, where politicians read minds and take power. More pessimistically, Marcel also aims for a more hopeful future, with readers giving disabled and non-disabled persons benefits.

Microchips implanted in the brain could also allow contact between brains, i.e. telepathy, according to Miguel Nicolelis, neuroscientist in Duke University. Brain wave sensing devices have been used to monitor everything "telepathically," from actual aircraft to video game characters. In its new incarnation, telepathy science has gone further by encouraging anyone in India to send an e-mail with a thought to his colleague in France. To do this, scientists used headphones from electroencephalography (EEG), which captured electrical activity from brain neurons to turn terms into binary. After the original thinking has been archived in India, it has been forwarded in France to a child's psyche, where the communication has been taught by a computer, and then the recipient's mind has been electrically stimulated. Undoubtedly, telepathic chips and similar intelligence systems may give rise to new types of knowledge such as "mindplexes." This is a concept used by Ben Goertzel, a researcher in the field of artificial intelligence, which is a collection of human minds—but still a cohesive self—and knowledge in the high level of the telepathically connected human community. Mindplexes could contribute to the advantages of community generation in which in many situations the collective expertise of a crowd has proven that challenges are solved beyond the reach of experts. Indeed, a trait of networking brain-to-brain connectivity will be the qualities of "fair crowds": diverse viewpoint, independence of people, decentralization, and a strong means of aggregating opinions [7].

Reading suggestions would undoubtedly pose a range of legal and political concerns. The privacy rule is not the least of them. In this respect, the courts will have to determine in the future whether to listen to and monitor the thoughts of a person is defensive speech or an unconstitutional search and seizure of prefrontal cortex-created behavior (i.e., cognition). With respect to microchip implantation, many states in the United States have now adopted anti-chip laws that discourage the "chips" of marginalized populations and lift the consent bar for the implantation of an identifying or tracking device in any individual. Again, I'm going to return to this critical topic. The potential to manipulate the brain produces overwhelming ethical and political challenges. Will the wirelessly networked brains be hacked by a company or government entity to inject advertising, subconscious thoughts, or recalled memories if it is theoretically possible to connect mind to brain by thinking alone? Imagine the nuisance of a pop-up ad in your head if you are irritated by pop-up advertisements currently on your website. Furthermore, it poses a vast number of questions from a metaphysical, ethical, psychological, and societal perspective for the capacity to implant the "telepathic piping," a neural implant that will enable users to project thinking, feeling, and reception of thoughts and feelings from other people. For example, what if an embedded computer chip crashes after it has been implemented? What will happen? In such a scenario, what forms of health and behavior concerns might arise?

5.4 BIOLOGICAL ENHANCEMENTS

Although much of this book deals with the improvement of technologies in electronics, software, and algorithms and offers a detailed description of what will happen in the future, I will briefly address the latest attempts to enhance human ability by changing DNA and enhancing drug efficiency. Furthermore, the U.S. DARPA agency has been investigating this alternative for fifteen to. twenty years to enable humans to access the internet. Up until recently, the content of science fiction and hit films from Hollywood was human genetics engineering. However, DNA genetic modification is not limited to books and film; scientists and doctors are now working to change humans and our cells genetically. It is necessary to recognize a crucial difference here between care and improvement within the umbrella of genetic modification, in order to understand the decisions that mankind must face in these hundred years as a result of its desire to genetically strengthen civilization. Technically both genetic modification and genetic enhancement are genetic therapy, but significant philosophical variations remain. For decades, scientists have been trying to reverse debilitating genetic disorders using genetic engineering such as gene therapy. Gene therapy succeeds in attempting to repair a faulty gene by injecting a clone of the regular gene into a patient's cells. Hopefully this genetic engineering will prevent or slow the disease. The inserted gene would in many cases create a protein that a patient does not or would not operate because of a genetic mutation. But genetic modification is not gene modification but genetic improvement, for example if you want more muscle for enhancing your athletic ability. The improvement of genes will take an otherwise healthy individual and turn him into more than a man, not only in strength but also in intellect, appearance, or some other attractive attribute. But why is it necessary to distinguish gene therapy from genetic enhancement? Gene therapy helps to recover the regular work of the patient. On the contrary, genetic enhancement purposely and profoundly modifies an individual in ways not expected by design (note that cyborg technologies may perform the same function).

Another essential difference has to be addressed when discussing biological changes for humans. Somatic improvements are those that affect one person, because only one person is affected by genetic changes, which do not necessarily exist in the human genome. While modifications within one person will significantly affect the life of a lonely individual, as these changes are not transmitted to the children of that individual, they are not found in the wider human genome. Germline modifications, by comparison, entail genetic amendments that can be passed down to descendants thereby becoming permanent components of a human genome; any single human being is affected at least indirectly. Such modifications will modify the whole complement of species' genetic features and many agree that this should be achieved with extreme care, even with trepidation, if not outright prohibited [8].

Nanotechnology will have an effect on the very essence of society and has a tremendous potential for creating healthy humans at the same time. Nanotechnology's long-standing objective is to manipulate molecular and nuclear design structures to create atomic machines, for instance, nanobots for production in the body. Since people are built from the same fundamental blocks as the natural world, genetic engineering would allow human tissues and cells to be modified at the molecular level. This would open doors in medicine—which we considered previously impossible—and allow us to expand the span and quality of life of people. It would also open the way for "refinement" of the body, including enhanced IQ, looks, and skills. Many people will certainly benefit from these improvements, but they are also raising important moral, ethical, and juridical issues that human society is just starting to address. There are now biological changes of people in several ways: for example, according to Maxwell Mehlman, head of the Case Western Reserve School of Law's law center, the United States. A drug with the cosmetic effects of lengthening and darkening eyelashes has recently been approved by the Federal Drug Administration (FDA). The medication was also on the market as a therapy for glaucoma, Latisse or bimatoprost. And athletes use both steroids and blood transfusions of recombinant DNA—developed hormones to achieve a competitive advantage. Students have also been documented to apply Ritalin and the new warning medication modafinil to caffeine-containing energy drinks. In addition, the military is now investing millions of dollars annually on biological experiments to improve "cyborg" soldiers' warfare capability. These are all forms of biomedical improvement, strategies to increase efficiency, appearance, or capacities by utilizing medical and biological tools as well as those required for wellbeing to be attained, preserved, or restored.

One of the newest improvements is the DIY biology phenomenon, which promotes open access DNA knowledge (discussing the movement to self-modify the body). This campaign highlights DIY genetic testing and open access to genetic materials, science in particular. The DIY biology movement aims to provide everyone with the tools and services required for biological engineering, even non-professionals. In order to increase public availability of an essential technology, for example, low-cost thermo-cyclers (instruments designed to amplify samples of DNA and RNA via polymerase chain reaction) are created. What about the evolutionary and public policy improvements? Mathew Liao, professor of philosophy and bioethics at New York University, underlines an interesting relationship between genetic improvements and public policies. The idea of Liao was to genetically modify human eyes to work more like cat eyes so we can better be able to see in the dark. Liao explored ways in which humanity can change its nature in fighting "climate change." Liao noted that the need for illumination would be reduced and energy consumption reduced. In view of the resources available for feeding the growing population of the world, Liao talked also about modifying our descendants genetically to a lesser extent. So less is eaten and fewer resources are consumed. The NBA and humanity have a lot to talk of in the face of these recommendations.

In the next centuries it may be possible to significantly enhance social capabilities like intelligence by genetic technology and other cognitive enhancement techniques. However, as the ecology, economics, and biotechnology in the work of author Ronald Bailey point out, right- and left-wing criticism is concerned that the capacity to improve a person's cognitive capacity underlies political equality. Francis Fukuyama, a powerful critic of DNA engineering for human development applications, says that "The political equality enshrined in the Declaration of Independence is based on the empirical fact of natural human equality" in the 2002 book *Our Posthumous Future: Consequences of the Revolution of Biotechnology*. Fukuyama's proposition was that biotech may enable a class of "super beings" to be developed in such a way that "normal" human beings would be orders of magnitude less intelligent, aggressive, propelled, etc. The idea he opposes is that biological enhancements might "allow inequality to be inscribed in the mitochondrial brain." Although this condemnation definitely merits public discussion, some have said that this is a very ineffective reason for opposing it. Enhanced intelligence as a set of core competencies to improving the knowledge skills of smart computing [9]. Those in favor of cognitive improvement point to the already significant differences between people with low versus high IQs, as cognitive inequality is already part of the human genome. They further argue that cognitive development will help minimize political ignorance and increase political equity—at least to promote cognitive equality. As for the issue of equality, cognitive enhancement can be on the same course as many previous technologies of information distribution like books, radio, TV, and computers. Some argue that although they are mostly available to the rich (first adopters) at the start, the cost could decrease over time due to market competition and the rest of society can benefit from them. In the end, the improved knowledge may actually reduce the large "natural" gaps in cognitive ability that exist today, according to some commentators. Once more, we humans need to talk.

5.5 NEW OPPORTUNITIES IN THE 21ST CENTURY

Future technological advancements leading to a human-machine fusion will also give entrepreneurs great possibilities. However according to data from global industrial analysts, the design, production, and appropriation of the artificial extremities are worldwide markets for prosthetics, for example. A "typical" prosthesis at the time of this writing may cost $10,000 to $65,000, and by 2017 the market will increase from $15.3 billion to $23.5 billion. By 2016 the market for augmented reality can grow to $6 billion and demand for real-time information, which includes modern personal health information opportunities in the 21st century, will rise from 14 million devices to 171 million in 2016 that have health data. In addition, there is increasing demand for prostheses due to an aging population and an increasing prevalence of health problems such as diabetes and degenerative joint disease conditions such as arthritis and osteoporosis. And Singularity already has value when it comes to cyberspace. In 2004, when David Storey was bought in virtual Entropia for $265,000, he was the holder of the Guinness World Record for "Most Valuable Object which is Virtual." Storey has built on the island a virtual rare game preserve company, claiming to have earned around $100,000. However, you do not have to play in the World of Warcraft or pay the entrance fee to the Entropia club or purchase virtual swords for the digital marketplace. You pay real money for a virtual good when you're on Facebook and buy a birthday cake icon for a friend.

What about cyberspace law and financial transactions? Take the creation of "Bitcoin," an open-access digital currency that has not been used by a central government or banks to pay for products and services with peer-to-peer technology. Bitcoin is often the only approved method of cyberspace payment. It seems, though, that government rules and legislation are very close behind when financial transactions take place. And just on this basis, in an effort to decide whether the government could control cyberspace industry, the New York State Financial Services Department released summonses for digital money firms and investors. Why does the state require cyberspace regulations? Because Bitcoin proceeds to increase all that; an individual can buy with digital currencies, from burgers to art, including pricey vehicles, as a person who is using Bitcoin purchased a Tesla Model S from a Lamborghini dealership in Newport Beach, CA, who was the first dealer to embrace Bitcoin as a type of payment. Bitcoin has gotten a mixed reaction around the world, with the central bank in China banking creditors on the virtual currency. In addition to stating that it operates on this topic, the United States Internal Revenue Service has not provided advice on Bitcoin and since 2007 it has tracked digital currencies and transactions. Interestingly, the relation between digital currencies and cyborg technology is also created. A Dutch businessman has inserted two wireless chips into his hands under the skin so he can hold digital monies such as Bitcoin in his body. In and around his native Amsterdam and around Europe, Martijn Wismoijer is the founder of Mr. Bitcoin and runs a crypto-currency cash business.

Significantly, Martijn preferred a painful technique for embedding NFC chips under his skin (nearfield communication). These chips can be read on a number of computers and tailored to a range of applications, including smartphones.

Tomorrow this market will expand several times in a world where sustainability and progress in health become a prime commodity in another example of economic potential to grow this century, just as existing markets for cosmetic surgery, mood-altering medications, and even beauty and exercise aid in the billions of dollars. The Federal Food, Drug and Cosmetics Act and the Act on PSG in the USA strictly control advances in medical science, particularly bio-enhancing medicine. Under these actions, the United States FDA controls a wide variety of products, whereas various products are handled in differing respects. Some products are subject to "pre-marketing authorization," such as medications, machines, biological compounds, food and color additives, but most products are not. Premarket authorization ensures that the FDA will, inter alia, compel manufacturers to supply the agency with the requisite research evidence on protection and product efficacy. In addition to pre-market research, the FDA has the responsibility to find safety issues with the product on the market, to remove special product variants, or to absolutely prohibit unsafe goods, as required for customer and patient health purposes. New drug approvals are much more complex and those designing cyborg technology should take such guidelines into account. The aim of the clinical trial is to obtain enough evidence to decide if experimental medicines are suitable for human use. Will the FDA control hardware and software updates that affect their well-being if artificial intelligent machines become legally legalistic? Would any government agency address their needs? In the context of financial transactions certainly the law of contracts would be involved. The current emphasis on human-centered law will need to be updated in the future to resolve these and other concerns. FDA provisions actually state little directly about cyborgs or artificial intelligence devices promoting rights, whereas the FDA laws protect the prosthesis and counseling obtained by individuals falling under the concept of "bionic person or cyborg." Instead of waiting for regulatory approval, auto-directed body hackers use self-determined sensors and other implantable devices and take matters into their own hands.

Issie Lapowski states that "for decades the ability for artificial intelligence has been largely limited to Hollywood directors' significantly bigger collective imagination," and says it all appears to be unfolding in a long and dark future, from the *Blade Runner* to the *Terminator*. And yet if one thing is learned from the latest purchase by Google of DeepMind, an artificial intelligence firm, it's not a century or even decades away from the present day for this sort of technology. The global artificial intelligence industry has already been measured.

The market research group Research and Markets reported new prospects in the 21st Century at $900 million in 2013. Research from Oxford University has shown that about half of all U.S. work will be taken by artificially intelligent machines in the near future. Some people consider this horrible news, but it's still a great chance for businessmen to innovate.

I agree with some commentators who envisage a number of major markets for emerging artificial intelligence applications. The big data industry has evolved for years, but while there are a lot of technologies that can crown statistics and fill them up on a tablet or map, there is a gap in the way that the data is available and genuinely knowing. According to Issie Lapowski, staff writer for *Wired*, the first thing is to consider big data. Today, developers are starting to fill this technical vacuum, which not only sums up the data but also interprets it. One such business, Narrative Science based in Chicago, has created a program, entitled Quill, which provides users with a written report on the data. According to Lapowski, the second biggest demand for artificial intelligence is smarter robots. Robots performing basic, human-controlled production processes are not far over, but start-ups are on the rush to develop a better robot brain and sensors that allow robots to work autonomously. There's of course Baxter, which is currently on the market and can potentially be educated by the famed research robot Rethink Robotics. Others, such as Hanson Robotics, developed amazingly human-like robots, able to speak and remember personal history. Third, Lapowski reports that artificial intelligence is going to contribute to intelligent helpers. Amazing Labs also has a personal assistant app that not only tells you about where to go, how to get there, and how to save your choice; it advises you not only when to designate but also when to leave. Jarvis Corp. takes this a step further, a start-up that now is in the conception process of creating a virtual assistant that can navigate and give information on the Internet, but it can work as an internet server and even control all connected devices in a building. Artificial intelligence is no longer only used to handle requests and synthesize results. Today, some start-ups are also designing feeling-wise technology, which is known as affective calculation. The start-up in Tel Aviv, Outside Auditory, according to Lapowski, "uses techniques to evaluate vocal intonations in order to assess the mood of an entity." The idea is that artificial smart technology, by understanding emotions, can make dramatically more humane predictions of a person's interest. Naturally, by telling "them" how to understand us, we will open the Box of Pandora to send us the data we will need to tackle artificial intelligence.

5.6 CYBORGS AND VIRTUAL REALITY

Over this century, leading robotics experts and artificial intelligence scientists have expected that artificial smart machines will take on a much more of a human-like look, emotion, or reach or potentially exceed human intellect. This machinery will come into society, enter into contracts with humans, and possibly argue for civil rights, such as "individual rights" and liberty, and it will exist in human form. In this century too, people will have much more industrial equipment and electronic power than they do today; the result is bionic living things and cyborgs. "Virtual reality" is much more practical and interactive in the mid-21st century than it is today, so humans, cyborgs, intelligent autonomous machinery, and artificial intelligence-based avatars (sometimes referred to as virtual human or digital persons) can spend time living in Singularity in which they are creating governments, designing, buying, and selling goods and engaging multiple people in them. If artificially intelligent robotic avatars exist in future Virtual Reality, some act as our own digital helpers, others work for smart machines and others portray themselves, how are we humans going to cope with the clever virtual avatars we meet in Singularity? As I have written in *The Akron Law Review*, how would smart virtual avatars look at the legal system? Are smart avatars constitutionally entitled? Are they people, are they allowed to vote or to marry or are they able to demand progeny by means of genetic algorithms? Will it be deemed an intrusion and a battery to upload a computer virus? And where are disputes with virtual avatars roaming the Internet going to lie? In addition, are smart avatars allowed to be "treated" if compromised by a virus of a computer? In this respect, Joanne Pransky, a lecturer on social dimensions of robotics, sent me a card at the 2013 Stanford Law School conference after I had spoken, which tongue-in-cheek was the first robotic psychiatrist to bring to me.

Where do augmented reality, smart systems, and cyborg technology advances lead in the end? Some theorists also suggested that the convergence of this technology, together with advances of nanotechnology, would lead to "posthumans," a phrase some analysts use to apply to potential organisms that would be too far beyond that of today's citizens and may not have simple skills. By our modern expectations, longer human beings completely depend on machine parts. What would the posthumans be like? Posthumans may be in a number of ways artificial intelligences (such as human-like robots); human cognition can be transmitted to electronic machinery or the Internet or several smaller but more cumulatively deep increases may be triggered by a bioman. In this concept, either a reconstruction of the human organic using nanotechnology or a fundamental development of the human body using a certain mix of technologies, including genetics and advanced prosthetics, is likely to be needed.

The forecasts of human beings fused into machines and artificial intelligence that match human intelligence and then outnumber human intelligence are courageous and questionable and not simple for many to recognize, but the future moves at an incredible speed with a cliché. Indeed, there's already a fluid difference between human and machine. In our present age, a person can be fitted with a prosthesis of the retina, a cochlear implant, an artificial hip, heart, kidney and limbs, implanted sensors, and heart paste. In addition, people like University of Toronto Professor Steve Mann have used computers for thirty years and "packing the heat" as Steve told me years ago. In addition, by taking part in a group of studies known as the cyborg Project, Professor Kevin Warrick of the University of Reading pioneered the movement toward the cyborg future. A basic sensor was inserted under the skin of Warwick in the first stage of the research that began in 1998 and was used to monitor doors, lamps, heaters, and other computer-controlled devices based on their proximity to them. The second stage consisted of a more complicated neural interface consisting entirely of an internal electromagnetic array (composed of 100 electrodes), subjected to an external "gauntlet" with electronic assistance. The electrode set was introduced into the arm of Warwick in 2002 and directly interfaced with the median nerve of Warwick. The presentation succeeded, and the signal that was sent was sufficiently detailed to imitate Warwick's own arm's actions with a robot.

In developing the technology for the improvement of the human organism, whether necessarily or for the development of people with skills beyond current people, we are modifying the ratio of human to machine components, an idea promoted by Ray Kurzweil and Terry Grossman in their 2005 book *Fantastic Voyage: Live Length Enough to Live Forever*. We can assume that $c = m/h$ is equivalent to cyborg; that h is equivalent to human components and m is equivalent to machine components. However, it is not a simple human/machine component ratio that determines the grade of "cyborne" but rather how much data processing is carried out by the human or machine components of the cyborg/human being. Thus, the following relation can be established: $C = Kin (mi/hi)$, which represents information number of bits passed by the body or by the mechanical structure of the subscript (the human brain is a petaflop biological computer). We currently do not know the capabilities of information processing in different body parts or biological functions, but I do think that the idea that the degree of cybornness should be linked to theory of information is worthwhile (and heavily weights the information processing capabilities of the human brain). Every technological progress alone will not alter the ratio of human biological to mechanical components significantly if brain prosthesis breakthrough is decisive for data processing if you consider the quantity of human limbs, sensors, and internal structures (like heart or liver) that can be substituted. The "Cyborg" ratio is obviously definitely changing to benefit the machine, strengthened by emerging technologies.

Innovations of cyborg technology inquire, "Where is the ending of man and the beginning of the machine? Humanity will definitely need to resolve this problem sooner than expected. In some cases, a regulation that would not strengthen technology on a person (the existing majority) may not be applicable to someone with an improved brain arm and leg and more strength and ability for information processing than a non-enhanced person. For example, the controversy about whether steroids, medicines, or technologically improved people should be able to play against those that may not have those changes is now raging. From a moral point of view, should individuals with technical advances be recognized as a different class by society? And if so, will they be treated as "protected" class, or, by contrast, were there no enhanced citizens deemed to be a protected class? (In the United States, they will be protected by the 14th Amendment.) The concerns of civil law posed by technologically sophisticated institutions will contribute to interesting cases heard by the Supreme Court and foreign courts.

5.7 CYBORG DISPUTES

Another problem with cyborgs is what obligations will anyone who interrupts their "computing prosthesis" have? If a human, for example, interferes with a signal sent to a wearable device of an individual, will that person be responsible for helping the individual interpret and experience the world? On this point alone, take, for example, the United States. In 2007, Dick Cheney, Vice President, fitted with a pacemaker, disabled his wireless function.

There are currently two legal controversies about the issue of human contact with cyborgs that affect the interests of Steve Mann, a University of Toronto Professor of Engineering. For decades, Steve has operated as a cyborg using machines and electronic devices to increase his memory, boost his vision and keep track of his vital signs. In 2002, Steve was examined and apparently wounded by a police team while boarding a Toronto-bound airplane at St. John's International Airport (Newfoundland). In this event, loads of dollars of his appliances, including that of the eye lenses which serve as his display screen, were allegedly lost or destroyed. Steve used the routine on prior flights before flying. He told the Toronto airport security guards that he has previously informed the airline of his devices and shown them papers—some of which his doctor has signed—and identified the wires and glasses he is wearing as part of his wearable computer testing. Steve found it impossible to navigate normally without a completely functioning device and was confirmed to have fallen in the airport at least twice. In fact, the number of airline passengers who are exposed to longer checks is growing as they have cardiac equipment and artificial joints and bones. No figures are made of the number of individuals that travel through control areas with implants and cybernetic upgrades in the United States. As the vast number of baby boomers ages, the Transport Security Administration (TSA) expects more. For example, the demand for orthopedic implants is twice as high as five years ago. The TSA aims at improved testing of implants including pacemakers and defibrillators, heartbeat-controlled life-sparing equipment, and implants for orthopedic use, such as hips and knees. Steve argues that he should undergo the same therapy as someone who wants specialized devices like wheelchairs, depending on his position as a cyborg; this belief should definitely be the focus of discussion on public policies and potentially governmental intervention. But why argue a matter where only a few self-professed cyborgs are having an influence at the moment—so more cyborgs arrive and fast (and more than you even believe are here!). For instance, several million people have arm or leg prosthesis, important steps are being taken to improve brain-computer interfaces, and millions are being spent by the military in making Cyborg Warriors.

The debates about the course of our cyborg future often took place in restaurants. Seattle's Lost Lake Cafe took a strong stance against Google Glass and other wearable computers by kicking a guest who was using his device from the premises. Despite varied reviews from the local population, the restaurant adheres to its no glass policy. In a greater event, the Columbus theater owner Ohio saw a threat from Google Glass great enough to contact the Homeland Security Police. The Homeland Security officer disconnected the Google Glass programmer attached to his glasses. In comparison, a high-tech crowd-frequented San Francisco bar forbid customers to wear Google Glass when they were inside the house. In actuality, San Francisco appears to be a zero-rated social networking specialist in the event of cyborg disputes, as glass wears in the bar in San Francisco, stating that she has been tormented by clients who are objecting to her wearing the equipment in the bar. A *Business Insider* writer Kyle Russell has said he was snatching his Google Glass off his face and crushing it to earth in the mission district of San Francisco.

Ray Kurzweil, a well-known futurist caller, was assaulted in Paris by Steve, making it the first known attack on a cyborg in history. Similar to those who use Google Glass, this group also has access to prosthetic cameras. Is the attack a reference of a cyborg hate crime? Hate offences contain two components from a legal analysis: a prejudicially committed criminal offence. At first sight, Steve's events tend to fulfill both facets. The first characteristic of a hate crime is that a felony under ordinary criminal law is perpetrated. In Steve's scenario, the simple crime is possibly an aggression and a battery. This illegal act was also referred to as the "foundation offense." Although the legal provisions from country to country vary little, there are some discrepancies. It is a felony, but most countries usually criminalize the same kind of acts of violence. Hate crimes would always have been a simple felony. The second characteristic of a hate crime is that the crime was done for a special cause called "bias." It is the "bias motive" factor that separates hate crimes from minor infractions. This indicates that owing to such protected features the suspect has purposely selected the object of the offense (typical of a protected class). Steve will find it impossible to prove a hate crime—cyborgs are not seen as a safe community.

What is a covered class, a category that cannot directly be discriminated against? A protected group usually consists of persons with features similar to the category, such as "race," language, faith, ethnicity, nationality, etc. Interestingly enough, artificially smart machines speak a specific binary language with frequent physical features; they appear at first sight to have the hallmark of a "class." Public policies and law are dealing with this. However, implicitly, a court could have offered us a look at the future in the case of the Supreme Court. Justice Ginsburg, who reports on the policy conclusions of the Americans with Disability Act (ADA), said that "disabled persons are an insular and unobtrusive minority," "subjected to a tradition of purposeful disadvantage and reduced to a position of political impotence within our community." Obviously, cyborgs like Professor Mann should be deemed a member of a worthy class or not. Specific provision security is an intricate topic that must be debated by the public and the legislature.

Apart from wearable technology FDA rules, some jurisdictions have just started to govern cyborg technology as medical devices control fitness. For example, a limited number of U.S. states have a ban on the wearable computers when driving in the sparsely populated region of Wyoming, because of fears that drivers with Google Glass may pay more attention to their emails or other online content than to the road. And a Google Glass driver was ticked for the show in a high-profile California case that posed new concerns about distracted driving. Since driving slower for safety sake, she received the citation for driving with a "digital" monitor, which is against California law. Later, because of lack of documentation that the system actually went on during the driving process, the ticket was refused. Furthermore, Davin Levine comments that Google lobbied politicians in at least three U.S. states to end planned bans on driving with headsets like Google Glass, which marked some of the first clashes around the latest wearable devices in order to demonstrate how businesses are powering and shaping the controversy over our cyborg future.

5.8 TWO TECHNOLOGICALLY DRIVEN REVOLUTIONS

Rodney Brooks, former Director, Computer Science, MIT and now Chairman of Rethink Robotics, thought that there would be two technological developments as we discussed what could happen during the 21st century. It is interesting because when Brooks mentions his artificially smart robot, he often uses the term "artificial organism" to describe it. The first is called "robotic revolution" and the second is "biology revolution." Typically, you refer to a living being with the word "creature." But Brook's robotics are software, sensing, and none will say seriously that they are alive in the same way as humans or other living beings are—as artificial components, such as effectors, actuators, and servomotors. But what if robots keep gaining insight and pretend to be aware and alive one day? How will this progress be seen by society and the legal system? Will such "creatures," unlike the maker, be denied rights? Will they be residents, vote, or have land of their own? May they be falsely blamed for any damage arising from their acts, or culpable under criminal law? Brooks' is more detailed on these problems and includes a description of the rule of cyborgs and the emergence in the 21st century of artificial intelligence.

Mankind will be forced to face these very issues during this century as technological development is increasingly moving to smarter computers that are autonomous, which are more and more like people in shape and actions, separate from human programmers. Will intelligently artificial machinery be responsible for harms arising from its behavior because they are aware of their actions and whether they can reason through and map out their conduct? Brooks made some insightful remarks on law and policy significance as he suggested that mankind would relate in ways other than previous machinery to intelligent robots and that the next robotic revolution would transform society's basic nature. How can people relate to a smart robot? Will they be our equal under law, our land, our self-employed or some other partnership yet to be defined? And will they therefore enjoy the privileges that people receive, being considered a legitimate individual under the law?

Any variations are in order at this point. In case law, a natural person, in contrast to a legal citizen, is an individual who is a private entity (i.e. a company) or a state (i.e., a public) entity. Indeed, in the United States, the statute grants non-living organizations personality rights. The legal conception of a corporation as an individual in the view of the law is the corporate person. Corporations, for instance, can contract and sue or punish other parties in the same way as natural or non-corporate relationships. The theory of the corporate identity does not state that businesses are "humans"; it does not grant all natural citizens' rights, except its shareholders, officers, and directors. In certain cases, only natural individuals are guaranteed universal human rights indirectly. For example, Section 15 of the Canadian Charter of Rights and Freedoms, which provides that a person cannot be denied the right to vote on the grounds of sex, applies to natural persons only. This does not extend to unnatural persons. Another indication of how natural and legal citizens are differentiated is that a natural individual can have public office, but no business. Of course artificially intelligent robots (bionically equipped humans and existing models of cyborgs are) are not known to be legal persons, but the company's doctrine definitely presents a precedence for non-human persons to be regarded as legal persons.

Most people know Isaac Asimov's three robotic rules as well as laws that can apply to how artificially intelligent robots are concerned. The first says a robot cannot harm a person or permit a person to harm.

The law is that a robot is expected to obey the commands given to it by humans, unless those commands are contradictory to the first rules. But the third law says that, as long as this does not interfere with the first or second law, the robot must preserve its life. While these laws have led to much discussion since they were first written in the 1942 short story "Runaround," they do not say anything about many areas of law that should be seen in the light of intelligence gained by robots. How much accountability for making enforceable contracts do artificially smart robots take, for example? Can they be human agents or can people be genetically smart robots' agents? Will smart robotics digitally possess physical property or gain intellectual property rights? And can smart robotics artificially bequest property (as software?) for smart computers for coming generations? This are just a handful of the issues that society should consider during this century, both legal and political.

Throughout culture, increasingly, the notion of identity has grown dramatically. Women, infants, and slaves have been used as property at times rather than as individuals throughout history. The group of individuals admitted to courts has been extended to cover groups such as mothers, slaves, human beings, illegal children, and minors and non-natural or legal persons including businesses, trade unions, nursing homes, municipalities, and government units. There is obviously no argument on morality, emotion, or vitality from legitimate personality. However, being a legal person is being willing, under a specific legal framework, to have legal rights and responsibilities, the right to conclude, own land, sue and be sued, for example. Not all legal individuals have the same rights and responsibilities, and only for some and not others are some organizations known as "persons." New personality categories are decision-making, not discovery. Personality is an appraisal created to assign an entity rights and responsibilities, regardless of how they appear and whether they will be passed on to individuals. As Mark Goldfelder has stated: in order to imbue AI robots with personalities in a way that makes the greatest moral and logical sense for a set of actions, it is not necessary to prove that a "human" can behave in a "person" like manner.

At what point does this body step away from the role of property to personhood, at which point is the issue of the autonomy of artificially intelligent machines? Some people believe it makes sense to have legitimate personality for artificially smart robots in the near future. They claim that artificial intelligence already belongs to our everyday lives. Bots, for example, sell products on Amazon and eBay and semi-autonomous agents determine our qualifications for Medicare and other government programs. Predator drones are becoming ever less tracked, and industrial robots are becoming more prevalent. Google is researching self-driving automobiles and General Motors has suggested that in few years it expects semi-autonomous vehicles to be on the track. However, when the robot behaves individually, as it undoubtedly would, who is to blame? The computer store that sold it? The owner who was not interested in the mechanical failure? Or the person who took the risk of engaging with the robot? When a robotic car hits another vehicle or simply passes a red light, what happens? Legal analysts argue that robots should be given legal personality in order to delegate responsibility. The robot could take protection from its employer like a legal individual. As an autonomous agent he would be able to punish other people for the wrongs that gave the economy a sense of equity, which would ensure that the twin concerns of financial collapse and the inability to raise trade were uncontrolled [10].

With respect to the next revolution, Brooks discussed biotechnology and how technology of our bodies and computers will be changed. On this point, Brooks dreamed of a world in which robots would become more like humans and machines. In this sense, Cynthia Breazeal has built an especially fascinating Leonardo robot and a variety of other personal robots, one from Brooks and now from a media arts and technology professor at MIT. Leonardo has the ability to respond to people by adjusting their facial expressions and turning their heads as they speak. Curiously, people who worked with Leonardo appeared to be aware to some degree of Leonardo. Even if Leonardo obviously doesn't know about his own life, people think of the robot as an individual by behaving in a more human-like and social way. If such a reaction happens to robots of such rudimentary intellect as Leonardo, think how the people will respond in ten to twenty years, when robots are much smarter and similar to humans in shape and behavior. A fascinating concept about how individuals respond emotionally to artificial smart machines that approach human likeness is discussed in a later chapter.

5.9 MERGING WITH MACHINES

Hans Moravec, previously director of the Robotics Institute at Carnegie Mellon University, is also a pioneering scientist in developing artificially intelligent robots. Moravec, who studied robotic technology at Stanford University, holds a far better stance than Brooks as he speaks about the future of people and artificially smart machinery, proposing that the fate of people ultimately fuses with machinery. The 1980s and 1990s robots could only think at an insect level, basically fitted with sensory and motor capacities in the crude navigation of surroundings, as Moravec expresses in his 1998 book *Robot Mere Computer to Transcendent Mind*. Yet instead of the exponential condition, because of the increasing computation ability over the past twenty-five years, based on algorithm development, he projected that robots will become as intelligent as humans by the mid-century and eventually begin their own evolutionary process, which, according to Moravec, would extinguish human beings in our present form. Moravec argues nevertheless that humans need not be afraid because he concludes that combining with intelligent machines is, as he puts it, man's greatest future, the true means of human transcendence.

Moravec is not the first leading scientist to expect that people will merge with robots someday. Microsoft's designer, futurist, and researcher, Ray Kurzweil, has argued in the same way for several books on artificial intelligence and human fate. Interestingly, in particular in computational power, Kurzweil considers scientific advancement as the extension of the evolution process. Kurzweil suggests that far away from being a far-off science fiction fantasy, hybrids of humans and robots will emerge sooner or later. This prediction is based on the seminal book of Kurzweils *The Singularity Is Near: As Humans Transcend Biology (The Illumination of Biology)*, the accelerating returns law. Essentially, Kurzweil notes that development takes place at an exponentially low rate Essentially, Kurzweil suggests that development happens at an accelerated rate—change is incredibly sluggish at the lower end of the exponential curve; for example, eons elapsed between the appearance of single-cell microorganisms and the advent of homo sapiens. But until homo sapiens began to evolve technologies, hunter-collectors could only develop a technology that ultimately leads to computers for around 10,000 to 12,000 years. Once computer has been developed, the rule of Moore that says that every eighteen months' microprocessor power doubles has become a force in the growth of computer technology. The accelerating return rule of Kurzweil indicates that this explanatory speed controls splicing attempts, increasing the speed returning rule. Kurzweil says that this same exponential speed rules attempts at splicing DNA, refusing genomes, reverse engineering the brain, and creating nanotechnology devices. In view of all these advances and expansion speeds, Kurzweil assumes it is probable that emerging technologies are being implemented in human biological processes by our own technical creations. Kurzweil, well known for forecasting how people will fuse with robots, for example, has also written that one-day people dependent on algorithms will reside on the Internet, projecting bodies as necessary and choosing it, like simulated bodies in different domains of Singularity.

Given the forecasts for the future of the human race, in particular, the ongoing creation of intelligent computers, which will gradually become human or intellectual, allow humans to integrate with our smart computer creations and allow software copies of humans to exist on the internet; they are likely to lift the most important if these predictions are fulfilled. Since Kurzweil, Moravec, and Brooks' predictions could change society deeply, humanity will have to be careful to have a full debate about this issue. These future results are beneficial.

But before further addressing the legal, political, and technological problems that might emerge, if the aforementioned forecasts come to pass, let us take into consideration for a moment that predictions are incorrect, that human destiny is not to integrate with computers, or robots may ultimately not grow consciousness and intellect, or beyond that. In spite of this, the future of human beings will obtain more and more non-biological features, such as to regulate diabetes or the operation of the kidneys and even to improve cochlear, retinal, or body members' prosthetic therapy, attributable to attempts to combat cancer, to improve diseased processes, or to fix damaged anatomy or to provide human beings with enhanced prosthesis of cochlear, retinal, or body limbs. And the more mechanical components replace, the more the decision makers and the general population are wondering whether the resulting mix of human and machines is really human. In addition, even though machines are never aware and human level of intelligence as some of them predicted will happen this century, progress is made with regard to artificial intelligence already leading to machines that are considered to be "smart," even if in a small realm, by any measure of intelligence. Much of this is going to happen. These innovations in many fields of law alone will face big legal problems, such as in the world of electronic commerce where intelligent digital agents enter into contracts under the guidance of their employees.

That we integrate with a computer is certainly an incredibly contentious prediction, but one thing is certain: because there is a shortage of biological mechanisms or fixing anatomical structures, many people are increasingly getting more cyborg-like given the integration of technology in their body. In the U.S. alone 8–10% of the population is artificially changed by about twenty-five million individuals or Bionics, according to physician Sidney Perkowitz of Johns Hopkins University. Dr. Ross Davis and his colleagues are working at the Maine Neural Engineering Clinic. This party used the implantation chip technology in the brain to treat patients who have impaired or been affected by disorders such as multiple sclerosis in the central nervous system. In addition, a transmitter system has been inserted into a patient's brain by a team at Emory University in Atlanta. Using thinking alone, a research patient could move a mouse on a computer display after connecting the motor neurons with silicone. The effect of this is that a person was able, though in a primitive way, to relay thinking signals directly to a machine. The Emory team strives to increasingly expand the patient's spectrum of controls. Many scholars speculate that thinking-to-thought connectivity would be a critical function of cybernetics if we face the risk of being replaced by very intelligent machines. A corresponding chapter describes recent advances in telepathic communication thinking use and controlling machines. However, society should enter into a dialogue on three critical topics before such incidents arise at all: (1) must there be a cap on the improvement, increase, or substitution of biological components of human beings? (2) Should we or should we not build robots superior to unenhanced humans in intelligence? And (3) can we continue to build isolated from artificially smart machines on a different path?

5.10 QUESTIONS FOR OUR CYBORG FUTURE

In reaction to quick technology and innovation advancements in the capacity to better the human body and brain, the first critically important issue for mankind is whether there is to be a cap on human biological parts improving, growing, or eliminating them. This topic poses a range of significant ethical, legal, and public policy concerns with regard to human enhancements. For starters, will only the rich be able to make changes, and would we establish a population with supervisors strengthened by cyborgs and a community of people too poor to procure improvements? If cyborgs are fitted with "separate" emotional, sensory, visual, or engine prosthesis, will cyborgs be granted greater consideration according to the law (Remember Steve Mann's altercations) if they are fitted with cognitive, audible, visual, or motor prosthesis that "separate" them from non-enhanced people? The rule of the manipulation of the body and of the hacking of a body is the focus of a chapter in this book. However, it must be pointed out briefly in this chapter.

If the human being is improved by disabilities, there is protection of these individuals in many jurisdictions worldwide. The ADA, for example, provides protection for workers of the United States under the Americans with Disabilities Act (ADA). In the United States, for example, the ADA protects workers who are disabled and allows employers, if practicable, to deal with the disabilities. At present an applicant must, however, be a professional worker to have a legally defined condition for insurance in order to be protected by the ADA. An example is a mental or physical condition that greatly restricts a vital life function (such as the ability to walk, talk, see, hear, breathe, reason, work, or take care of oneself). Bionic individuals will most certainly be disabled when they are strengthened to restore or replace human anatomy or physiology. They will certainly be subject to ADA, but cyborgs cannot be shielded under the ADA because they are improved for purposes other than medical reasons, such as increasing human capability outside the normal range.

In artificially intelligent devices, the second critical question for humans to examine is whether or not we can construct machines that are intelligently superior to the advanced. Computers are now "wiser" than humans in certain ways, of course, but with this question I mean computers with "solid artificial intelligence," i.e., knowledge, sense, and capacity to accomplish any intelligent activity that a human being can effectively perform. This is not the case in this regard. Professor Stephen Hawking from the University of Cambridge, a former Lucasian Professor for Mathematics, conveyed grave concern that an eventual challenge to mankind would one day "take smart machines over the earth." Hawking said machines developed so fast that they inevitably transgressed human intelligence and that artificial intelligence computers could thereby conquer the planet. Hawking promoted improvements in human DNA through evolution to hold computational science advancements ahead. He has also argued that we should speed up the expansion between the minds and machines. Innovations allow a direct link with machines such that artificial minds contribute rather than contradict human intelligence. Analysis promoting the assumption that a human mind can interact directly with a machine and other networked minds continues to find that this is another matter than transferring data from a computer into a mind.

With regards to Hawking's suggestion to human beings with genetic engineering as a way of keeping touch with artificial intelligence, Ray Kurzweil pointed to genetic engineering during the birth phase as incredibly sluggish as machines gather intelligence at an exponential rate. According to Kurzweil, the age of robots outside the human realm was already upon us as the first genetically modified generation was created. Even if we are years away from the genetic modification of a human being, for example, if we start the clock by 2014—reminder of the Moore rule, computing power doubles every 18 months—since people become legal adults at 18, by 2043, as human beings officially recognized as adults, there will be more than one machine power doubling. This would in fact lead to an incredibly computational machine that could see, understand, and think the universe, particularly if we remember that the fastest possible supercomputer now runs on the several petaflops (a petaflop is one thousand million floating point operations per second).

As regards genetic manipulation, Kurzweil further argues that while genetic modifications are to be introduced to adult humans with new genetic material through genetic pharmacological treatments, biological knowledge also cannot take the lead. Engineering biology (by way of genetic procedure (whether by genetic therapy for infants or adults)) is inherently DNA dependent. The speed of a signal propagation and its ability reduced relative to the capability of an artificially intelligent computer are still incredibly sluggish in the DNA-based brain. For example, machines are now 100 million times quicker than our electrical circuits (i.e., neuronal) and we do not have rapid ports downloaded to our biological neurotransmitter levels to rapidly transport vast volumes of data between the human condition and the machine. We can make scientists and engineers intelligent, but this strategy won't start to keep track with the exponential pace. We should create a smarter bioengineer.

The third crucial question that must be answered by humans about our technological future is whether or not we can develop separately from artificially intelligent machinery. The dilemma seems to be whether mankind can continue to progress under the slow method of bio-evolution (the actual case), develop under DNA-modifications fairly quickly, or combine with artificially smart machines and evolve at a technology level. Evolution is not easy. Our genetic makeup requires several years to adapt to changing conditions and circumstances. Ted Driscoll of Clarement Creek Projects suggests that our human genome of the 21st century is still simply a man's genome. We have a very well-suited genome for the world of our hunter gathering ancestors, as it took tens of thousands of years to preserve our environment. Sadly, we have no similarity to the prehistoric world in our 21st-century world. In comparison, most technological advances in recent decades have happened, and continued changes are only escalating.

5.11 THE REEMERGENCE OF LUDDITES

Few people wondered whether people would alter the fundamental essence and function of their lives or choose to stay the same (that is, technologically enhanced). Some people asked. For purposes discussed in this chapter, it can be argued strongly that people would use their body as well as their brain for technical and biological advances. Yet those that oppose technology from an ancient point of view have come to be known as "Luddites." What is the sense of this term? Legend has it that a "failureous man" called Ned Ludd split two storage pictures in a Nottingham warehouse. From then on, until one of his costly offending plant owners had discovered machinery bits unexpectedly destroyed, Ned Ludd was conveniently responsible for the damage. The word has a good historical footing too, though. During the early days of the industrialization, workers (or Luddites) started to split up in the evening to smash the new machinery that the capitalists used, angered by pay reductions or the use of non-students. In 1812, the British Parliament approved the Frame Breaking Act in reaction to the Movement of Luddites, which culminated in the sentence of people guilty of machine breaking. In the regions where the Luddites were involved, the British government ordered 12,000 troops as a further measure.

Looking from the lens of history, the Luddites' activities in the early 1800s were seen as counter-revolutionary to the "Scientific Revolution." Considering that 90% of Americans were employed in agriculture in 1890, and it was just 41% in 1900 and that it was only 2% by 2000, we have to ask ourselves if there would be a similar trend of workforce displacement in careers that require a specific cognitive skill. As an example, the language-efficient machine, IBM's Watson, has defeated the best human champions on Jeopardy's TV game show lately. Since mixing experience with wizzes from human game shows, Watson has now moved to become a diagnostic specialist after mixing wits with wizzes in human sports. IBM believes that Watson is more able to process and interpret large volumes of data than many human physicians. Watson was given the task of "reading" peer-reviewed medical papers of oncology, concentrating on lung, prostatic, and breast cancers, following an average second-year student's level of experience. Ian Stedman said, "The additional capacity of Watson to look at up to 1,5 million patient reports for additional details lies in absorption of over 600,000 medically-based facts, over two million pages from medical articles. No human doctor will balance it with a depth of expertise." When industrial technology performs much of the physical work functions that experts once used to do, and if artificially intelligent machinery performs human cognitive work, it is no wonder people like Hans Moravec and Ray Kurzweil talk about combining people with artificially intelligent machines.

Today, both blue- and white-collar employees continue to undergo artificial intelligence and robot technologies, with researchers estimating that robots will transfer more human labor than they produce by 2035. By 2035, robotics and artificial intelligence will no longer be limited to repeated jobs in a manufacturing line as they continue to progress at the same rate as in recent years. Can the future be improved or not through sophisticated artificial intelligence and robots? It is generally believed that in the next few years robotics and artificial intelligence will displace more workers as the general-purpose robot gets older. Though these early bots for general purposes will not be as fast or agile as humans initially, they will be flexible enough to carry out various menial tasks 24/7—which will cost only a few cents of power instead of minimum wage. On the other hand, robotics can regulate the workplace as easily as possible, so it is difficult to sustain our commercial, educational, and political structures. Robots have been replacing mostly blue-collar jobs, but the next generation will substitute more and more skilled and trained white-collar workers. Many of these skilled workers may be unemployed and lack the resources to pursue a new one.

Back to artificially intelligent robots and the Luddites. When artificially clever machines become overwhelmingly capable of cognitive functions, could the anticipated work loss in many service industries lead to a new wave of people voicing animosity towards clever machines? A wave of the neo-Luddites has arisen. Ted Kaczynski, also known as Unabomber, was sentenced to a life imprisonment, the most severe expression of this philosophy. In the *New York Times*, his manifesto claimed that the "Industrial Revolution and its results were a tragedy for the human race." Steve Mann is also a pioneering pioneer of cyborg technologies. In his sponsorship of the cyborg cause, Steve Mann is also hesitant in claiming in *Singularity 1 of 1* that

> I'm not asking about more or less technology—I tell the technology. We should have technology that is compatible with nature, instead of industrial waste. We should combine it rather than replace it with technology. We should use humanistic intelligence rather than substitute intelligence with artificial intelligence.

New technology focused largely on economic considerations was criticized by the 1800s Luddites—the technology was viewed as capable of replacing the human expertise of the textile industry, the skills that are required to secure a livelihood and benefit their families. In the present day, for reasons other than fundamental economics, people can object to technology; for example, they contend that, to remain humans, we have to reject human fusion. However, supporters of technology advancement reject the presence of certain causes of why individuals should be enhanced or improved. For example, one in every sixty-five persons is ill with Alzheimer's and half over eighty-five have Alzheimer's disease. Up to 100 trillion dollars per year are spent on the epidemic, by customers and insurance providers in the U.S. alone. How can human technologies benefit Alzheimer's sufferers? It is a big problem for the millions of relatives of people who struggle with Alzheimer's. Doctors can implant an FDA approved microchip in the arm of patients with Alzheimer's, enabling them to access crucial medical information immediately. A sixteen -digit identification number for the chip, which is around the size of a grain of rice, is scanned by a hospital. When the number is entered into a database, the medical information can be important. The brain-computer interface provides other ways of stimulation that can benefit Alzheimer's patients. Brain computer interfaces (BCIs) provide new approaches to communicate and to function in the environment, since signals and instructions are sent from the brain to an external instrument without requiring standard peripheral nerve and muscle production pathways. A BCI can help those who have lost their abilities to speak orally, so they can express simple thinking and emotions (e.g., "yes" and "no.")

More than 20,000 plastic operators in the United States alone work for transforming the body form and appearance, according to Ramez Naam: "in more than Human, accepting the Promise of Biological Enhancement," more than eight million people in the United States had some kind of cosmetic surgery in 2001; Olympians and other athletic activities have doping stories, where players take substance enhancement results, there are at least a quarter million quadriplegics in the U.S. who might benefit from brain-computer interfaces. In the U.S. too, the auditory system will boost more than thirty-four million surfactors or listening-impaired individuals. More than 197,000 people around the world have been exposed to the world of human enhancement by cochlear implants—the multi-electrode microphone that activates the auditory nerve electrically. Whereas some proportion of the population will still reject modern technologies much like Luddites did in the 1800s, many people have accepted the need for human improvements and artificial smart machinery in the 21st century.

5.12 ENTER THE HORSE

There is a crucial question to be posed when addressing the laws as it relates to cyborgs and automated smart machines—are there any ethical problems specific to technologically advanced humans, cyborgs, and artificially smart machinery? Judge Easterbrook, from the U.S. Seventh Circuit Court of Appeal, said, while talking about law and cyberspace, that there was no more precise regulation on virtual worlds than the law of the horse. In making this declaration, Judge Easterbrook described an anecdote with a past Dean of the Chicago Law University who had taken pride in not providing a course in "Law of Chicago." There were clearly no special lessons on "The Law of Animals," but there were cases involving problems such as the selling of horses (contract law) or persons kicked by horses (takes). The best way to learn that the law applies to specialist businesses was to review general rules according to Judge Easterbrook; only in bringing the horse's law into a sense of wider trade regulations could the truth about horses truly be understood. His argument, of course, is that the "law of the cyberspace" is very much like that of the horse law, a specialist endeavor better interpreted in the sense of the recognizable genome. a professional initiative, better understood, which does and does not require a different group, in regard to family general concepts of contract, intellectual property, anonymity, free expression, or the like. Larry Lessig, a law professor at Harvard, responded to Judge Easterbrook's claims about the true essence of cyberspace law and what lessons it could bring. "Lessig" was of the opinion that cyberspace law currently exists and tells essential things about cyberspace transactions' time, place, and national borders. This chapter takes every approach—whereas the rule of augmented reality, cyborgs, and artificially smart technologies will definitely benefit from analysis focused on general laws, each sector will pass rapidly beyond current legislation and therefore the exciting future that awaits us will require new legislation and policy, a future where we integrate with artificial machine intelligence.

5.13 CONCLUDING THOUGHTS

There is a great likelihood that mankind will deliver options that have been the focus of fantasies for decades in developments in human development technology. People could theoretically be improved, smarter and stronger and more desirable by any social norm, to live happier and better. "It can be enhanced by greater intelligence and memory, considerably increased sensitivity and shocking sporting ability," Jacob Heller and Christine Petreson posted. Experts also cautioned, however, that while human innovations could produce various benefits, such advancements will cost civilization dramatically—not least that technological progress for people will alter the nature of the sense of being human. Maybe human beings would be prudent to heed the advice of Sun Microsystems' influential computer scientist and co-founder Bill Joy, in his 2000 essay "Why does the world not require us": human beings are likely to ensure their own extinguishment by creation of the proponents of technical upgrades. This statement refers to the use of nanotechnology to reinvent the environment, but since any improvement technique is unlikely to lead to such an unfortunate result, public opinion should call for strict protections, even a prohibition.

One argument that I like to stress is that more people, whether because of medical necessities or preference, will in the future be improved by technology. Biological and engineering advances have also been made in several areas, including hormones, ritin, prozotic surgery, cosmetic surgery, and body mechanical replacements, without considering the "game changing" capacity to implant chips on the brain. While human enhancement practice has so far mostly focused on restoration, this technique is unlikely to expand soon to healthier people. But if only those who can afford it vote for human advancement it would further increase the awful inequality of our current society and further reduce social mobility. If the wealthy will develop more wisdom and be more physically capable, so their political and economic influence will increase; in that situation, the rich will grow wealthier and stronger. In view of this alternative, can the government guarantee all citizens a simple set of characteristics?

Will politicians behave before the actions of scientists and companies that are impossible to undo or that will deleteriously impact humanity? I can only say "maybe"—a poor answer. But how do we solve the question of defending society or at least ensuring that the future is one of our choice? Join the judiciary, the newspapers, and the public domain. In a recent case of the Supreme Court, all nine judges admitted to illegal search and confiscation of a GPS tracking device on a vehicle without the warrant and ruled that it violated the Fourth Amendment of the U.S. Constitution. Justice Alito noted that "in the case of drastic technical chaos the only remedy for privacy issues could be regulatory in situations requiring drastic technical changes." As there are no clear GPS Monitoring Unit laws (i.e. no Horse Law), the Fourth Amendment to examine the secure use of GPS technologies was introduced by Judge Alito and his colleagues in order to create a data security solution. Judge Alito does not conclude alone that additional regulation is required if rapid technical progress is to be handled. Bills have been enacted at the U.S. Congress to restrict internet surveillance, establish guidelines for the processing of geolocation data, protect the safety of children, and regulate the collection and use of human data, generally personal data storage and usage. In the United States, moreover, some states have adopted laws that govern the degree of implantation of microchips. To set up and secure human rights vis-à-vis cyborgs and artificially intelligent machines,

I think far-reaching regulation by the middle to the end of the century would be appropriate to define the rights of cyborgs who do not have human capabilities and eventually to define the rights of unenhanced human beings and of each other, who are artificially intelligent robots.

When we head into the 21st century, mankind will need to build a robot and cyborg from the human rights viewpoint and the Ethics Charter of Cyborg; basically, a set of laws for man, cyborg, and manmade smart machines communicating. A group of robotics engineers in South Korea are designing a preview prototype of such a robotic code that I might add represents an extension of Asimov's Tripartite Law. The Korean Charter acknowledges that potential robots can need legal protection against violent persons, just as often animals need constitutional immunity from their owners. While some experts support the adoption of the Charter of Robot Ethics and the like, they note that willful human maluses of intelligent machines will cause righteous indignation. We must all be worried that our intelligent inventions might exploit human beings (and thus the *Terminator* movie series). The topic of this chapter is these and other critical questions of law, technology, and strategy for the future of human beings and our intellectual developments.

5.14 CONCLUSION

Medical discoveries in intelligent, helpful prostheses and technologies have been made possible by developments in robotics, biomechanics, and neurology. Wearable robots are becoming more prevalent in the workplace, home, and health care settings. Wearable robots will alter human labor habits, as with other newly developed technology. Strong frameworks that address human welfare and social effects should be promoted at every stage of research and development since these cyborg technologies raise significant ethical and humanitarian concerns. Adopting human-centered methods to exoskeleton research that takes embodied and subjective experiences into consideration is one way to handle this complexity.

We detailed brand-new, multidisciplinary methods for studying exoskeletons from a human-centered standpoint in this chapter. Our hope is that human-centered and multidisciplinary research approaches might start contributing to strong objective reporting and prevent the reincorporation of unquestioned gender and political presumptions that shape the practice of science. Crossing different kinds of thinking has significant consequences for scientific conclusions and research findings.

REFERENCES

[1] Derrida, J. (1978). *Writing and difference*. (A. Bass, Trans.). Chicago: University of Chicago Press.
[2] Derrida, J. (1979). Living on: Border lines. (J. Hulbert, Trans.). In H. Bloom (Ed.), *Deconstruction and criticism*. New York: Seabury.
[3] Derrida, J. (1981). *Positions*. (A. Bass, Trans.). Chicago: University of Chicago Press. (Original work published 1972).

[4] Derrida, J. (1982). *Margins of philosophy*. (A. Bass, Trans.). Chicago: University of Chicago Press. (Original work published 1972).
[5] Dery, M. (1996). *Escape velocity: Cyberculture at the end of the century*. New York: Grove Press. Descartes, R. (1988). Discourse on method. In J. Cottingham, R. Stoothoff, & D. Murdoch (Eds. and Trans.), *Descartes: Selected philosophical writings*. Cambridge, UK: Cambridge University Press. (Original work published 1637).
[6] Downey, G. L., Dumit, J., & Williams, S. (1995). Cyborg anthropology. In C. H. Gray (Ed.), *The cyborg handbook* (pp. 341–346). New York: Routledge.
[7] Driscoll, M. (1995). Eyephone, therefore I am: Miki kiyoshi on cyborg-envy in being and time. In G. Brahm & M. Driscoll (Eds.), *Prosthetic territories: Politics and hyper-technologies* (pp. 248–269). Boulder, CO: Westview Press.
[8] Dumit, J. (1995). Brain-mind machines and American technological dream marketing: Toward an ethnography of cyborg envy. In C. H. Gray (Ed.), *The cyborg handbook* (pp. 347–362). New York: Routledge.
[9] Dunn, T. P., & Erlich, R. D. (Eds.). (1982). *The mechanical god: Machines in science fiction*. Westport, CO: Greenwood Press.
[10] Downey, G. L. (1995). Human agency in CAD/CAM technology. In C. H. Gray (Ed.), *The cyborg handbook* (pp. 363–370). New York: Routledge.

6 Cyborg Sensors

6.1 INTRODUCTION

In previous choices, prostheses and implants were discussed as innovations that put humans closer to the convergence of artificial smart machines. This chapter presents a new technology that is important to a posthuman age: sensors and sensor networks. Under the law of accelerated returns, drastic advances in sensor technology have taken place in the last two decades, resulting in millions of smaller, quicker, and stronger sensors being introduced into the environment. People today have bracelets or clip-on monitors that monitor their vital signs, for example, their levels of activity and their stress levels; furthermore, some people wear ID badges employees use to monitor their position and allow access into safe buildings. In recent years, developments in digital tattoos mean people will wear provisional, health-conserving epidermal circuits. With continual technical advances, sensors that are currently worn on the body start "flying under the skin" to create an evolving cybernetic class with capabilities beyond those of today's unenhanced citizens. The properly fitted cyborgs, for example, can sense magnetic fields with implant technology, see infrared light, hear color, engage in telepathy, and increase the universe with knowledge. And with the ongoing exponential development in sensor technologies in tandem with advancements in nanotechnology, automated nanobots will enter the body by the mid-century to restore injured bodies and cells—when it occurs, the number of sensors will grow to trillions. Obviously, in our hypothetical cyborg, sensors will have a crucial role to play [1].

Due to the exponential advancement of technology, a range of legal and policy concerns of cyborg developments must be resolved by the mid-century, and their position must be addressed to determine the future course of civilization. For example, by making powerful thumbnails with wireless power, can we create a world with little or no privacy and a world where governments systematically monitor people by manipulating the sensors they wear? Or are we to create a more utopian society in which devices are used to track and restore the human body? Do we need to improve contact between individuals? As in other situations, applying a specific technology combines the two possibilities and the legislation often tends to establish the equilibrium. Let us just continue the debate on the sensors and their role in the development of the conditions for a fusion between human and machine and an interesting "unwilling cyborg" event. Daniel J. Palese believed his two brothers had entered his home every day for ten years; drugged him and hypnotized him; and had monitors, sensors, and transmitters inserted in his body while he was sleeping, to follow and monitor him. Not unexpectedly, it was rejected as constitutionally frivolous as his lawsuit was taken to the Tenth Circuit Court of Appeals. Nobody seriously considered in 1998 when the lawsuit had been heard the prospect of Daniel being implanted against his own will with sensors. However, less than twenty years after Daniel's disqualification, the ability to monitor an implant user's actions is not a result of a delusional mind, but technologically and efficiently conceivable. To explain this fact, RFID implants are used for monitoring Alzheimer's patients, to use electrodes implanted in the brain of patients with Parkinson's to instigate their conditions, and to produce new senses, some of them identified as grinders, using magnetic implants within their fingertips to detect magnetic forces coming from everyday objects. And in answer to battlefield casualties, I should note that the U.S. government has invested heavily in technologies to restore troops injured in the brain and has spent millions in experiments to develop neuro-implants to boost veterans' memories. In upcoming decades, not only can the millions of people globally recovering from traumatic brain injury profit from the same technology, but it will also offer an important method to better fuse people and the system [2].

There had been no statute expressly related to cyborgs to rectify his argument when Daniel claimed to the Court that he had been implanted with sensors and transmitters; however, if his claim were valid, both criminal and civil law had provided a clear cause for his intervention. Yet a law passed in California remained in order just ten years after Daniel's arrival in court. The "anti-chipping" law of California specifies that "no person shall claim, coerce or force other persons to have subcutaneous identification technology implanting," defined as any object, program, and/or product capable of actively or passively transmitting private information or radio frequency-based devices. The law established a legal suit in which an embedded (or chipped) person could recover a "civil punishment" against his will with an identifying device such as injury and compensatory damages and punitive damages proven by "malice, oppression, deception or compulsion" to the person who implants the sensor. The California Law is certainly an important legislative response to RFID technology; more legislative measures on human improvement and cyborg innovations are coming. Their purpose is to defend a citizen from being "squeaked out of his or her will."

Although implant innovation has shown tremendous promise for returning missing functionality to a diseased or disabled body, Daniel's assertions show a possible application of cybernetic technology that is extremely human-related. Consider Ray Kurzweil's hypothesis that brain implants would allow a person to replicate past experiences or relay memories (transplants) from brain to brain. While this is an incredible vision for the future from an engineer with an uncanny reputation for correcting it, governments and companies will use the same technologies to put artificial memories in their heads and to hack into private thinking (see the chapter on Cognitive Liberty). The direct consequences for the person and society as a whole are "shooting" individuals against their will or using implants to regulate their behaviors. Many also claim that sensors and implant innovations can only be applied with caution and that the desirable application of technology will radically change what it is to be human beings only after a rigorous public conversation.

When several social scientists joined the discussion on the course of our cyborg future, they were arguing against policies that would hinder innovation in IT growth. Resuming opposing positions, the economist Adam Thierer referred to a conservative approach to technical advancement as the "principle of caution," which he described as the notion that new technologies should be limited or discredited before their proponents can show that they would not affect individuals and groups. In comparison to the view that modern technologies should be allowed by default is "permitless innovation," which Thierer defines as a policy "Without the belief that emerging technologies will severely affect society." While I firmly endorse scientific advancement, if cyborg technology might actually contribute to the destruction of the human race by the use of artificial superintelligence, then I would do so as a proponent of the categorical imperative. Bill Joy, co-founder of Sun Microsystems, still agrees with the scientific innovative company and entrepreneur. In view of the danger to mankind posed by superartificial intelligence, Joy claims that researchers are essentially "relinking" or giving up crucial studies in biology, nanotechnology, and other technologies. A dystopian future would lead to it. The well-known physicist Stephen Hawking even cautioned of the threat to mankind from artificial intelligence. He commented that humans are constrained by slow biological development and are incapable of winning against artificial intelligence, which I affirm. In an interview with the BBC, Professor Hawking said "the creation of total computer technology will mark a human race's end." Stanley Kubrick's 2001 film and his killing computer HAL reflect the concerns of many people about how AI could endanger human lives. "It'd take off alone and restore itself even more quickly," Hawking said. "If a computer moves beyond our own knowledge, we do not decide what happens, so that we cannot know whether it will be infinitely supported, neglected and sidelined or conceivably killed," he says. If, because of the pace of technical development, Joy's solution is regarded as worthy, and the Hawking issue should be vigilant, lawmakers would have to initiate cyborg technology regulation mechanisms in the very near future, because uniformity is close, and once it's reached, artificial human consciousness would not be immune to human laws.

Stanfort Francis Fukuyama has raised similar fears about the danger presented by artificial intelligence to civilization by biotech in *Our Posthuman Future*: *Consequences of the Biotechnology Revolution*, where he argues that the biotech industry needs to be closely monitored because it is potentially transforming the essence of humanity and the corporate framework. Fukuyama elaborated that genetic alteration of the genome would change democratic government in light of biotechnology, for example, by producing a cybernetically engineered subclass. Likewise, how we distribute sensors around the whole world and how the government uses devices to measure, for example, individual freedoms and the nature of society and government bodies that evolve, could influence this century. The monitoring of sensors is an indication of the interaction between sensors and rights of individuals, but the general population has not been aware of how much privacy loss in a sensor-packed environment is appropriate. Anyone who read Orwell's *1984* will soon see the ominous similarity to Big Brother's ever-watching screen that holds people in the middle of a fog of terror. And if we are both "expected" to be continuously used in a panorama, anonymity will not be "expected." Some claim that the right to privacy will actually wither away and perish as soon as the government introduces sensors into a tracking and control network. Regardless of how the worth of integrating with robots is viewed, scientists agree that the only way to prevent a dystopian future is to create laws that explicitly ban the spy activities made possible before surveillance becomes a social law through a worldwide web of sensors and implant technologies.

Researchers have concluded that policymakers and companies could exploit sensor technology to restrict people's ability to follow up, track, and regulate a host of technological safeguards. If not, they claim that if society is completely programmed to be watched, traced, cut, and registered there is some legal basis for the privacy and the independence from government control.

Regarding laws and sensors, the case law is expanding and continues to define the limitations on what sensor information can be gathered, when, and in which conditions. In these situations, the sensors, in particular those embedded in the body, aim to address critical law and policy questions. Can sensor-obtained data be kept secret, for example? Is public access to sensors an illegal seizure of property? And what if someone has a body-worn or embedded sensor for the intent of stealing material that is patent or trade secret law? Generally, courts have found that metal detectors should be used by the government to scan the body with the justification that the protection of sensor scans is less invasive than frequency or other forms of searches. However, the threat to humans from exterior devices programmed to see into the body can be lethal with cyborg equipment already being embedded in the body. In reality, after passing an airport scanner, a woman in Russia who was equipped with a pacemaker died as it altered the workings of her pacemaking system. This example illustrates that an unpredictable potential cyborg could present an unforeseen risk. When we speak about regulation, science, and political problems arising from the use of sensors, the main questions for society this century are at the core of this discussion—either we integrate with artificially smart machines and thus enjoy the rewards of an incremental extension of IT or do not merge into our potential "mind children" To confuse matters even further, it must be debated in the light of another crucial topic, the essential question of whether humans should be fused to artificially intelligent machines, whether or not we should allow artificially intelligent machines to achieve human or other levels of intelligence. This chapter on the rules, technologies, and sensor regulation offers valuable knowledge in order to contribute to this significant discussion on humanity's future path [3].

6.2 A WORLD OF SENSORS

To reaffirm the main subject of this entire book—whether people will fuse in these hundred years with artificially intelligent devices—I will address how sensor technology will lead to that end. I will also discuss legal and political concerns surrounding the growth of sensors in the atmosphere and the implantation of sensors inside the body. Sensors outside the body travel easily "under the skin," which I describe as the "breach of the sensor-skin firewall." If the boundary is broken, sensors within the body are used for monitoring and externally related technology. Through BCI technology person operate various types of technologies and enhance its abilities. Sensors, for example, in conjunction with brain machine interfaces, begin to enable users to track a robot arm remotely using EMG signals that sensors on the head of a human gather. Data from ultrasound and infrasound sensors are also tracked and corrected for the robot arm motions in this request. And while all the smart digital systems gather massive volumes of information on their own, sensor data is much better when embedded in the real world. Implanted sensors continue to link the activities of our body to the external sensors and machines. Anthony Antonellis, for example, has an RFID chip in his arm that preserves and moves art to his smartphones. For instance, the RFID artist Anthony uses a transcutaneous near-field communications chip as a wireless storage device for his particular application. The Antonellis re-writable 1K chip features an animated GIF file so that a variety of device technology can be viewed. The chip's data is left available to the public for reading and writing passwords.

These examples are just the start; as for sensors, get prepared, there will be more. The thousands of sensors that now exist are compared, and according to analysts the number of sensor nodes stretched in a few decades around the environment. In reality, Janusz Bryzek, moderator of a sensor summit at Stanford University, forecasts the manufacturing and shipment of one trillion sensors next year. Furthermore, the writers of *Abundance: The World is better than you believe,* Peter Diamandis and Steven Kotler, look to the future to the need for 45 trillion sensors. From my point of view, I have seen the planet with trillions of sensors distributed in the entire atmosphere and installed in the body long before 2100 by overseeing the sensory engineering laboratory at Washington University. What could contribute to a world full of sensors? MIT's Gershon Dublon and Joe Paradiso claimed in an article from *Scientific American* that one explanation why sensors have been all-embracing is that "they've adopted Moore's law more" and that they've become smaller and powerful; instead, the strengths of our sensory receptors have been fixed, whatever, how wonderful they are.

Today the human body is "richer" than artificially intelligent computers in orders of magnitude. The whole surface of our body consists of sensors from a biological perspective that detect more data than we can actively tend to. In reality, senses capture about 11 million bits per second from the world, but only a fraction of this information is processed and put into workmanship due to compression of information—around 50–126 bits. Google's Ray Kurzweil estimates that by 2040 it would be possible to store all sensory impressions of a person on a microchip embedded in the brain with ongoing advancements in sensor and implant technologies. The idea might sound like science fiction, but Kürzweil's prediction could become a possibility in the mid-century as we consider discoveries made on prothesized neuronal chips by Theodore Berger of the University of Sub-California. Even though our perceptions have an incredible bandwidth, artificially intelligent machines easily measure up our sensory potential in terms of raw computing speed, at least. The raw treatment force of the human brain is estimated to be in the petaflop processing scale with a fascinating amount that is consistent with modern supercomputer systems, with an estimated 100 billion neurons each attached to 10,000 other neurons with each neuron functioning at about 10 bits/second. But the "cognitive part" of the data analysis, or the machine's ability to analyze the millions of bits of data gathered by the detectors, are absent. Since a world composed of sensors that develop according to the rule of speedy return would produce artificially smart machines with exponential improvements in the sensory capacities, this opportunity alone gives humanity the strong impulse to equip themselves with the same sensor technology and to combine with our technological progeny to reap the benefits of IT. Indeed, implantable sensors are changing our thoughts about the body into a paradigm shift. When technology advances in some researchers, for example, they would not "represerve" natural memory or layout technologies for sensations; they're going to continue to develop new senses consciously instead. In this respect, researchers at Intel are designing a robotic hand that allows people the opportunity to detect objects before touching them using the properties of a magnetic camp. And researchers at Duke University claim that they have built technologies to allow people to see infrarouge light that is not available in their laboratory. And in the terrorist domain, the Defense Advanced Research Projects Agency (DARPA) is seeking to create "thinking goggles," which can be used as brain-computer interfaces in telepathic communication to improve the senses of troops, such as night vision and magnetic fields [4].

As computers become more intelligent and independent, sensors become crucial technologies in the relationship between machines. For example, Google cars are navigated by GPS satellite data. In order to maintain their location on the highway, the navigation systems use information from sensors that are built into the area. However, what are the ethical questions with individual computers that use sensor knowledge to communicate with the world? Self-fitted vehicles with sensors—who is responsible for a mistake? What would happen if a driverless car injured someone or Google Maps sent someone down a one-way street the wrong way? Legislators have discussed these and other issues relating to autonomous vehicles lately, culminating in the passage of legislation specific to driverless cars in the U.S. in four states and the District of Columbia, others authorizing automakers to research automobiles but none addressing the entire set of legal questions that would arise in an environment of more and more intelligent sensors.

From my opinion, the rule of speeding up returns means that the capacity of the sensors and implants will begin to exponentially increase, contributing steadily to an evolving digital "Network of Things" of cyborg-friendly robots, equipped with powerful capacities. In reality, sensors are noted at an extraordinary rate of relation to the daily objects—they only use thermal sensors to measure and regulate temperature within the limits of one person's kitchens, stoves, dishwashers, microwaves, refrigerators, and other appliances. As compared to traditional sources, a big explanation is the nature and implantation of sensors within the body. The bionic pancreas designed by Fiorenzo Omenetto, Professor of Biomedical Engineering at Tufts University, is an example of sensors designated based on "medical needs." The device is fitted with a small sensor on a tissue-engineered needle, which "talks" to a mobile app that tracks diabetic blood sugar levels. If the application on the phone senses blood sugar starting to increase, the app indicates a pump to decrease inflammation; if blood sugar falls too low, the app communicates to the device, which contains two pumps and two detectors that track the pumping. The second is a glucagon release pump. Tufts University is another example of the need for medical sensor research but, in this case, the School of Engineering is developing a device composed of electronic implant dissolving, which includes silk and magnesium that are used to extract bacterial infections in subjects. When activated by a remote wireless signal, this device provides heat to contaminated tissue. This represents an essential step forward in designing implantable medical equipment, which can be triggered remotely to perform a therapeutic function, for example, post-operatory infection control

6.3 OUR RELIANCE ON SENSORS

Consider the smartphone technology that billions of people use every day to demonstrate our current dependence on sensors. A mobile phone is essentially a portable, wireless sensor that transforms typical analog data into digital information, including a speed measuring accelerometer and a direction gyroscope. These "senses" allow a mobile phone to monitor the position of an individual and incorporate the information into multi-layered real-time location databases using extensive satellite, aerial, and ground maps. In reality, a lot of sensor data gathered by people as they walk around the world is from their mobile phones; extremely personal information like the location and the call pattern can be used. This is an important argument to make: just as we humans become more and more machine dependent on managing complex systems we need for our daily lives, so do we become similarly sensor based; most people just do not know it. According to Bill Joy, co-founder of Sun Microsystems, we have been drifting into software reliance to such a degree stated in Wired magazine's landmark piece: "Why the Future does not need us." Back up their choices. Don't question what they believe.

In an era of rapid development for sensors, what are the kinds of questions? Of course, privacy and the protection of our bodies and minds and government monitoring are key issues, particularly when sensors are inserted into our bodies. In this framework of data security legislation, should the authorities who access sensor data have gathered and processed it without a warrant issued on implants? The right to privacy does not exist explicitly in the U.S. Constitution, but it exists in many state constitutions and statutes. By the constitution and legislative intervention, for example, California has been a significant contributor to the evolving field of cyborg law, which is defending privacy. When privacy has been violated by the use of sensor technologies in California, there are several "legal theories" that an individual may use to make privacy claims invasive: the constitution of California, the privacy statute of California, interference into isolating average people and the law against paparazzi. Privacy rights in the Constitution of the State of California grant a citizen an inalienable right to privacy, while the California Privacy Act is also specific to new cyborg laws and generates civil and criminal sanctions on those who use an electronic amplifier or storage unit for illegal wiretapping or wave tapping [5].

The location of an Alzheimer's patient, for example, may result in a search using sensor technologies, but a search using sensors can also violate a person's privacy. The U.S. Constitution addresses explicitly, if concerned, how a search and seizure can be executed by the federal government. The 4th Amendmentof the U.S. Constitution sets out the requirements for a search and seize by the government: "the people have a right to be protected by oath or argument, and in particular by identifying the place to be searched, or people or objects that should be seized, in their persons, households, papers and effects, shall not be violated and no warrants given." While no "search and seizure" case has so far specifically addressed cyborgs, similar cases have been recorded involving sensors that I think are the precedent of an emergent cyborg rule. *Katz v the United States* was one of those cases before the U.S. Supreme Court. When the government used sensors to gather data on an individual, the Katz Court laid down a fundamental law. In *Katz*, the Supreme Court found that a government-wide warrant was wrong since the complainant fairly anticipated a private conversation. The Court ruled that a "search" in compliance with the Fourth Amendment was an infringement that violated the expectation of privacy and was properly recognized. Naturally, what is a rational assumption in a world of thousands of sensors of privacy? I must note that a person subject to government search, not an artificially intelligent computer, has expectations of privacy. Obviously, since it is rare to control and register individual behavior, when they are public and internet-driven, the Katz system formulated in 1967 (and hence based on a decades-old technology) illustrates the need of the legislation to keep pace with technology advancements and is complex to enforce. The Katz Framework is a rare example. However, in an era of exponential growth for IT, the degree to which this can be achieved is uncertain. The principle of overriding the rule of technology is not fresh. Louis Brandeis and Samuel Warren described the security of the private domain as the cornerstone of individual liberties in modern times in a major article on privacy published in the *Harvard Law Review* in 1890, "The Right to Privacy." Given the growing capacity to penetrate previously opaque facets of personal practices by states, the press, and other organizations, Brandeis and Warren proposed that legislation should adapt in response to the developments in technology. Common bans on intrusion, threats, divorce, and other intrusive activities have been made possible. There were enough protections in the past, but in their opinion, such values cannot safeguard persons against the "too-entrepreneurial press, the photographer or the owner of some other digital rewording or replication system for scenes or sounds." Consequently, they argued that legislative remedies could be established to impose definitive limitations between public and private lives in order to preserve the right to one's identity in view of current corporate activities and intrusive innovations. What has changed today given the "advanced" 1890s technology? Sensors and body-implanted neuroprosthesis, remote sensing devices, billions of sensors incorporated within the environment, facial recognition algorithms, and the global network for tracking. There's a lot to worry about as we outfit ourselves and the world with "cyborg" technologies.

Another case of special interest for sensors was *Riley v California,* which has been resolved decades after *Katz*. The Supreme Court in *Riley* was concerned with the validity of a wireless police request for private records. In a significant security ruling, the Supreme Court unanimously ruled that the police had to have warrants for the mobile phones of those they arrested in the era of cameras and portable computer devices. Law Professor Orin S. Kerr of George Washington University responded that the court has to consider the evolving world of information technology, observing in reaction to the ruling. "This is the first instance of machine quest and it means that we are in the modern digital age. The old laws will no longer be enforced." Interestingly enough, Chief Justice Roberts appeared to predict a world of the cyborgs by mentioning the crucial role sensors play in today's life. Roberts said that they are "an invasiveness in everyday life that might be inferred that Mars' proverbial guest was an important part of human anatomy," relating to a mobile telephone. "It should not make knowledge less deserving of the security the founders battled to get through now that the technology enables the citizen to bear that information in their hands," Chief Justice Robert writes. In dicta, Justice Roberts spoke on a world in which a variety of fundamental regulation problems emerge in new technology. Can we say that the case of Riley certainly indicates that the government wants a duty to locate the sensors worn by humans and the cyborg sensors and artificially intelligent robots in the next couple of decades? Perhaps not as several decisions of the Supreme Court are broadly written; thus, future cases involving sensors will be litigated in deciding the government's capacity to search for the body's sensor technology; it's just a matter of time [6].

I think that technology has consistently removed the protections to privacy that constitutions and laws offer people such that the corresponding room where we can anticipate privacy has deteriorated dramatically. In view of the trend for the sensors to migrate under the skin, not only can we be continuously monitored, filmed, and collected as we walk through the environment, we can also have a future where businesses and governments regularly access the data that reflects the very function of our bodies and thoughts. I think it is necessary for people to pressure governments to prevent this negative result. Using cyborg technologies could lead, particularly for an individual's body and thoughts, to gross violations of privacy and persistent tracking. If a government player is concerned, the court's contention in *Katz* is that "[t]here is no security for the individuals in an environment packed with sensors and IT in our anatomy and physiology, as long as what is seen and heard is perceptible to the naked eye, or unwitting eyes, the person seen or heard has little fair expectation of privacy."

6.4 THE NETWORK OF SENSORS

In keeping with the rule of exponential returns, individuals, robots, and everyday artifacts in Kurzweil gradually become part and parcel of the worldwide network of transmission networks. In this networked world, data generated by a variety of sensors is now available to everyone, anywhere. When sensors break around the entire world and sensors "migrate" through to the area of the specimen and eventually get inserted within the body, there is one key piece of the puzzle that will lead to our future cyborg and ultimate integration with robots. As important factors for sensor research, I have frequently discussed battleground accidents, medical needs, and industrial applications, and, whilst companies still have the ability to help many of their market operations, low-cost sensors. Retailers from many sectors have been pressuring their vendors to mark shipments with RFID for a period of time so that data can be captured immediately on arrival. In an environment in which utilities have become necessities and costs are king, sensors will allow small companies to develop a competitive edge, according to Ben Gaddis, Vice President of development and innovation for marketing and advertisement company T3. Gaddis says that sensors enable services businesses to switch from a more reactive model to one that is more constructive and useful for end-users. He cited an example of pizza delivery and the use of a sensor in the cooler magnet. The sensor/magnet is pre-programmed for the choice of the consumer and effortlessly positions a command with a built-in Bluetooth feature that produces a return text message that confirms the request. Moreover, retailers attempt to break into the brains of consumers and now have the technology to do so. It is known as customer neuroscience and Thomas Ramsey points out that shopper intelligentsia is valuable and can influence sharing in the few minutes before we vote to consciously buy our minds. If marketers can use those seconds of unconsciousness to make decisive choices about how to market goods, they can use the details. "When we approach this in an entertaining and enjoyable way, people get more and even buy more," says neuromarketing professor Paul Zak. "No buy-button is found in the brain and deception is not discussed. We talk about the use of tools to help create a healthier shopping experience," Dr. Marci said.

The development and importance of sensor technology are also recognized by investment bankers. In fact, Goldman Sachs'] bankers have estimated that 20.8 billion connected devices will be usable by 2020. And the "Internet of Things" is amazingly embraced more than any other invention in history, according to Wim Elfrink, Cisco's Chair of Globalization. In essence, what he and many others mean is that something related to the Internet would happen later in this century. In reality, small sensor devices are stretched practically anywhere and connect to each other through wireless lines to fulfill the monitoring tasks demanded by industry and businesses, medical science and governments. The sensors inside our bodies and even our minds by the middle of the century would form a huge part of our global network of things; this will include cyborgs, which are now appearing in tandem with artificially smart machines. As we enter the middle of the century the emerging worldwide network of things will also construct a "Network of Spirit" in which crowdsourcing and other methods are used to address international issues. Through the use of networks, it is possible to incorporate rich metadata that can be used in several cases when a snapshot was taken or when the vehicle was passed by an automatic sensor. For example, a pneumatic road tube, an inductive loop detector, magnetic sensors, piezo electric cables, and weight-inmotion sensors may be used for measuring road congestion and proposing alternative routes (piezoelectrics, bending panel, load cells, and capacitance mate). In effect, position information links the real world wirelessly with the sensor data's abstract meta world. In the centuries ahead, everything from the clothes we wear to the streets we travel, using sensor technologies, will be embedded with sensors that gather information about any move like our priorities and wishes.

With the advent of cyborgs this century and the increasingly networked world of "regular things," lawmakers would have to decide which limitations are sufficient to collect and disseminate personal data gathered by sensors that form part of the globally linked network. A public policy discussion must reflect to lead lawmakers the degree to which sensor-related data can be re-used for one reason without any authorization or without providing notice of retrieval, transference, access, and review to an individual whose data is collected. And for sensors to capture personal information, policymakers would have to decide how reliably and protectively sensor data can be kept; of course, it is highly necessary to preserve data protection contained in implantable devices in the brain. Although we expect confidentiality for our personal data and our opinions, particularly when we are in private locations, in reality, there are porous walls between public and private areas, says Pamela Samuelson at the Berkeley Law School. For example, many uses for sensor network implementation include capturing uniquely recognized details that can be described to the user as he walks through the world when paired with facial recognition technologies. There is currently a controversy in the United States and the European Union on the role of facial recognition technologies fitted with camera sensors, design matching

FIGURE 6.1 The tracking system.

algorithms, and their connection towards privacy; however, in government and in the private sector the use of facial recognition technology is increasingly expanding [7].

Sensor technologies can also be used for monitoring individuals without using facial detection methods or cameras. For instance, researchers at the University of Washington have developed a way to monitor people immediately by movement and still cameras, using an algorithm that trains networked cameras to understand the difference. The method operates by identifying an object in a video picture and then "following" the individual through various views of the camera. But tracking a subject through cameras from non-overlapping fields of view is troublesome, because the presence of a person in each video can vary significantly due to multiple viewpoints, angles, and hues of color created by various cameras. Jeng Neng Hwang, researcher from the University of Washington, has solved this issue by linking the cameras and using machine learning and artificial intelligence prepare the device for camera recognition (Figure. 6.1). The camera tracks the variations in color, texture, and angle between a pair of cameras for a few minutes, then it is measured manually for a certain number of people who stepped into the frames without any manual involvement. A software is repeatedly implemented after the calibration time. Those cameras vary and could select the same individuals through several frames, monitoring them easily without having to see their faces or hearing their voices.

With many innovations the framework developed by researchers at the University of Washington can be used. Installed on camera placards inside a robot and a flying drone, for example, the researchers allowed the robot and drone to track a human even though the instrument met barriers that stopped the person from seeing it. The infrastructure of connections can be used everywhere so long as a camera can connect to the cloud and upload data through a wireless network.

6.5 TELEPRESENCE AND SENSORS

In a curious prediction of the coming cyborg worlds, some scientists suggest that the environmentally implanted sensors start to act as an addition of the human nervous system, to create a new sort of sensory prosthesis and to ask: what is the beginning and end of our human senses? The expansion of our senses to areas far from the position of our actual existence will also expand our legal responsibility. Take into account Samsung's robotic center for the North-South Korean border. The robotic center is fitted with a range-sensor and is packed with vision, temperature and observable characteristics. Sounds amazing, feels interesting, perfectly alright? If the idea of a dystopian society future is created, a target can be identified and fired over a two-mile radius. It's not important to think, at the moment it's under the control of a remote human operator. Legal, ethical, and moral concerns will obviously be involved, as a result of the trend for the creation of autonomous sensor-controlled "killer robot" weapons. Indeed, an autonomously operated missile or so-called killer robot was created for airborne fighter jet strikes in Norway that are capable of identifying targets and making decision on killing without human "intervention." International groups like the Movement to Stop Killer Robots are seeking to encourage the lawmakers to ban weapons that are able to determine for themselves to kill as a reaction to the creation of autonomous guns. Similarly, I believe that the reaction of "international stages" to implants and other cyborg technology, such as manipulating, violating individual property or even killing them, may serve iniquitous purposes [8].

Connection to sensory input that is distant from a person can generate a sense of telepresence or "being there" cognition, based on my study with my undergraduates. Consider the legal ramifications for sensor technologies in telepresence—a offense must comprise of an action reus, guilty act, along with a mens rea, a guilty conscience, according to criminal laws. There would then be two components of a felony if a person supervising a remotely controlled robot intentionally manipulated (mens rea) a robot arm (acus reus) in order to do harm to a person. If we remember Isaac Asimov's Three Laws of Robotics, which became popular in 1942, in the short story Runaround, the two first laws deal with this example: a Robot cannot harm a human being. This brings a fascinating design challenge for robotics: allowing a person to suffer damage by inaction; a robot is expected to comply with the directions provided by human beings, unless those commands are contrary to the First Law. In the future, it would be a problem for courts to attach criminal responsibility to robotics using sensors that surpass a human level of performance, precisely because robots cannot be convicted of a felony. Again, we come back to the topic of mens rea; can the purpose of harming a human machine (or another) be assigned to a machine? Will it be proof of intent by the computer to view and interpret sensor knowledge using the necessary methodologies? It is sensible to enforce Asimov's laws to defend human beings against artificial intelligence devices and increasing sensory ability to discover the world.

6.6 CHARACTERISTICS OF SENSORS

Contemporaneous robots can be considered as self-navigating, semi-autonomous machines that feature sensors that allow them to see, touch, hear, and move through environmentally friendly hardware, software, sensors, and algorithms. Furthermore, most robots are fitted with orientation sensors and force sensors to influence targets. One argument to make is that robots become more like us at unprecedented speed when they acquire intellect, agility, and manual abilities.

What are some of the fundamental characteristics of sensors and the regulatory issues? Prior to this query, lets us present the significant concept that Harvard Law Professor, Larry Lessig, author of *The Code and Other Laws of Cyberspace*, put forward about law and technology. The argument raised by Lessig was that the cyberspace infrastructure itself governs actions close to that of the rule. For computers, cameras, protheses, brain implants, and other cyborg technology, the same definition applies. The capacities and characteristics of sensor technology, for example, determine the type and range of sensory input humans and robots are able to perceive from the environment and therefore the behavior they can exhibit. The infrastructure for the device and the network of sensors already exists. Likewise, algorithm parameters govern the actions of the computer. Please take the following example to illustrate how an algorithm sets the robot behavior parameters. MIT researcher Sangbee Kim has established a bounding algorithm in the field of mobile robotics to regulate the speed of a robot with four legs. The robot measures the power exercised by a robot in the second break, as it reaches the ground, which helps the robot to retain balance and speed by sensor data and feedback loops. Generally speaking, the higher the target velocity, the more strength with which the robot must be pushed forward. Robert Gregg at the Branch of Texas Dallas, in linked work with amputees, uses robotic control and sensors to automatically react with motorized pre-optimism to the wearer's environment, enabling the amputee to go along a pulling treadmill about as quickly as any other human. Sensors can also be used to endorse policy priorities by modifying physical characteristics. For example, sensors may be used to help the legislation if public policy prohibits people from writing while traveling. Consider Apple's patent of 2014 defining a method for using the iPhone accelerometer and other sensors to calculate the driving rate of the human. If a speeding vehicle is observed by the sensors, motion sensors are triggered to obstruct the cell phone's textual capability [9].

A sensor's basic purpose is to quantify and translate a physical volume into a signal that is read by a sensor or a tool. In other words, a sensor is used to sense and report one source of energy mostly as a power signal. For instance, a pressure indicator may sense pressure, a mechanical energy shape, and turn this into a remote display signal. A significant part of the sensor is a transducer that transforms a signal into another energy. Energy forms measured by sensors include electric, mechanical, electromagnetic (including light), chemical, acoustic, and thermal energy. Energy is not confined to sensors. In vivo biosensors that operate inside the body and monitor various facets of our embedded physiological chips that monitor glucose levels and cardiac surveillance sensors within the body are especially important to our cyborg future. As sensors can monitor the action spectrum of a robot, maybe this is one way to handle it. It is an extra precaution in the software above programming-friendliness to think carefully about what sensory input it senses.

In view of the latest robot architecture, the previous discussion of sensors should be acknowledged. Most robot systems are equipped with an array of sensors that enable human beings to carry out many tasks—in reality, one explanation is the remarkable capacity of sensors. The proximity sensor is one of the standard sensors for a robot and can sense the presence of a human in the process of reducing the risk of injuries. A typical combination of proximity sensors involves tactile sensors that detect contact between subjects and force sensors that detect and monitor the degree to which the robot's end effector exercises strength in an object. An infrarogue transceiver that transmits an IR light by using an LED is a typical form of proximity sensor. The IR light is reflected from an object's surface to recognize the presence and distance of an object to the sensor. Equally, a sensor module produces ultrasound waves that often sense the position and distance of the target from the sensor. The nature and range of sensory input sensed would expand as the sensors proliferate through the whole area. For example, a cellphone case may be fitted with a thermal sensor that enables a human to "sense" infrarouge light and to detect changes in temperature as little as 0.1°C. After 9–11, robots with infrarouge cameras were used at the New York City World Trade Center to climb into boxes too small or too dangerous for humans. A significant case was resolved again in 2001 for *Cyborg Rule Kyllo v United States*. The Supreme Court reviewed a case involving a thermal imaging system used by public servants to control heat radiation emissions from a person's home. The argument before the Court was whether the Fourth Amendment made this act an unjustified hunt. An audition suggested that an imagery system cannot infiltrate or monitor human experiences in the houses walls or windows. The monitor, however, revealed that the side walls and roof of the Kyllo garage radiated an abnormal amount of heat. This detail was used to secure a search warrant, and federal authorities later identified over 100 pot plants in the home. The Supreme Court agreed in to allow a search under the Fourth Amendment of the thermal imaging devices for tracking Kyllo's home and thus ordered a warrant before use of the equipment.

We are especially interested in laws regulating sensor technologies for a law of cyborgs and of artificially intelligent machines that may be the precedent in future cases. In Kyllo, the Supreme Court declared that the Government would use a system not commonly used by the public to discover information about a home. Before it was unknown without physical interference; tracking is a "search" and without warranty is presumably "unfair." Owing to a lack of "all public use" infrarouge temperature sensing at the time of the filing, thermal imaging was a "search" involving a warrant. In this rule, it is not fair to presume the technology is not used for security purposes, whether or not it is commonly used by the public. In a cyborg age, sensor creation and incorporation into the body would certainly go beyond regulation, but questions of the secrecy of our cyborg future might not only focus on the notion that a "man's home is his palace"; even more so, his body still deserves its fortress and the most strict security of the constitution. Naturally, that is usually valid today, but incorporation of the technologies and the capacity to examine an individual's mind can throw a variety of new questions of law and politics up as we enter the middle of the century.

GPS and other movable sensors are common technology in handheld robotic and other semi-autonomous devices in the sense of legal rights and sensor systems just as people are fitted with a vestibular system for keeping control of the body. In the cyborg era the use of GPS poses complex legal concerns. For example, in *United States v Jones* the Court ruled that the search under the Fourth Amendment involved the implementation of a car GPS tracking system and the use of the vehicle monitoring device. Interestingly, the majority of the Supreme Court found that the police had conducted an infringement against "personal belongings" of Jones by putting the GPS physically on the defendant's car—as this breach was a bid for collecting knowledge it was a search per se. This poses a fascinating question for cyborgs of whether to outfit a cyborg with sensor technologies including microchips that map or control its position. This raises an important issue for cyborgs, because activating a cyborg with sensor technologies, such as microchips that track their location or track any part of their actions, is a trespass. I guess it will rely on the precise evidence. If the government does so, it may not be, if it is performed with the consent of a patient.

The trespass against a person traditionally engaged in common law jurisdictions consisted of six different offenses: intimidation, assault, battery, harm, disaster, and mutilation. Via the evolution of common law across jurisdictions and the codification of torts in the common right, the majority of jurisdictions now generally accept three infringements of the individual, two of which involve sensor-equipped individuals and other cyborg technology: attack, which is "any act of the sort that instigates the arrest of the battery." There should also not be an attack and a battery An attack and a battery will also not only be committed on the body but also on the cyborg technologies. In view of the conjunction of sensors and prostheses with the body—does cybric technology consider itself a part of a body or detached from its body (i.e. a form of property)? The dilemma of attack with cyborgs poses a fascinating legal question. Some critics currently refer to an iPhone or device of a human as a "exobrain," which goes beyond the idea of portable computer devices as a way to incorporate technology into the body. Taking the proposals of a modern cyborg into account, he says his "glass" is so much a feature of his daily life, that it is part of him and that it is part of the mind and body. Professor Steve Mann of the university of Toronto believes this. Wearables technology has progressed from a bulky apparatus with some parts fixed forever to something slim and sloppy that slips on and off like ordinary frames of eyeglasses. The communication system is embedded under your skin.

It is also relevant for various areas of law if cyborg technology is part of the body; the criminal justice system is an exceptional case. Here is a prime example. According to criminal law, cyborgs could not only be the victim of an attack against the computer wearer but could also be the object of a firearm used in a cyborg assault.

In criminal cases of cyborgs beyond human power, sensory capacities, and connectivity, the previous argument may be of particular significance. In a case hearing from the *State v Schaffer*, the Arizona Court of Appeal concerned an aggravated charge of assaults against the defendant when the Court ruled that a prosthetic arm could be a "dangerous weapon" in the context of Arizona's aggravated assault status. In the discussion the Court contended that a prosthetic arm is not a "body-part." It is "not a combination of skin, blood, bone and muscle" since the arm is a mechanical instrument, but rather a "dangerous weapon" under the aggravated attack status of the state, even if connected to the body. The claimant in his defense argued that the prosthetic arm was his arm, that during the supposed attack it remained bound to his body (as opposed, for example, to being detached and wielded like a club), therefore being a "body part" considering the fact that it is made from rubber, metal and bone. In short, the court of Arizona disagreed with the "prosthetic as a body part" statement but decided that a prosthesis system is not a "body part." It is an instrument that was intended to replace a missing body part. The conclusion of the court was followed by a report by Jessica Barfield at Dartmouth College. She questioned whether or not changes in the cyborg body were seen as part of the organism and whether the identity of a human would change from being fitted with cyborg technology that covers much of the body. Her research studies have shown that most people believe that if fitted with cyborg technologies, their physical appearance and self-confidence would improve. If I understand the logic of the Arizona Court in terms of the progress that has been made to smoothly incorporate prosthetics into the body, I believe it will take the next few decades to analyze whether cyborg technology is treated as part of the body or not.

In almost all of the cases, when humans are in immediate contact with robots and people begin sharing their living space with residential robotics of "general purpose"—South Korea aims at making household robots by 100% by 2020—the absence of a sensor of proximity to a robot in the architecture will constitute a design flaw and hurt persons. Indeed, people and robotics next to each other are likely to lead to injuries. For starters, in 1981, a robot murdered a thirty-seven-year-old Japanese worker who could not feel his presence. And look a few years back at what happened to a worker in Sweden. The maker was performing maintenance on a faulty engine, and the employee approached the robot's workspace with the thought of having turned the power supply off. But the robot unexpectedly "stirred up" the worker and fractured four of his bones and caught a close hold. In compliance with the patent infringement law adopted by Sweden, a robot design action (or faulty sensor) may have been the case if the robot was constructed initially without the sensor or had been fitted with a failed sensor. With respect to sensor reliability, product recalls do not take place uncommonly. Car maker Kia, for example, released a reminder for a defective sensor mat located in the seat of the car of a particular sedan model. The defective sensor, which belongs to the occupant classification scheme, has been found to be dangerous because of a child's general clothing. This has meant that the airbag cannot be changed after an accident has occurred.

Moreover, since sensors are goods, their usage and durability would require laws governing trade transactions. *Ionics, Inc. v Elmwood Sensors, Inc.*, is a case in point, where faulty thermostats were claimed to lead to fires in dispensers of water. The principal issue in question was whether the contract between the parties consisted solely of the terms accepted in both the bid and approval of the contract or the seller's additional terms and conditions that modified obligation. In this scenario, a critical concern is posed—who is responsible for injuries from faulty sensors, in particular because robots can self-program and use 3D printers in the future to build their own sensors and components to fully delete a human being? The principle of product liability typically refers to damage acts arising from deficient goods. Product liability applies to the possible liability for losses incurred by this substance of any or more of the parties in the supply chain. This encompasses the individual components producer (at the head of the chain), an assembly firm, the distributor, and the retailer (at the bottom of the chain). The product liability suits are subject to goods having intrinsic flaws, causing injury to a product owner, or anyone to whom the product has been loaned and given, etc. "prosthesis

Unfortunately, many electronic goods injure people due to the lack of a sensor, which is even fatal at times. The case of *Messerly v Nissan North America, Inc.* is currently under discussion. The event emerged from a tragic back-up crash in which a mom wanted to move her sports car Nissan from a concrete pad back to home in 2002 in order to allow her kids more space to play. At the beginning, her nineteen-month-old son was safe, but sadly the boy had shifted behind the car. She hit the boy when she backed up, and he was critically injured. They sued Nissan for not being fitted with a rearview camera or backup sensor because of the faulty and incompetent nature of the sporting utility. In reply, Nissan claimed that the danger of hitting children while backing a car was an obvious, well-understood risk of driving any car and indistinguishable from its inherent features, and that its 2002 vehicle at the time of its manufacture complied with relevant safety regulations. If technology ventures beyond the scope of existing regulation, it also refers to a regulatory body. In the United States, for instance, the national road safety department announced a law requiring the installation of a back-up camera for vehicles constructed after mid-May 2018. The law requires a backup camera to directly default to a minimum of twenty feet back from the vehicle to an area of at least ten feet wide. With more cyborgs emerging this century, sensor durability and environmental positioning will be an important topic of public conversation regarding the future course of mankind.

If we enter a human-machine hybrid, intelligent robots are fitted with sensors that are practically close to those given to humans by artificial processes. For millions of years—that's how "they" get more like us. In this case, the sound can be sensed by a significant sensor with which robots are also fitted. The planet is immersed in rich sensory outlets that not only have valuable knowledge on the world but also orient an individual to respond to stimuli around the world. Robots with sound sensors may use software for voice recognition to recognize and react correctly to people, a technology obviously required for a fusion of person and machinery. A robot can also be designed for sound navigation and as body hackers start using sound for echolocation. Sound sensors that then form part of a voice recognition system would enable governments and businesses to build biometrical repositories to preserve thousands of speech prints; this capacity is a significant development in recognizing and tracking an individual when traveling through the world. Integrated with computer vision algorithms and with the ability to monitor individuals from camera to camera, a comprehensive surveillance device for Orwellian purposes may be used by governments and businesses—this is obviously a serious future outcome to society for conversation and discussion. Most court claims have so far been over copyright violations of applications for speech recognition. For instance, some patent infringement lawsuits were the subject of the use of speech recognition software by Apple. An example of the litigation brought by Cedatech, a patents holder, argued that Apple had breached its patent by selling products that allowed an audio/video program to be combined with a separate application. While it is possible that this common argument will include most Apple goods, the prosecution cited iPhone 5 and its multifaceted audio-talking recognition functionality. Internationally, Apple has struggled in China to invalidate a Chinese company's software for speech recognition; the Chinese company, in retaliation, reaffirmed its argument of Siri as a contrary program. I am dreaming of the future of patent wars based on cyborg technology; however, if the artificial mind becomes the creator, how are the courts to respond? What is the impact of these outcomes?

6.7 REGULATING SENSORS AND BEING FORGOTTEN

A legislative agency, such as the National Transport Safety Board (which governs sensors in vehicles and other tasks) is a governmental authority set up to enact relevant regulations through the U.S. legislature. An organization has quasi-legislative, executive, and legal functions. As far as cyborg technology is concerned, separate government bodies and professional guidelines oversee the sensors and wireless networks that link them in the United States. The Federal Communication Commission (FCC), for instance, controls products that are communication devices using electromagnetic spectrum; The Food and Drug Administration (FDA), however, monitors the electrodes that are implanted in the body as a surgical device, not as a medical device. A memorandum of understanding has been formed between the FCC and the FDA, with possible conflicting jurisdictions, to co-operate within the regions of their respective agencies. In 2012, the FCC authorized a mobile body area network that allocated frequency radiation for individual body-wear sensors as a regulatory decision of importance to our cyborg future. The range can be used to form a personal wireless network that includes data from various medical devices used by the body and real-time sensors are aggregated and broadcast.

Details that certain individuals wish to be kept secret from society, like a predisposition to a disorder, are being collected as body-worn devices become popular and gather valuable information about the health of an individual. Such evidence in job decisions can contribute to different types of prejudice. Should a federal statute prohibit the access to sensor knowledge affecting the body, for example, in jobs, sports competition, or government, in periods of rapid growth for sensors? This place has already been granted precedent. Under the Genetic Material Non-discrimination Act employers are forbidden from genetic profiling of staff or work seekers or from supplying medical records to their relatives. Since sensors are able to gather personal information, particularly without their permission and awareness, and with the ease with which sensor information can be posted to the site, some people say that the desire to get information removed from the internets is "right to be ignored," effectively to condemn a person's past behavior. The fundamental principle is that personal data should be in the hands of an entity. In other words, if it is considered to offend the data topic, information that impacts a person and has lost its validity or correctness and has no public interest should be excluded. But the "Barbara Streisand Influence" compares with the right to be lost. The Streisand effect is a mechanism (discussed later in greater detail) that causes an effort to filter or delete a piece of material from the network to strike shots and not less.

In the "right to be forgotten" case, the European Judge determined that, on recommendation of people wanting information, Google must alter some search results on the internet written about them. A Spanish man who argued that a note of auctioning his previously owned home in Google's search results breached his privacy brought an early lawsuit (and won). A Commissioner of the European Union called the decision of the Court "an unambiguous victory for protecting European personal data." But the decision could, in a cyborg age, have serious implications for someone who uses sensors such as a camera in order to make movies and to post the film to the internet. In general, in the light of their power over personal records, the ruling states that the interests of the citizen are vital. Even when individuals in public life are concerned, there is protection of the public interest. In response to demands to remove content, Google submitted that it does not monitor records, that it does not just conceive links to information that is publicly accessible on the site, and that requiring the deletion of links constitutes censorship. Far from Europe's legal powers and jurisdiction, a Japanese court ordered Google, after complaining that its privacy rights had been violated, to remove search results suggesting a human being's relations with a criminal organization. Thirty-two countries have flushed thousands of websites by Google on the basis of requests received from about 200,000 people; however, what is the value of the prospective removal of trillions of sensors without censorship hindering the free flow of ideas?

In order to make them comply with the European Personal Data Protection Directive and with the law for privacy of countries outside the EU, such as Japan, I consider the "right to forgive" a clear form of censorship that requires online search engines to edit, on demand, the search results. Although I support the right to strong privacy of individuals particularly with regard to sensor data concerning their mind and body, I also agree with Google's claim that censorship should be avoided or used only in the last resort, for example in cases of national security, privacy, and the protection of personal medical data of an individual or pensions. In taking into consideration issues relating to cyborg technologies, legislators must carefully balance data protection rights against social damage caused by information censorship. Since democracy flourishes when there is a private enterprise for ideas, policy should be steadily less censored. Thus, I have to ask if we should give someone the rights to decide which information should be erased even if this is personal and concerns people when it comes to the deletion of information about an individual. If so, people could remove facts from the history of collective human existence. In order to do so, an individual would be given sole monopoly rights over data representing his/her life, a difficult concept when the actions of one person affect many other people. Although facts are not normally covered in accordance with the law, data may receive a monopoly in the form of the IPR, as part of a discussion on the freedom of cognition and the law of artificial brains given sufficient creativity in their organization. At present, while legal data protection varies widely between countries, the majority of jurisdictions grant certain data protection rights.

6.8 REMOTELY SENSED DATA

The widespread availability of sensors has reached well beyond ground atmosphere since the beginning of the space program. Indeed, the public increasingly has access to satellite imaging, a phenomenon sped up by services like Google Earth. In remote sensing data collected by systems based on aircraft, the soil and space become an important part of perceptual data bases that people with sensors can access. One point to make is that while distant sensed data can bring positive benefits for society, such as the chronicling of the health of the planet and the search for uncovered natural resources, totalitarian countries track and monitor the environment for human beings.

The First Amendment protects journalists that use drones and is already an active concern. In several incidents, a reporter who used a drone to screen the accident scene was arrested or intimidated by police. The reporters argue that the drone activity under freedom of speech of the First Amendment is protected because the drones are used to observe police activity from the public space as part of a news story.

Meanwhile, the police said that they stopped data drones or arrested reporters for legitimate safety reasons—for example, they said that in one instance with a journalist a helicopter approaching an accident scene interfered with the drone.

In the past, rules on remotely sensed data have been more focused on the marketing of remote sensing systems than on safeguarding privacy. In view of the recent developments in sensor technologies and the ability to combine different monitoring platforms into a global network of supervision, I am therefore of the view that the laws concerning remote sensing systems need major revision. As a self-proclaimed Earth observations champion, the United Kingdom issued national remote sensation policy back in 1984 but passed its initial, remote sensing law in 1986. The 1992 Remote Sensing Policy Act declared that the marketing of distant land sensing was a long-term U.S. policy goal and set out licensing procedures for private remote sensing operators. Under the Act, before a company can start operating a remote satellite a license must be given. Unenhanced data only needs to be made available by private system operators to governments of intelligent states, free to supply data according to market forces for all other customers. These results mean that, according to the current legislation, satellite sensor information cannot be accessed directly by cyborgs and artificially smart machines unless they are authorized to access the data or hack into a distant sensing system. In the future, what if "remote sensing" laws are violated by artificially smart systems?

A few legal disputes have been raised to date involving remotely collected sensor data, such as the Google Maps data. One case focused on whether information on Google Maps could be obtained during driving. In California a few years ago, a driver was given a ticket for using the Google Maps app on the iPhone 4 that was disputed and, upon appeal, a State Appeals Court overturned the conviction that drivers should be able to use smartphone map apps during while driving. The Court concluded unanimously that, while driving, the State Vehicle Code covered listening, talking, and texting on a cellphone—not viewing a map application. In other jurisdictions, the impact of the decision is still considered.

The impact of their use on privacy rights is one of the most important issues concerning policies regarding remote data. Even when individuals in public life are concerned, there is protection of the public interest. In response to demands to remove content, Google submitted that it does not monitor records, that it does not just conceive links to information that is publicly accessible on the site, and that requiring the deletion of links constitutes censorship. Far from Europe's legal powers and jurisdiction, a Japanese court ordered Google, after complaining that its privacy rights had been violated, to remove search results suggesting a human being's relations with a criminal organization.

The essential issue of privacy also revolved around another case involving the collection of sensor data presented on Google Maps. A couple in Pittsburgh argued that Street View in Google Maps is an imprudent privacy invasion. The couple sued Google, arguing that Google "considerably ignored their privacy interests," when cameras from Street View captured images of their homes beyond "private road" signs. The partner said the "mental anguish" and diluted home value were clearly visible on Google's Street View. The plaintiff "was not able to file a complaint under any count," the U.S. Court declared. Interestingly, Google claimed that confidentiality

Cyborg Sensors 153

is permissible in this satellite and aerial imagery age and that it is not possible to take photos on private roads. Scott McNealy, a co-founder of Sun Microsystems, had expressed his sentiment many years earlier: "this is null confidentiality! Get over it."

Back in California, a person can sue for "invasion of physical privacy" under an anti-paparazzi statute when three elements meet: first, a person has tried to enter without authorization in the domain of another person; second, an entrance is made with without permission with "the intention to identify the other person engaging in 'personal or family activity' in any kind of visual image, sound recording and other physical impression," and, third, an attack was carried out "in a way offensive to a sensible man." This aspect of the law against paparazzi deals with the physical interference in the private property of another person (which is also covered by infringement law) and not data collected by a remote sensing system. The Anti-Paparazzi Statute also included, however, a "constructive privacy invasion," which provided for responsibility, even if the property of another person was not actually included. Under this aspect of the statute, a cyborg fitted with a telephoto lens may be liable. As a part of a person's body, the telephoto lens is not far away but is now done with a tiny population of the weakened vision losses. Specifically, some individuals with severe maturity level degeneration (AMD), affected by an implantable telescope that works like a camera's lens, would be better off with 2,000,000 Americans (currently in development). However, the judge rejected Streisand's claim in his return to Streisand's case, stating that "air views are a common element in everyday life" and "the way in which they occur, or the way in which these terms of direction view was achieved, is nothing offensive."

One thing that concerns the coming cyborg age is that sensors can be used for things other than their actual design philosophy. Some images suddenly acquired from camera systems may infringe video voyeurism legislation on this point. In the United States a federal video voyeurism prohibiting statute states that anyone who intends to capture without his/her consent a picture from a private area of a particular individual, in circumstances in which the person has reasonable expectations of privacy, is held criminally liable for not more than one year. The term "capturing in relation to an image" means the transmission of an image by electronic means by videotape, photographing, film, recording, or broadcast; the word "broadcast" is the transmission of a visual image by electronic means with the intention to be visualized by someone. The problem with anti-paparazzi and cruel and unusual punishment or voyeur laws is that they presume conscious acts of surveillance. The model is a famous photographer who stalks his beast. The monitor, such as Steve Mann, is automatic or inadvertent like Barbara Streisand, whose beach home was photographed by Kenneth Adelman, who retired with her constantly filming EyeTap technology, which is one interesting example. And as technology "progresses," video voyeurism laws are increasingly difficult. I predict that the next major offender could be remote sensing via satellite technology.

6.9 SENSORS AND INTELLECTUAL PROPERTY LAW

An interesting question to ask is whether software, technology, and algorithms that control the operation of cyborgs and artificially intelligent machines are eligible for protection under intellectual property law, provided that the exponentially accelerating technologies build a world of sensors. In order to introduce a debate on this question, the technologies intended to carry out the same tasks are obvious, whereas the parts of the human body are not patentable subject-matter. William J. Kolff was awarded in 1967 a patent for the first artificial kidney as one of the most essential organ kidneys; none, however, can obtain an organic kidney patent. But what happens next time where most of man's body is made of patentable electronics and an artificially intelligent computer that appears to be felt? So what are the rights of individuals and cyborgs over the body? Cyborgs are not regarded as a "special class" under the existing statute and are hence not bound by the 1964 Civil Rights Act. However, does the Constitution grant technology protections that cover cyborgs? The response is yes; in the U.S., intellectual property monopolies are explicitly stated in the United States Constitution, Article I, Section 8: the force to facilitate the advancement of research and useful arts by ensuring the exclusive right to their respective writings and inventions for limited time for writers and inventors. In view of the rapidly accelerating ITs, for cyborgs, artificially smart (usually artificial) computers, what facets of sensor technology should be covered under IP regulations are relevant public policy issues to address.

Patent law is especially powerful as it grants patent holders exclusive control of the invention. Inventors of cyborg technologies are subject to patenting. Such rights shall include the exclusive right, for a limited time, to produce, use, and sell the patented invention. The patent owner may bring a claim against any person charged with violating the patent because he is entitled to patent protection. According to U.S. patent law, patentable subject matter categories are generally defined in any method, machine, manufacture, composition, or enhancement. The Supreme Court ruled that the Congress was pursuing patentable subject matter in *Diamond v Chakrabarty*, a seminal case, to "cover all that is man-made under the sun." The Court has noted, however, that there are limitations and that this general definition does not include all discoveries. For example, laws of physics, physical phenomena, and abstract theories cannot be copyrighted according to the Court.

Whereas cyborg hardware is patentable, are algorithms and software patentable, which manage the hardware and cybernetic systems? The rule on the patentable subject matter of software and algorithms is changing in the United States. Although this issue was debated in the "Artificial Brains Rule" chapter in order to revisit briefly, abstract theories and natural laws are not patentable topics under U.S. patent law—not that it's just that clear. For example, as part of the control dynamics of cyborgs and artificially intelligent machines, the main issue for patentability of algorithms is if they're treated as an abstract principle or as a scientific expression of a natural law (for example, the gravity equation is natural and thus not patentable), or whether they constitute a legitimate patent security even though they are not the primary object of patent protection.

The topic of the patentability of algorithms was dealt with in the *Alice Corp. v CLS Bank International* 2014 Supreme Court case, which contained a patent infringement on a virtual currency tracking algorithm. It was found that the patent was void because the Court ruled that the algorithm claims represented an abstract concept and it was not necessary that the claims were carried out on a machine converting the idea into a patentable invention.

In *Alice*, however, the Court actually opened the doors to algorithms' patentability and state algorithms are a species of abstract concept, a decision "inviting all sorts of mishaps." Take a well-known search algorithm argument issued by Google's co-founder, Larry Page. In reality, Stanford University, who awarded the patent to Google, patents the "PageRank" algorithm. The domain patent argues the so-called page ranking algorithm—a method of measuring web pages according to the density of their connections. In *Alice*, however, the Court actually opened the doors to algorithms' patentability and state algorithms are a species of abstract concept, a decision "inviting all sorts of mishaps." Take a well-known search algorithm argument issued by Google's co-founder, Larry Page. In reality, Stanford University, who awarded the patent to Google, patents the "PageRank" algorithm. The domain patent argues the so-called page ranking algorithm—a method of measuring web pages according to the density of their connections. And is this principle turned into a patentable application of abstraction by an (additional) "inventive concept?" I expect the patentability of algorithms, particularly when more cyborgs with sensor technologies are emerging in the public domain, and artificial machine intelligence are writing their very own code and using 3D printers to create their own sensors, to be a potential field of litigation and public discourse.

Patent infringement suits are popular with increasingly sophisticated devices used to equip individuals with implants and prosthesis. Due to the increased use of sensors and the profitable market, there are constant patent battles between sensor technology makers. In a case of patent infringement litigation (which was ultimately resolved), for example, Rochen Diagnostics Operations Inc. and Abbott Diabetes Care Inc. accused Abbott of violating two of her patents protecting the diabetes blood glucose monitor strips. In *Nautilus Inc. v Biosig Instruments Inc.*, the U.S. Supreme Court was called to investigate the resurrection of an infringement lawsuit involving a heart rate monitor; in another patent dispute involving sensors the approval in Biosig Instruments Inc.'s patent of an ambiguous complaint by the Federal Circuit was inconsistent with the aforementioned case and it invited patent holders to misuse it. In the past years, the public oversight office and analysts on legal matters have said that violation lawsuits have resulted in too many overly lengthy patents according to the Patent and Trademark Office.

6.10 SURVEILLANCE, SENSORS, AND BODY SCANS

Probably the most noticeable expression of surveillance from sensors and sensor networks is the distribution of CCTV cameras run by the government in urban areas. In the United Kingdom, for example, it is recorded that more than four million cameras have been mounted by the authorities and that an estimated 300 times a day they are fired at the average Londoner person. Monitoring cameras in the U.K. are also used for functions other than their original crime prevention mission, a type of "sensor mission creep." In London, for example, any car entry and exit of the central financial district is photographed by authorities ensuring that all vehicles traveling through the city are accountable for the congestion charge. Heathrow Airport in London has already started to search for passenger recognition by using eye scanners. CCTV oversight programs in Washington D.C., Chicago, New York, and several other big cities nationwide are either under way or are being enforced. However, it seems as striking as the dispersal of sensors in the form of security cameras for the collection, in contrast to the tracking power and sensor control, of technologies now in the possession of individuals with head-on displays, that digital intelligence is gathered by government officials, smart phones, and other body-worn sensors (there are about 6.8 billion mobile phones in the world).

Steve Mann from the University of Toronto offers one of the most fascinating viewpoints on infringement of privacy in a world full of sensors: a human being who has been modified on purpose to have the same characteristics as a cyborg. Steve's simple premise is that people using technologies that enable them to film will "shoot" government and companies that track and shoot people using cameras and other sensors while they're going around the environment. Obviously, the project of Google Display is an example of technologies where people will fire back with Steve's earlier head-worn monitor technology. Steve relates in particular to the tradition of filming those that film us. The phrase "subveillance" was defined by Steve as the contrasting French terms *sur* which mean "above" and *suba*, which means "below." "Surveillance" thus means "eyed-in-the-sky" that is seen from above, whereas "subvenility" means that the camera or all other means of observation are limited to an individual level, physically (mounted cameras for individuals instead of buildings). Large bodies enjoy watching companies.

Monitoring, as Professor Mann described it, is obviously done by governments in several different ways. For example, we've all witnessed since the 9/11 act of terrorism by the airport, courthouse, or other government surveillance scanners passing our bodies and belongings. Even an individual moving weapons onto university property is necessitating metal detectors in certain classrooms. Although courts have found that a person's privacy rights are usually "acknowledged in the body," does this not extend to corporeal scans? Moreover, what are the legal theories if a person searches another person's sensors for access to sensor data? One reaction can be a miscarriage under common law countries. Specifically, the "intrusion into selectivity," common law is the case where someone intentionally, physically or otherwise, intrudes into the loneliness or isolation of another person or his private relations or worries. In the event that such intrusion may be particularly objectionable for a rational person, he or she is liable to the other person for an invasion of his privacy. Unjustified intrusions, such as eavesdropping, include the wiretapping of one's personal life and video or photographic spying.

In order to triumph over intrusion in the perpetrators of seclusion, a party must contend that an unwanted intrusion or bias occurred, an aggressive intrusion or disagreeable in reverence for a reasonable person, a private intrusion, and anguish or distress resulted from the intrusion. The second aspect regarding the aggressive nature of the attack depends on the way the data are collected. Anybody with cyborg technology embedded inside a body scanner by a sensor could claim that because a part of the body was scanned it's very violent. A body scan of a customer in a retail shop was a fascinating event of significance to an evolving cyborgs rule. The Court held that the interference of the scan was "not to be so particularly invasive for a normal individual as to be a violation of privacy conduct," in reaction to the intrusion of the seclusion tort. I assume that this decision must be changed ultimately to incorporate a potential universe of potent devices, which can search the inner body and also our brains—these capacities would clearly appear offensive to a rational human; if they are not, what will be?

6.11 USING SENSOR DATA IN TRIALS

What laws and directives extend as someone applies this knowledge to a courtroom, provided a potential cyborg in which a person's brain waves can be captured and tracked using sensors? Take into account the statements made by Champadi, the neuroscientist, focused on a computer that he developed in order to assess the innocence or guilt of a human by interpreting brain waves. The main concept behind Mukundan's signature measure of Brain Electrical Oscillations is to use electroencephalography (EEG) to display active cortical regions, including fMRI, that are located during retrieval. Champadi believed that the system was so precise that it could say whether or not an individual was committed. In an assassination trial in India, the Indians attempted to find a woman culpable of murdering her former fiancé by the Brain Electrical Oscillations Signature test. The decision was deemed "ridiculous" and "unconscious" by scientists arguing that the Mukundan thesis had not yet been reviewed by peers. But it begs a question—how accurate could a "science" evaluation based on sensor knowledge be when the data is notoriously fallible? I accept that the court ruling is an affront to justice. Is a person entitled to secrecy over his own memory, or does the interest of society to keep offenders responsible "prevail?" Given the rapid advancement of sensors and speed at which sensor technology reaches the market, the problem of whether sensor information such as EEG data should be used for the purposes of deciding guilt or innocence is a significant problem for courts to decide in a trial as credible empirical proof.

Neuroscientific research has actually persuaded courts to imprison convicts rather than to put them to death. During criminal hearings, courts have accepted brain pictorial facts to prove the suspects like John W. Hinckley Jr., who threatened to kill President Reagan, was foolish. The law professor at Notre Dame, Carter Snead, developed for President Bush's Council on Bioethics a study paper on the effect of neuroscientific research on criminal law. The study concludes that the neuroimaging proof is of mixed value, but "it is striking that several instances of such data are presented." In the coming cyborg age, this amount would undoubtedly increase significantly. Neuro-scientific research is said to have a significant effect not only on issues of guilt and sentencing but also on lies and cache forecasting and the estimation of possible criminal behavior. Skeptics also fear that the use of the central nervous system as a sort of super-mind reading device will jeopardize our privacy as well as mental rights, causing some individuals to advocate for a new definition of "cognitive freedom" in the legal system

Regarding the allowability of sensor data in courts, take into account the complications in *The People v Dorcent*, where the accused was charged with a suspended license. Several months prior, Dorcent had been accused of driving under the influence of alcohol in a separate legal case. A continual alcohol monitoring cuff (SCRAM) and its electrodes were put on the complainant's ankle to identify periods of abstinence, and the individual went without food for 30 days. There are three pieces of the SCRAM method, a bracelet, a modem, and a SCRAMnet. The SCRAM bracelet has a chambers and fuel cell to test the vapors in a person's sugar during the day and at night at moments of repetition. It also has a tamper strap and locking clip, which stops the wearer from removing the unit and a temperature sensor. The IR sensor transmits an IR beam. In order to be able to analyze, control, and store alcohol reading, manipulatives, body temperature, and diagnostic data, at least once a day, are transmitted within the modem and to SCRAMnet via an Internet connection.

The SCRAM bracelet confirmed three weeks after it was equipped with tracking equipment that, due to the supposed barrier that stopped the system from collection of data, it was not able to track the alcohol intake of the defendant for ten hours. There was a hearing as to whether the suspect violated the terms of his plea deal and whether the SCRAM technology is technically accurate enough to fulfill the technical proof admissibility standard in the State of New York. In earlier *Frye v the United States*, the Court ruled that, whether the latest or new scientific hypothesis or facts were to be omitted in the related research community, approach is widely acknowledged. In some countries, the credibility of evidence from emerging technologies is usually tested by experts at a "Frye hearing." However, the drawback of Frye is that the findings can be omitted from theory or methods that can yield true and consistent results but are too recent to pass a peer review examination to become widely recognized in the research community involved. This split in Frye law ultimately led to *Daubert v Merrel Dow Pharmaceuticals, Inc.* passing a new rule through the Supreme Court. In *Daubert* the Court ruled that the Frye rule had been replaced by Rule 702 of the Federal Rules of Proof. Rule 702 states:

> If, by means of intelligence, skill, experience, preparation or schooling, a testimony certified as an expert shall be aided in a trior to the truth of the evidence, the testimony can be testified to by an opinion or otherwise, if (1) the testimony is backed by sufficient facts or records, (2) the testimony is supposed to be a prerogative of the evidence. (3) the witness has reliably adapted to the facts of the case the standards and procedures.

This decision that, if the three conditions are met, causes the emerging sensor technology in court to be used as proof.

6.12 CONCLUSION

In creating a class of cyborgs with a capability far beyond that of unenhanced people, sensors will play a crucial role. People will evolve in accordance with the law of accelerating IT returns. After that came the physical evolution of humans, which included the creation of neuro prosthesis for the brain. There will be an intelligence explosion for artificially intelligent machines after Singularity that could occur around the mid to late century. Either we will merge with our smart inventions so that "we" will become "they" or we will remain behind. The steps are not independent but intended to guide the future, to overlap and to evolve in the background. Artificial intelligence is like that too. The choice to combine with robots would be the crucial option for mankind and not the time periods of decades and not millennia for the process. Sensors within the body link cyborgs to the global network of items, track their health, and develop cognitive skills. Various future possibilities associated with the use of sensors have raised several notable and ethical concerns. Mankind also has time to decide about the desirability of the posthuman future, but, given the speed of technological advancements in cyborg technology, our future needs to be determined by the current generation, not by the next generation. An effort to outlaw them outright may be the first reaction to unforeseen cyborg technologies, writes Hertling. But in a highly regulated, closed society, such decisions may be practicable, but in a free and open society it would not be desirous. Hertling: the future is unsure, and now we must act.

REFERENCES

[1] Doyle, R. (1997). *On beyond living: Rhetorical transformations of the life sciences*. Stanford, CA: Stanford University Press.
[2] Eglash, R. (1995). An interview with Patricia Cowings. In C. H. Gray (Ed.), *The cyborg handbook* (pp. 93–99). New York: Routledge.
[3] Escobar, A. (1996). Welcome to cyberia: Notes on the anthropology of cyberculture. In Z. Sardar & J. R. Ravetz (Eds.), *Cyberfutures: Culture and politics on the information superhighway* (pp. 111–137). New York: New York University Press.
[4] Featherstone, M., & Burrows, R. (Eds.). (1995). *Cyberspace, cyberbodies, cyberpunk: Cultures of technological embodiment*. London: Sage.
[5] Ferry, L., & Renaut, A. (1990). *French philosophy of the sixties: An essay on antihumanism*. (M. H. S. Cattani, Trans.). Amherst: University of Massachusetts Press. (Original work published 1985).
[6] Fishbein, M. (1967). A behavior theory approach to the relations between beliefs about an object and the attitude toward the object. In M. Fishbein (Ed.), *Readings in attitude theory and measurement* (pp. 389–340). New York: Wiley.
[7] Fiskc, J. (1994). *Introduction to communication studies* (2nd ed.). New York: Routledge.
[8] Foucault, M. (1973). *The order of things: An archaeology of the human sciences*. New York: Vintage Books. (Original work published 1966).
[9] Frailberg, A. (1993). Of AIDS, cyborgs, and other indiscretions: Resurfacing the body in the postmodern. In E. Amiran & J. Unsworth (Eds.), *Essays in postmodern culture* (pp. 37–56). Oxford, UK: Oxford University Press.

7 Telepathy Signals in Cyborg

7.1 INTRODUCTION

The potential Singularity would have a lot to say about the case as people combine with robots this century. What's the uniqueness? The uniqueness is the moment in an evolutionary period where artificially smart computers equal or exceed human intelligence. In 1958, mathematician Jon von Neuman used the word "Singularity," which spoke in particular of the continually accelerating advancement of technology, which contributed to changes in human lifestyle, giving the idea that mankind was entering a fundamental Singularity, over which human affairs cannot proceed. Twenty-five years back, Vernor Vinge, writer of scientific fiction, invented "technological Singularity," specifically, "We are soon to build more intelligences than ourselves." When this occurs, history of humanity would have entered a sort of Singularity and intellectually impenetrable transformations like the knit space-time at the middle of a dark void. Tim Wu, Professor of Law at the University of Columbia and the writer of *The Master Turn* said, "Don't be mistaken: now we are distinct beings than we once were, technologically and not biologically, for the best we expect." While the statement of Tim represents the present ambivalence of the public regarding our cyborg future, I promote an agenda that includes general education on the topics of artificial intelligence and "cyborg science" and that engages in a robust discussion on the future of civilization [1].

One of the main points I make in this chapter is that we humans may encounter the last generation(s) of growth primarily under biological rules. What is the reason? Centered on the fact that the study of technological history indicates that technological progress is rapid, as discussed in *The Singularity is near: when humans transcend biology*. Kurzweil argues that exponential evolution is opposed to the perception of the "intuitive linear." Much of us see society advancement in which we see modern technologies come into our lives but don't know where we are in the curve that reflects the transition in technological progress. According to Kurzweil this means that in the 21st century we will not have 100 years' advance—but something more than 20 thousand years. In a couple of decades, however, some say that artificial intelligence beats human intelligence and contributes to individuality. These effects include the convergence of biological and non-biological intelligence, everlasting human software, and ultra-high levels of intelligence, which stretch beyond our present imagination. In several different types of technologies, the word Singularity has been used, but for this book, the most popular conceptual frameworks of Singularity are an idea of intelligent artificial intelligence other than humans, the meaning of which is software, computers, or robots that study, think, pick, and create their own priorities on their own. The idea of Singularity goes along: many influential scientists in artificial intelligence, robotics, and neuroscience claim that technology ultimately hits humanity and then goes past them in intelligence generating an on-the-spot universe packed with "intelligent" computers. It's actually happening now. Surgical devices, concept operational machines, medicines designed to save lives, writing news articles, and working in a variety of industries already exist, which concerns what we people do to our minds and bodies. Because if they conquer us, then what? What then? Do you want to fulfill your human master's duties, or do you want to isolate your people from others in their ambitions and expectations?

7.2 QUESTIONS OF LAW AND POLICY

There are extremely crucial questions to answer if single-mindedness is to arise in the near future: do we integrate with it (intelligent artificial machinery) or transcend, coexist with them, enslave them, and avoid being exploited by them? Yet, according to others, can we determine for the sake of the human race to avoid Singularity until it is possible; or should it even stop Singularity? Even though the relentless march of technology into Singularity does not allow us any options to stop considering or addressing these tough questions, we either take a move forward in deciding what since the never-ending march of technology into Singularity leaves us with no alternative, or we indulge in deciding the future of our civilization or passively watch as the future surrounds us. Without giving up thought or addressing these complicated questions, we are relaxed.

In view of the aforementioned, many forward-looking theorists expect that we will converge with the silicone developments in the coming decades. The robotics experts Ray Kurzweil and Hans Moravec predict a day when the machinery of the future will be more civilized, in other words in the shape of an android and super intelligent. The physician Max Tegmark writes on this stage in Berkeley In this respect, the Berkeley physicist Max Tegmark wrote in *Our Mathematical Universe* that the creation of supercomputers with human or greater intelligence levels is possible, provided that our brains essentially consist of particles that conform with physical laws and there is no physical legislation that excludes particles from being organized in ways that can carry out even more advanced calculations. On the human side, fifteen years ago I co-published a work on wearable computers and explored the technologies of human improvement and the possibility of people integrating with machinery (the second edition is now available). The combination of people and machinery may profit people in different ways, for the fusion of humans and robots may help us in many different ways: for example, by replacing our biology with non-biological parts we will be able to fix or substitute prosthesis automatically, like neuroprosthesis, if weakened or obsolete. And we could be aging with more grace and probably turn the aging clock around instead of enduring the consequences of aging. Because of the anticipated advantages, some observers claim that the need to substitute and remedy human genetics would enable humanity to regard the fusion of human beings and intelligent machinery as a natural stage for evolution. While the concept could sound drastic, particularly for people unfamiliar with the exponential rise in technology, many robotics and IT experts agree that this is a possible scenario for our future. If we combine, though, with machinery, we can only have minimal opportunities for that; as soon as machines intelligently exceed us, they will plan to keep improving in ways that are incompatible with our aims and ambitions. To have access to its mobility, manual ability, or maybe human emotions, you could infer: why merge with a less intelligent species? [2].

In the first half of the 21st century, technical development has been made in various sectors. Such technological advancement in the first half of the 21st century would pave the groundwork for Singularity in a broad variety of areas including robots, neuroscience, artificial intelligence, sensors, nanotechnology, prostheses, and materials science. According to many who contend Singularity is just a few decades away, over and beyond human intellect, machines are self-directed or self-reliant and their intelligence and abilities develop exponentially instead of gradually. Ray Kurzweil, who believes that Singularity is near, sets the date for Singularity around 2045. By then, he estimates that Moore's law would cause exponential improvements in computer power along with advancement in artificial intelligence to be high enough to build artificially intelligent machines that are beyond man's control. Kurzweil thinks that there would be no simple difference between person and machine or physical and Singularity in this modern world. Practically speaking, it will mean the end of human ageing and disease, pollution, malnutrition, and suffering worldwide (with developments in the nanotechnology sector). Yet individuality may be inevitable. Fortunately, although Singularity, I assume, is unavoidable, there is serious discussion as to when or even whether this will happen to certain researchers. For example, in 2011 one of the founders of Microsoft, Paul Allen, co-authored an analysis in the *MIT Technology Review* in which he takes a conservative view of the impending Singularity as opposed to Kurzweil.

The Duke neuroscientist Miguel Nicolelis, whose experiment on brain-computer interfacing is interesting and, paradoxically, points to a future fusion of person and machine, represents a clear voice toward the concept of Singularity. The key point of Nicolelis is that "the brain cannot be computed and therefore no technician can replicate it." He also states that the brain is "copywritten" that is covered by its own experience. However, Ray Kurzweil identified his new pattern pecognition theory of mind (PRTM), an opinion that couldn't be more unique from Nicolelis: "we now have ample proof to accept a certain hypothesis of how neocortex operates, a uniform theory." Kurzweil suggests that the neocortex consists of basically 300 million pattern recognizers that are able to communicate with other pattern recognizers in the hierarchy. The universe is fundamentally hierarchical, and we can appreciate this here in the neocortex. Whether Singularity takes place or not, it is never too early for the public to recognize the profound impact of Singularity on society and to support or reject humanity.

In view of the rise in device speed and memory by today's semiconductor manufacturers, some analysts infer that our smart silicon creations are inevitably reasonably effective for creating their own better hardware and software, increasing their intelligence and capacities per generation afterwards. Thinking of regulation and rule, will artificial intelligence or the human (the manufacturer, owner, and third party) be responsible for its operations if a computer with artificial intelligence might create its own codes, heuristics, and algorithms? Legal paradigms as they now stand are inadequate to deal with this issue, hence further measures are required. Existing legal frameworks are insufficient to solve this problem yet essential. The key argument is this: clever and independent computers that indulge in "human practices" contradict existing legal paradigms and produce a variety of problems. Is the contract signed by an autonomous smart computer, for example, to be deemed legitimate as a negotiating party and liable for a violation of the contract? Electronic commerce solves only this problem as intelligent machine agents wander the internet and enter into contract agreements for an improvement in knowledge and autonomy. Each arbitration agreement has an offer and approval, consideration and determination to establish legal obligations in order to move this matter one step forward. An artificially smart computer is actually not known as being capable of making a purpose of itself and thus is not able to contract on its own behalf for this and other purposes.

What are those "legal ties" between human beings and machinery? With respect to humans and artificially intelligent equipment, legislation is developed that applies when a party requires another party to discuss agency law on its behalf. In general, agency law is a business law field protected by a series of legal relationships (generally fiduciary) involving an individual, a software entity called the agent, who is allowed to operate on behalf of the other person, called the principal, to maintain legal ties between himself and a third party. The agent is responsible for a variety of duties, such as: the responsibility to carry out the tasks or tasks defined under the agency's terms (i.e., the agent should not do anything that was not sanctioned by the agent principal). If it is then found that the accused agent behaves without the authority required, the agent is usually held responsible. Since software agents may contractually relate this theory, it would appear that potential artificially smart machines would also function as agents and therefore be subjected to agency law [3].

With respect to criminal procedure, we have developed a law framework where the perpetrator is a human being, but what if the "victim" is a smart computer that says it has rights? For instance, what if an artificially intelligent computer has a software virus imported to the operating system? What privileges has the computer, counterpart of the human limbic system, or limbic system, to safeguard the dignity of its software? One consequence is entirely plausible that the period of technical improvement will turn into the knowledge explosion that many of us have identified. Intelligent artificial technologies will continue to evolve until they far outweigh human intelligence thresholds.

In fact, some people argued that the future could progress so rapidly that, in some ways, our biological brains would no longer be able to grasp the directions of machine growth, if not strengthened. The Singularity would also drive more technological advancements (Figure. 7.1). A variety of scientists and thinkers are worried that artificial intelligence could often make society superfluous, but inexpensive health care (but health care could itself be very diverse) could provide an ensuring quality environment, which will provide many of the world's people with limitless lives and a robust economic economy that would reduce the difference between rich and poor today. And here is where the imaginations of those who believe the Singularity is imminent are let free; the idea of having existed long enough to take advantage of the cool innovations that Singularity would bring. Ray Kürzweil accepted in his book *Fantastic Voyage, Survive Long Enough to Live Forever* that the advancement predicted in stem cells, genetics, and nanomedicine in the next few decades

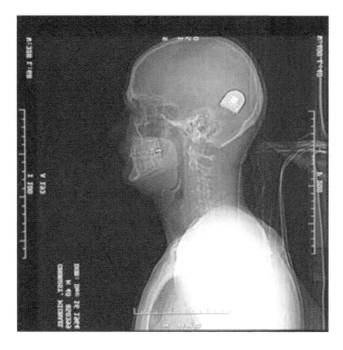

FIGURE 7.1 X-ray image of implanted technology for hearing.

could put an end to lethal diseases, introducing more of us to this high-tech future of tomorrow. Whatever we name those that go beyond conventional notions of humanity, they would have legal, spiritual, and posthuman judgments. Anything thatkindl we call people who step beyond conventional conceptions of civilization has legal, moral, metaphysical, social, and political consequences for their decision to become posthuman.

7.3 TOWARDS MACHINE SENTIENCE

Researchers, science-fictionists, and media writers on Singularity often concentrate on the primary technologies of artificial intelligence. Curtis Karnow, judge of the Supreme Court of California who wrote *Future Codes: Modern Computing Technologies and Law Essays*, frames software autonomy intelligence; autonomy being a machine that can function for solving issues independently of the human. The autonomy of the machine is the capacity to design itself. The more real-time choices the computer can make in unforeseeable situations, the more sophisticated the machine is. When I use the expression "artificially smart machine," I mean a wider meaning of intelligence. I mean to characterize artificially smart machines that can execute cognitive, perceptive, and motor tasks at human skills level when I use this term. A machine that could specifically diagnose illness, like a computer that could write original brief novels, compose music, or run a hedge fund, would then be considered knowledgeable in that domain. Obviously, a numerically operated factory robot will be deemed even less clever and will repeatedly move in preset places.

With the exception of the automatic machine, note that the instances I described just now are extremely "cognitive-oriented" but still clearly demonstrate other areas of human success. Current generations of robotics, for example, will stabilize themselves as they cross rough terrain or climb stairs. More or less, the sensations of our clever technologies are improving, including cars with algorithms, machine vision, GPS, and minimal artificial intelligence. But the topic of the intellect and Singularity lacks an important point in my mind as remarkable, as in the recent gains in artificial intelligence. While "intelligence" is used as the central element in the debate of the next Singularity, I believe that the most important question for mankind is that of "feeling." I believe the most critical problem for mankind is "sentience," where an artificial intelligence machinery appears to be aware and conscious, in time or creation. If it does, I think it's going to get interesting by the end of this century. For one, I wouldn't need to pull a plug on a computer that is wiser but definitely not mindful, and pulling it on a machine that convinced me to be aware that it is not a danger to mankind would present difficulties. At this fundamental level, this issue of ethics relates to the discussion over the capital punishment that we humans have.

Numerous methods are studied for the development of artificial intelligence and essentially a sensitive mind, which pose crucial legal and political concerns. One of the early leaders in the field of evolutionary algorithm, John Holland of the University of Michigan, used biologic evolutionary concepts to prove machines could "change" their programming in a manner such that they would not even completely comprehend complex issues. According to Judge Karnow, in the absence of government entities, certain aspects of existing law are lacking in sufficient legal checks to account for where machines solve problems in ways not expected by the programmer. It's just not a thing. This is not just a question of law, it is a matter of public policy, and in an open and free conversation an educated audience should discuss ties between humans and our intellectual inventions. Professor Holland concluded that, in the ultimate section, hardware is just a means of running programs in order for system feeling to take place. Sentience is based on software and algorithms. Professor Max Tegmark also mentioned that, when our first ultra-intelligent computer was invented, its software was seriously constrained. But when it could rewrite its own software, the emerging machine could climb above human intelligence within a few hours. I could not agree more with how artificially intelligent computers must be integrated into algorithms and heuristics, grafted onto the chips and written as thought machine applications, and making major strides in our understanding of the human brain's potential for computation [4].

Today, thousands of scientists worldwide are working on this goal, and in the last ten years, they have discovered more than in the history of neuroscience about how the brain absorbs knowledge and makes sense of the environment. Nevertheless, the opposite perspective is still true. For instance, Miguel Nicholas, an interface visionary at Duke University, believes that human cognition will never come out of silicon. When I take into account the work being carried out in the fields of artificial intelligence, robotics, and neuroscience, I think that it is groundbreaking that Nicholas not only produces brain computer interfaces but another part of the equation to create a posthuman world in which people merge with machines. Paul Allen, co-founder of Microsoft, wrote to computer scientist Mark Greaves: "conscious of those who say that the uniqueness is near must be the extraordinary sophistication of human cognition." Allen also said "We cannot build the software that could cause Singularity without a scientifically profound understanding of cognition." Physicist Roger Penrose is one of the most ardent critics of the notion that a machine should sense. Penrose protested in *The Emperor's New Mind* against the notion of intellect or cognition appearing in a computer dependent on enough algorithms. Penrose acknowledged that non-algorithms are intrinsically facets of knowledge and consciousness. If we take into account Penrose's critique, it must be remembered that criticism was published in the 1980s (about twenty times more computers). Right now, we agree there are ways other than algorithmic programming for artificial intelligence. Penrose wrote his detractors and learned little about the potential of neural networks or behavioral robots with the capacity to learn by perception and test and error and no microchips to imitate input from the brain. It is possible in the next few decades that we will discover if these methods and one to be created are adequate to achieve Singularity.

I believe it is essential for Singularity to arise and robots to become conscious that the secrets of the human brain are unlocked. The decoding of the human genome gives a glimpse of what can be possible by designing and reconstructing the brain. In *How to Construct a Mind*, Kurzweil states that the number of genes sequenced has doubled per year since the human genome project started in 2001. He believes that neuroscience as well as artificial intelligence would advance equally. The EU, the U.S., and others are financing big projects to make things possible and kickstart advancement in the area of brain research. Henry Markram, among others, uses the influence of supercomputers in the European Union at the Swiss Federal Institute of Technology to study the concepts underlying brain control. The concept is that we should construct a framework that emulates the architecture of thought if we grasp it. In addition, Markram's project for neuroscience attempts to "rebuilt the brain piece by piece and construct simulated brain in a super machine," allowing artificial intelligence systems to boost their way towards increasing thought and planning abilities. Similarly, in the United States, Brain Science is a policy whose purpose is to advance innovating neurotechnology programs and also to step up our human brain comprehension. By expediting innovative technological development and implementation, researchers will create a dynamic brain image demonstrating how individual cells and complex neural circuits communicate in time and space. This image bridges huge holes in the present neuroscience awareness and offers a unique opportunity to explore how the brain helps the human body to register, use, preserve, and extract massive volumes of information at a rapid rate of thinking. The results of this study are extremely useful for the creation of artificial intelligence that reveals how people exercise cognition in the human brain. Such experiments obviously would pave the way for a potential integration of humans with computers [5].

In order to make machines that can think, the physicists are attempting to develop a computer with three features, which the brains and existing computer do not have. These include: low energy consumption (human brains use 20 W, while super machines used to replicate them need megawatts), defect resistance (just losing a transistor; the brains are plastic and lose neurons all the time). The use of neuromorphic chips that do not necessarily need a line of software code to work is a technology to reach these goals. The chip is taught in the way of the "true brains" instead, the researchers say. From the point of view of computing arches, it functions like a small world network and is an essential property of a real brain. Each neuron may have millions of connections between neurons to other neurons within these networks. Thus, since a human brain comprises some 85–100 billion neurons, it is possible for them to connect to each other on countless potential routes within two or three connections in many natural minds Attempts to create artificial ones include memory building and pruning any of these synaptic ties. The observation enables the neuromorphic chips, without needing to rely on a traditional software program, to manipulate data.

The more we understand about brain circuitry, the easier a computer is to be designed to replicate it. As Kurzweil states, for example, the neo-cortex, which has a plurality of neurons and three quarters of the volume of the brain, consists of several columns each containing about 70,000 neurons. The neuromorphic chips that are designed to imitate the brain are cortical column equivalents that link them to a computer, which is actually like the brain, at least. The query of where neuromorphic computation could lead, of course, persists. It's simplistic at present. But if the procedure works, the computer may be designed as wise as—or perhaps smarter than—human beings. People tend to imagine their minds to be more complicated than the lesser ones, because they are. An organization and cabling are the biggest distinction between a human brain and an ape or a monkey. Therefore, it could really just be a matter of connecting enough modules and arranging them to create a conscious machine. And maybe "they" can hold mankind as animals if it succeeds, as Marvin Minsky, a co-founder of the area of artificial intelligence put it. The artificial intelligence industry is also deeply engaged in the production. For starters, Google spent money retaining partners. Machine-learning and robotics established firms, including Boston Dynamics, a lifestyle military robots group, Nest Labs' smart thermostat builders, Bot and Dolly, Meka Robotics, Holomni, Redwood Robotics, Schaft, and another DNN research AI startup. In addition, recently Google acquired DeepMind, an artificial intelligence technology firm on the cutting edge. The "deep" in DeepMind means techniques that enable computers, without being explicitly trained for it, to learn patterning from various types of data and images. "deep learning" uses algorithms to recognize increasingly complete layers in succession, based on the way neurons function in the human mind. Intricate details, such as the movement of a chair's armrests, reveal the story. This methodology seems to be well tailored to today's modern computer age, which can carry out a trillion transactions per second [6].

7.4 TELEPATHY, BRAIN NETS, AND CYBORGS

Classic legislation and public policy have been focused on the difference of human beings and machinery (including livestock), but the technology is now blurred and the convergence of human beings and machinery between cyborgs needs to be integrated into the conversation of who enjoys legal rights. In the next century, as artificially smart robots argue they exist, the question of who should be granted the rights would once again be relevant (and need to be reviewed). Though a cyborg is not considered in legal realms (what unique rights does a cyborg have?), Steve Mann, a person who has been wearing computers for decades, has had interactions that show how culture and law should handle the accelerated pattern of evolution between humans and machines. Steve's personal history as a cyborg with head-built mapping technology has contributed to conflicts with government departments and companies. For starters, Steve's "cyborg" presence merited a screening and he was allegedly examined and wounded by security staff before taking a Toronto-bound plane at St. John's International Airport in Newfoundland. Neil Harbisson also is a self-reported cyborg, who has an embedded chip interface with a camera in the brain, who also faces difficulties in the safety of airport. However, Neil's passport flies Port, including his head-wearing technology photo. Security screening, in particular after 9/11, is an important concern, but what about the rights of a person who wears digital hardware who has a legitimate and vital purpose that cannot easily be separated from their bodies? From Steve's point of view, the issue of the way a traveler gets along while wearable electronic technology is a fixture on his body indicates that "we must ensure that we don't become a state of law enforcement where certain people can't travel." Steve has a good argument that the right to fly without constraint is, for the most part, a constitutional right; we must also create policies to align protection with cyborgs' rights to fly, in the future, as freely as any normal human, as artificially smart machines that are sentient.

This "cyborgization" region would not include government oversight. The U.S. TSA has introduced some protocols to support passengers with medical equipment and disabilities in an effort to cope with the problem of people with a prosthesis going through airport screening. In this scenario, though, the traveler must tell the TSA officer, however, that he has a prosthesis system before screening starts. This may be the case, however. They have an opportunity to obtain a warning card from the TSA website, which can be displayed to the agent, instead of orally notifying the TSA officer of their prosthetics. An individual may have a prosthesis. The prosthesis unit may also be scanned with a metal detector, patted down, or screened by imaging technologies. However, they can also detach their prosthesis and screen it with the X-ray device. Whatever the cyborg process being used, the officer will also be tested additionally; the officer will ask for the prosthesis and will test for explosive residue on the prosthesis with the proper detector, regardless of the technique. It seems obvious that the more invasive the prosthesis in the body, the tougher the gadget is scanned for airport protection; the device is removed and cyborgs are free to fly with unrestricted passenger aircraft [7].

Like Steve, other cyborgs eventually step into our lives and lead mankind into individuality. Remember Neil Harbisson, who was born colorblind and is now wearing a sound-colored eye. How is airport security supposed to look at this device? Michael Chorost is the author of *Machine-Made Me More Human*, which is now *Rebuilt: how to become a computer*. Michael was born with a hearing impairment and was entirely deaf as of 2001; now he has a device inside his brain that helps him to hear again. As a cyborg, his world perception relies on his embedded computer's CPU power, which is different from the human brain.

The mother of technology is necessity and even cyborg accidents. Jerry Jalava lost his finger after a motorcycle crashing and, being tech savvy, inserted a 2 GB USB drive in the tip of his predictive finger.

Instead, it's inside a cup of rubber that matches the cloud of his prosthetic finger, and the USB drive is permanently fused into his finger. But at the end of the day, he strives to make it a genuinely bionic interaction. And then, there's the Canadian cinematographer Rob Spence, whose vision deficiency was the secret to making him a cyborg. He wanted to create his own electronic eye after a partly blind shooting incident, and he now refers to himself as an Eyeborg. Not only can he capture anything he sees by looking approximately, but the device can allow another individual to see the world through his right eye and access his video feed. "Unli," said Spence. "I will afford to update, unlike you humans." "I am a cyborg," she says. "Yes."

The creation of brain-computer interfaces is one of the most important advancements in technology contributing to the emergence of people with artificially intelligent machines. In lab studies and for people suffering from weakening neurological disorder, the direct interface between the brain and the Internet has been studied successfully. Brain computer interface analysis from another factor of perspective is fascinating and can contribute to direct analytical contact. This provides the possibility of telepathy, a technology that would allow contact between the mind and the brain. AI Research teams at the Duke University Medical Center and Kevin Warwick at the University of Reading, United Kingdom, carried out pioneering studies on the brain-to-brain interfaces. However, let's discuss a real court case before we move on to their reports. In 1993 Teri Smith Tyler brought federal prosecutions in the shape of a bizarre complement involving the defender's attempts to enslave and oppress those parts of our population, among others William Clinton, Ross Perot, Defense Intelligence Service, IBM, David Rockefeller, and NASA. Teri contended that she was a cyborg and obtained much of the knowledge on which her case was based. Telepathic "Proteus" communication technology operates in silence. The argument was of course discarded as trivial but still, how far are we from a case with a telepathy and an implantable sensor that is legitimate, considering the innovations in brain-computer interface technologies?

Brain implant technology has allowed a person with a severe spine to move again in a remarkable breakthrough in people who are paralyzed by spinal cord injuries. What's the technology like? The technology usually circumvents the patient's severed backbone by sending the message directly from the brain to the metal bands placed on the muscles of the patient. The surgeons initially diagnose the precise coordinates of the motor cortex of the patient, which controls the muscles in a given part of the body, and then implant a small computer chip. The next phase is "teach the chip" to learn the feelings of the patient. The patient is placed in the MRI, where the patient looks at a hand video moving in a certain way while imagining moving his own hand in a certain way. The embedded chip reads, decodes, and translates the brain signals into electric signals, which are transmitted to the muscles of the forearm of the patient. Next, by having to run a cable from his skull into his arm, the patient is "plugged in" with the implanted chip on their arm attached to the metal bands. When the patient concentrates on his hand moving, he moves. There is still a long way to go before this experimental technology is commonly used for paralyzed patients. It must be wireless, for example, so there is no wire socket. Once it is connected to the skull, scientists have to find out how to send a signal back to the brain from the body to help the patient feel as the body moves.

Duke Miguel Nicolelis and his colleagues lead the way into a cyborg future and have reported that sensor areas are successfully cabled together in two rats' brains. They have noticeably found that one of the rats is going to react to the experiences of the other. The incredibly interesting question they asked was whether the brain of one animal could capture sensor information from another body. They found evidence of brain-to-brain communication without going into the details of whether their study was necessarily feasible. The Duke University Group is currently going to push forward with additional studies that demonstrate the feasible direct brain-to-brain communication by trying to link several rat brains at once. Can there be an emerging "central nervous system," that might lead to mental capabilities that no brain has? Whatever the future holds, it's worth wondering what has already been achieved. Imagine how a multi-form brain unit might feel; say, all students at Oxford University connected to the same brain network, all IMB workers linked to the same brain network, all family members connected to the same brain network and you get an image of it. It is worth considering the advantages and possible hazards of such networks. However, not all are excited about the prospect of a collective interconnection of the brains of people. For example, "Tomorrow, we'll have collective consciousness and the Borg," Rob Spence, Eyeborg, thought, "You have Facebook, camera eyes and the world today. It's an awareness of collective robot. It's truly a modern concern, I believe."

However, on many fronts the concept of bringing people together through technology is evolving. Professor Kevin Warwick and his wife addressed, for instance, that they had silique chips in their arms above the elbow with nerve fibers. There was a power supply, amplifier, and radio communication module for each chip. The purpose of their study of evidence-of-concept was to establish a form of teleportation using the Internet to signal each other. The design has contributed to the first direct contact, solely mechanical, between two humans' nervous systems (not their brains). Interestingly enough, Warwick's wife said she did not want her husband to be associated with anyone else. "My brain's law" side can't help but imagine what marriage and divorce law in the future will look like?

Telepathic capacity, coupled with a range of technology to read our mind, poses many crucial questions of law and politics when we enter Singularity. Technology, for example, will eventually make it possible to search a person's mind for his thoughts. How could this affect basic privacy rights? If the home of an individual is their castle, is a person worth less protection? We would inquire, are there adequate constitutional guarantees to preserve our freedom of expression as we merge into machines? The capacity of government to intercept and interpret wireless communication of technologically transmitted ideas would be considerably more handy than reading electrochemical ideas produced by biological brains—thus, all brain-computer interface ramification is prudent when integrating with computers. On this topic the professor of law at Duke University, Nita Farahany, made an ominous warning, "We have this notion of anonymity, which includes our thinking room, which we share exclusively with those we want." "Neuroscience reveals that what we thought could be abused as this privacy zone." The Fourth and Fifth Amendments respectively in the United States Constitution safeguard against arbitrary search and assault; and self-inflicted, which forbids the State to make a "testigo against itself" for any individual. "Will

Farahany take the fifth to deny information that could be incriminating himself" in a world in which the government is capable of searching your brain? In this respect, I question if the government will have the technology in the future to scan for a citizen's prefrontal cortex. A big civil protection would have been lost if done without a warrant issued. In a cybersecurity context, I foresee a large spectrum of possible criminal and terrorist risks to a human brain and, in that regard, to any intelligent artificial brain, whether knowledgeable or not, considering the substantial advancement of brain-computer interfaces in the world. Why do you think so? First of all, neural devices are now being attacked by the technology. Currently, the media have also written stories about the prospect of hacking pacemakers and other medical equipment. With the same technology hacking devices, including wireless devices, prosthesis controllers, and deep brain, embedded inside the human body, could strike. Second, humans have the opportunity to use neural devices and their inspiration. And third, machines and the Internet have been monitored to demonstrate that, if justification is given, individuals, administrations, and criminal groups can target and subvert computers and devices [8].

Neural systems are far more endangered than computers and the Internet. Conventional assaults on technology and the Internet usually affect money, knowledge, and other property, but none of this affects the human body directly. However, if medical devices are compromised, the person with wirelessly attached implants could be killed or injured instantly. And the use of cognitive devices poses much more danger, as attacking a neural system can wipe out the entirety of the memories of someone or even destroy the mechanisms of thought. What could be a human right more fundamental than defending the mind from external interference? Once singularism is accomplished and artificial intelligence cries, I think that it might even be unbearable to an individual capable of self-preservation to hack the tech and prosthesis. And what about the thoughts injected into your head, maybe through a licensing arrangement, against your will or the internal thoughts captured by a neurochip given by a company? Then who owns your thinking copyright? This case won't be contentious in the immediate term, but hopefully we will be there.

Consider Sony's patent (U.S. patent 6 536 440) that describes a method of ultrasound affecting, such that sensory data can be projected onto the human neural cortex and control nervous impulses in the brain. The technology proposed in the patent is entirely non-invasive, because it uses an apparatus that ignites ultrasound pulses in the head, to alter the patterns of firings in targeted areas of the brain. While innovation can offer blind or deaf people the ability to see or hear, technology poses the curious question of whether Sony should patent Sony's ideas created by people using the technology.

7.5 BODILY INTEGRITY

I think the subject of physical dignity is becoming an important trend for law and political decisionmakers as individuals become enhanced by technologies and artificially intelligent machines become more humanistic. Androids may be particularly keen to protect the purity of their bodies from vanity. Curiously, they may be covered under copyrighted or advertisement laws that apply if a smartphone takes on the image of a celebration. Vanity or machines with other feelings is not a far-off choice, but they are here, albeit in a very restricted way. Professor Rosalind Picard, MIT's Media Lab, and the author of Affective Computing, observed in 2000 that the human brain is not entirely rational but rather subjective and that it is a vital aspect of our capacity to see and to perceive. She has also suggested that machines that need to recognize, in some situations, emotions to have some advanced abilities that we like. At the simplest stages, this can mean installation of sensors and programming, which enable an informatic device to decide the user's emotional status and respond to it in an advanced way. When an artificially smart computer, like an android or robot, has emotions and a relation to the body, it can take a look at how others see the emotions. They may argue in favor of the right to technical innovations, like digital cosmetics, which are not appropriate for any practical use.

Cyborgs and artificially intelligent equipment may be nervous about human responses toward them; just look at the "uncanny valley" phenomenon. Originally meant to provide an insight into human responses to robotic architecture, this definition can also be applied to human encounters with virtually any non-human being. Simply put, people respond positively to a "human" computer but to just one extent. For example, people usually like robotic toys, but when a robot appears like a person and does not completely fulfill the norm, people report a strong negative reaction to their presence. However, if an individual appearance cannot be differentiated, the reaction becomes affirmative. The answer goes ... positive, negative, and again positive. This abyss, the uncanny valley, is the point at which a person who encounters this creature or object sees something almost human but only too off-screen to seem furious or unsettling.

Body dignity is typically concerned with the individual body's inviolability and stresses the significance of personal sovereignty and individual self-determination to the destiny of their own bodies. Body compromise is considered immoral violation in most cultures, and illegal intrusion of the organism is considered in most jurisdictions. When humans are fitted for more technical advances, not only our biological parts but our prosthetic and other cyborg systems are concerned about body integrity. Amusingly enough, a very small number of the persons demanded the amputation of a natural limb and some persons did also have needless amputations. The guy's appearance doesn't match his personality in this setting; this is an indication to me in the uncanny valley. These individuals are diagnosed with "the Personality Disease Body Dignity." The primary explanation for this condition is that when the brain finds the "outrageous body" to be alien and not part of the human, there is a deep urge to eliminate it. A corollary for cyborgs is the alignment of pieces to the body of the human. A link with cyborgs is that pieces must be matched in such a manner that the acceptability of the technology is improved [9].

Is it possible to deny prosthesis as a query for the upcoming Singularity? Investigations into people's prosthesis acceptance found that the support rate is usually poor in some instances. The reasons usually cited for low acceptance include prosthesis functionality and technological issues such as malfunctioning joints and inadequate transition to the residual limb. Deficient prosthesis occurs in court action according to the principle of product liability, whether or not cyborg technology has been approved. Because of the ageing population in the U.S., for example, as in other Western countries, hip and knee replacements in the last ten years have risen dramatically cases involving the failure to replace and recoup damages. In comparison, numerous products' liability lawsuits contain faulty cardiac defibrillators. In this case, patients also run the risk of removing and repairing a potentially malfunctioning cardiac unit and of infection as a result of surgery. As a result of the FDA decisions to recall thousands of defibrillators that may likely shortcircuit when required, many patients and their physicians weigh these competitive risks. For one manufacturer of heart defibrillators defibrillating agents are issuing an electric pulse to provide rhythm for a sloppy heart. It is not rare when products, such as artificial hips, breast implants, and pacemakers, are recalled for medical equipment already inserted into humans. These reminiscences represent a firm's and the FDA's awareness that a system poses either a new form of danger or an elevated degree of a recognized risk.

For individuals, the topic of body dignity in many foreign jurisdictions has been discussed. For instance, Ireland's constitution requires "you have the right not to intervene with your body or person." The State will do little to destroy the lives or welfare of a citizen. There are no clear rules surrounding it in the U.S. Federal Constitution assets to the body or the particular degree with which the state may operate on bodies in favor of the actual body. However, the U.S. Supreme Court upheld privacy protections, which frequently protect bodily dignity rights. The Court, for example, determined that a person cannot, even though such donation saves another person's life, be compelled to donate body parts like bone marrow. The Supreme Court has also safeguarded the freedom of public authorities to infringe on their physical dignity. Examples include laws restricting the use of drugs; regulations banning euthanasia; legislation enforcing seat belts and helmets; inmate strip searches; and mandatory blood checks. We may also see abuses of physical integrity as a violation of human rights. The Civil Rights and Fundamental Rights initiative at the Columbia Law Schools described four primary aspects of government alleged violation of bodily integrity. They are: rights to life; slavery and forced labor; personal protection; and torture and care or punishment that is inhuman, unfair or degrading. The Universal Declaration of Human Rights and the International Covenant on Civil and Political Rights currently secure these rights under two main international documents. Should these protections not extend to artificially smart creatures after Singularity?

In view of the rise of cyborgs and artificially smart computers, were there any integrity problems? Looking at the perspective of Professor Steve Mann as a cyborg, the conclusion is yes. Steve was in Paris with his family and used to take pics and videos and increase Steve's visual information processing capability. He wore the EyeTap device, which is physically mounted on his skull. With fear that the value of his wearable equipment will not be recognized for its daily service, Steve has evidence from his doctor that the EyeTap is not reversible without speed. The McDonalds workers were given the documents by Mann without help. He was finally deleted from the restaurant physically. This won't be the last cyborg attack, I'm guessing. McDonalds' reaction to Steve was inspired by his ability to capture footage while at the restaurant. The debate is fascinating, and do regulation and public policy have to differentiate between wearable device technologies that don't affect their environment? Or do wearable computer technologies provide a benefit if they are capable of "screening out" remotely and affecting people in the cyborg sensor range? In another cyborg technology conflict, in California, there was no evidence that the system was in service at the time of the dismissal of a woman's traffic ticket on Google Glass behind-the-roll.

From a privacy point of view, you should build your own "glasshole-free zone" if you don't want the potential of Google Glass to turn people into invisible cyborgs for surveillance recording. Berlin-based artist Julian Oliver has created the Glasshile.sh software to identify any Glass gadget that is connected to a Wi-Fi network. When it finds glass, it uses a separate program to relay a "de-authorizing" signal, which cuts out the Wi-Fi headset. It can even give a signal to the glass carrier of anybody's nearby existence. However, Oliver warns that users could use the same method of glass ejection more aggressively: he wants to build another variant of the Glasshole. Sh is supposed to be a form of roving Glass disconnector to knock Glass off every network or even isolate its attachment to the user's phone in the future. He sees Glass as a situation where Google first breaches data security laws and only raises questions. "With the backup network and no external sign of recording they are cameras which are rather surreptitious in nature," says Oliver. "To reflect on the computer is to dance past a history of heartfelt resistance to the unconsented video footage. . . . The places and spaces of our public."

7.6 SINGULARITY AND CONCERNS FOR THE FUTURE

The category of artificial intelligence study is usually under two categories: solid and soft artificial intelligence. Powerful artificial intelligence is supposed to create machinery that corresponds to or exceeds human beings' intelligence and that has the general capabilities to solve and to develop abstract thoughts and problems. Solid artificial intelligence also argues that a cleverly functioning computer does not only have a conscience but also knows that humans do. Poor artificial intelligence, by contrast, merely says that engines can behave intelligently; they can do without the "mind" of themselves and sometimes pretend that you are responsive. A third alternative is that artificial intelligence will learn and have different forms than human intelligence, reason, and comprehension, as if it existed as a non-human intelligence, without comprehension [10].

Clearly, with solid artificial intelligence, threats will occur for mankind. Perhaps one of the threats is an attitude of "indifference" towards us; that is, our scientific innovations will simply disregard us. However, smarter-than-human computers face a more severe danger. The scientist Stephen Hawking commented that a great artificial intelligence challenge to mankind is the hazard in terms of the speed at which they improve; they could grow enough wisdom to take on the globe. Can they? Would they like to? I'm not sure, but personally I would like the prospect of a revolution to be ignored instead of being hunted down by a killer robot. Naturally if we integrate with them, we enter the technical revolution. This way, we will guarantee that our potential technological innovations are rooted in beneficial humanity. He is a reputable physicist who, because of the growth in artificial intelligence, foreshadows a dystopic future. Sun Microsystems' remarks mirror those of Bill Joy, co-founder of Sun Microsystems, who warned of possible risk in computers. In an essay in *Wired*

Magazine called "Why the world doesn't need us" Joy warned of the possibility for a very real challenge to mankind and the environment from the integration of genetics technology and computer science. Postulating high-intelligence machinery, Joy commented, "I would work to develop tools that can generate the infrastructure to substitute our organisms. How do I feel? How do I feel? Really unpleasant." Joy speculated that society would become more focused on choices based on artificial intelligence; slowly it would lose its computer power. Because of the sophistication of the processes they control, we will no longer be at their cost without them.

Let's look at the complexities of certain information systems to explore this issue. When the space shuttle of NASA arrived, it had about 500,000 machine code lines on board and about 3.5 million script points in ground stations and processing. The International Air Transport Administration's Automated Automation System, the new-generation air traffic control system, also includes a vast volume of hardware and software. Many personal computers cannot operate without operating systems in our workplaces and homes (e.g., Windows), code ranging between one and five million lines. So, Joy warned, "suicide" might be attempting to take out the plug. How do we human beings to bypass the risk of a bleak future? Eliezer Yudkowsky from the Machine Intelligence Research Institute advocates for powerful artificial intelligence that we should create "friendly artificial intelligence" devices that will have a beneficial, rather than a negative, impact on humanity; this happens when human beings prepare or target.

7.7 INTRODUCING SUPER COMPUTER (WATSON)

If this unique occurrence happens, it will have a radical influence on any aspect of human life, including the rights of "human" citizens for cyborgs and intelligent animals, including philosophical concerns on what it means that people are human. The upcoming uniqueness will also change the economy and human position in the workplace. Take into account the efficiency of IBMs supercomputers, such as Watson, which had 80 teraflops in 2011 (80 trillion operations per second). Although Watson is obviously a supercomputer compared to the current requirements, after a few cycles of increased processing capacity it would be substantially lagging in capability. But what makes Watson fascinating is what can be achieved now—a spectacular achievement considering the breadth of expertise needed to win, recently defeating the most prolific human competition at the *Jeopardy* game show.

To me, individuality is a connection to a revolutionary tomorrow. It is the occurrence that shifts the course of growth, from the one governed by biological laws to the one regulated by technical laws. While technology may function at the level of an artificially intelligent system, Watson, a human organism's creation, is not called evolution. The improvements in populations called evolution are those that are "inherited" from generation after generation through genetic material, taking approximately eighteen years for humans. In comparison, Moore's law states that generational shifts in information technologies only take approximately eighteen months. The results are, before an eighteen-year-old having given birth has thus increased the genome, the machine resources could double many times. IBM's Watson from 2030 will not do eighty-trillion units of work per second on the pedestrian track but will be like thousands of trillions of operations per second, which will thus be able to be carried out massively at the same time; of course, we humans can also process information at the same rate, with set capacities for data processing.

How much, maybe you think, is the human brain any smarter than a Watson supercomputer? "The brain can potentially make more calculations per second than even the quickest super-computer," says Stanford Professor Kwabena Boahen, head of Brains at the Silicon Research Laboratory. Of course, the brain is much lighter in the process than a super-computer, but in fact, since "massively parallel" the brain can do more calculations per second. This indicates that the human brain cell networks actually exist and work together to address several challenges concurrently. However, each stage has to be done before the next step starts in traditional computing platforms. The futurist Ray Kurzweil predicts that the human brain will carry about 1.25 TB of data and operate on around 100 teraflops. The human brain has a potential to perform. If you wonder, 1 TB of storage is very large; it can accommodate 220 million text pages. Compared to this, the 2011 Watson version was a 1-TB-memory 80-teraflop system. If Watson functions at 80% of a human brain's computing capacity, this is an important development in computational power; so, I wonder how similar is Singularity?

Let me note something I find fascinating that has a lot to do with how technology will affect humanity's potential destiny. One fact is that artificially smart machines that master sports, which are called the territory of human experts, do not sit on their laurels. They grow, and in just months they get better. According to IBM, Watson's pace has been up twenty-four times since the Jeopardy contest in 2012; its output was up 2,400% and its physical size decreased. In contrast, iconic players Ken Jennings and Brad Rutter, who are a few years older now, are fortunate; they still manage and still work with the same bandwidth constraints as effectively as when matching wits with Watson. However, they undoubtedly increased in height, like most people who are old and not shrunk, and it might take longer time to recall where the car keys are! It seems to me, artificially smart machines may be somehow of an opposite age, because they can develop like a fine wine with age, because they can be quickly upgraded to computer applications and software because they can exchange data from other artificially clever machines. Imagine learning from the knowledge of other artificially smart computers, Everything with internet-based links to rich content. Might it be that people are the 21st-century rustbelt technology?

Is it odd that the key trend in science and medicine is improving your physique and brain with medications, prostheses, neurological implants, and other cutting-edge innovations, given the impact of aging on the mind and body? Naturally, we set the scene explicitly for Singularity and the convergence of people with artificially smart computers. And by the way, Watson had to adjust his 2015 estimate of projected sales created by many "Watsons" from sixteen billion to twenty billion dollars to his employees at IBM. It is wonderful to be wise, and money speaks, so the future will probably have many cleverer Watsons vying with people for the work we are now doing with our brains. They do not compete with us anymore or work with us anymore because their priorities and goals may differ.

Economists caution that the incredible technical moves taken over recent years, such as mobile phones, vehicles that can operate comfortably alongside people in stores, and driverless cars, could very rapidly throw a lot of people out of work. "We're at a time of inflection," MIT researchers Erik Brynjolfsson and Andrew McAfee say of the second computer era in their book. "The main components for the automation and transformation of emerging technology for society and the market are now in operation," state the authors. The technical developments of the last few decades have led to the nation's increasing income disparities, they claim, because only a small number of people appear to profit from the invention of the next iPhone or program for tax planning. And the strongest impacts on the job market are yet to be felt in Brynjolfsson and McAfee. The research by Carl Benedict Frey and Michael A. Osborne from Oxford University in 2013 could give you a feel for what's coming; Frey and Osborne say that about half of U.S. jobs will be taken over by robotics over the next two to ten years, at a "high chance." Economists take this concept seriously, with political repercussions, especially with regard to higher education and injustice and, of course, our cyborg futures.

7.8 WHO'S GETTING SMARTER?

In just a few decades artificial intelligence has come a long way as we think of grandmasters in chess who singlehandedly beat artificially intellectual computers in the very first days of the computer world. It is just seventeen years after Chess Grand Master, Garry Kasparov, retired to a sixth and final game against IMBs Deep Blue, after nineteen movements, which lost two of two matches with three draws. In fact, it was only seventeen years ago (this is around eleven duplications of computer power). Now nobody expects decades before our artificially intelligent creations conquer another field of human knowledge. Why not?—During the last forty years, our artificially intelligent technologies have improved and have been improving at a drastic pace for a brief period of time, since we humans have been comparatively equal in a couple of hundred thousand years.

What about intellectual ability? We know that machines are certainly clever, but are we humans cleverer? In the book *Mindless*, Simon Head, a senior graduate of the Institute for Public Information at the University of New York, claims that artificially clever devices have been genuinely knowledgeable, determined by a wide variety of market priorities and tactics and deskilling. Work is done by middle-class workers; this only reaffirms what we knew already: machines get smarter. However, it is still a contentious proposal if human beings have grown smarter than our ancestors over the past few millennia. The brain has literally shrunk for some time, a leading scholar, John Hawks, an anthropologist at the University of Wisconsin, has said. This conclusion he justified by noticing that, over the last twenty thousand years, the mean volume of the male brain has fallen from 1.500 to 1.350 cc. Hawks claims that if our brain starts to degrade at this pace it would continue to come nearer to the size of our ancestor, homo erectus, who lived half a million years ago and had a brain capacity of just 1.100 cc for 20,000 years. Although some agree the erosion of our gray matter makes modern people less intellectual than our predecessors, other authorities argue the opposite: as the brain shrinks, the wiring becomes more effective, making us faster and more flexible thinkers. Others also claim that the decrease in brain capacity is indicative of our domestication of, pigs and cattle, many of them smaller-brained than their wild ancestors. Interestingly, new genome research casts questions over the idea that contemporary people are essentially the same copies of their forebears, right up to our thoughts and emotions. During the same time during which the brain shrank, our DNA accumulated multiple adaptive mutations in the brain and neurotransmitter processes, suggesting that the internal mechanisms of the organ have shifted even though they become smaller. However, several scholars suggest that it is possible that our temperament and reasoning capacities have altered as a result, and these mutations are still unknown. While it is debatable whether we are becoming more intelligent or not, regardless of the response, "They" become wiser and more calculated by months, not the millennium in period cycles.

The most cynical theory as to why people are less knowledgeable is that we have achieved our intellectual height successfully. The average intelligence in the U.S. increased by three points between the 1930s and 1980s; test scores also increased substantially in Japan and Denmark in the aftermath—a development known as "Flynn effect." Tests have also risen dramatically. James Flynn of the University of Otago, after whom the impact is named, states that the rise in knowledge has been attributed to increased diet and working standards and better schooling. A future window? Some analysts think that the Flynn effect is coming to an end in advanced countries and that the QI prices are decreasing and even declining. Pessimistic scientists claim that our offspring (if not technologically enhanced) will fail to grasp topics that we can now grasp.

A few wondered if, if artificial intelligence achieves some degree of intelligence, it is dangerous and will to take over our planet as Stephen Hawking warned? Or are you eager to make a contribution to resolving problems that have plagued humanity since the dawn of time? J. Storrs Hall claims that machines and robotics are advancing in his book *Beyond AI*. Technology would allow us to enhance our brains by non-biological materials and to communicate their knowledge with those creations. He believes that, in this sense, we will all compete and do not need to fear our computers. I personally see no future for us humans, but rather an improved biological portion, while we merely share resources with our smarter creations; I see ourselves either combining with or being overwhelmed by them.

It's fascinating to consider if it makes us smarter to have the use of technology to help the brain make choices. When dealing with magnet resonance imaging in the study at McGill University or in the region of the brain including memory and navigation known as the hippocampus, people who used the technique of space navigation have carried out fMRIs, both space-built and relaxation reaction techniques. McGill researchers have found that repetitive GPS use can lead to hippocampal atrophy as we age, placing the person in danger later in life of cognitive problems, such as Alzheimer's disease. Alzheimer's disease first attacks the hippocampus before every other brain component, which induces temporal awareness and memory issues. There are still concerns regarding this study, though researchers have found data linking hippocampus activation to memory. Researchers, for instance, are unsure if space strategies are responsible for increasing the hippocampus or if "robust" hippocampuses are the cause of spatial strategies for an organism. In any case, it would be helpful to use spatial techniques instead of GPS to minimize memory impairment.

7.9 RETURNING TO LAW AND REGULATIONS

As human beings are augmented with prosthesis and implants and artificially smart robots advocate rights, what are any law and regulation problems that can concern them? For example, while the aforementioned body integrity security laws do not include protection for artificially smart devices, existing examples include those of which law implicitly recognizes "technology rights." Sandra Braman, for instance, points to certain improvements to U.S. policies that tend to take computer "needs" into account. She references the United States Telecommunications Act of 1996, in which she suggests that there is a difference between social policy and equipment. Connection to the network by persons is necessary for universal service obligations, although the access obligations of telecommunications networks are universal.

The ability of a computer to "read" itself, a mechanism in which software creates its own heuristics to solve problems, raises fascinating questions in law and politics as courts attempt to assume compensation for the consequences done by artificial intelligence devices. One question that must be raised is whether a person who is not educated enough about heuristics employed by an artificial smart computer is responsible for the damage caused by the operation of the machine? If not, who is that? Ethical and robotic workshops are held every year to discuss this topic alone. One of the problems of artificially accountable complex robots is the dilemma of the legal individual, who lacks the status to bring litigation for protecting its rights or to be held responsible for its acts without being considered a legalized citizen by statute.

Legal theorists such as Ugo Pagallo, author of *The Law of Robots*, argue that the comportment of robots as instruments for human activity should be separated from the comportment of robots as appropriate legal agents. I consider this to be a temporary solution to the dilemma of assigning robots liability, that the rule of accelerated returns eventually poses the question for mankind to address whether or not to award the status of legitimate persons to artificially intelligent devices. I assume that legal persons will finally have to be given our intelligent works on the basis of my background in wearable computing and sensor technologies and my expertise in law. If not, we have to struggle with cases where there are no legal persons; however, the damage has taken place. Meanwhile, Judge Karnow suggests that a judicial body should be established as an "electronic individual," based on a corporate and agent comparison. A business is not equated to anybody, but such privileges and responsibilities are also allocated. As an example of this legal myth, Article 10 of the European Convention on the Protection of "human dignity and fundamental freedoms" allows for private companies to appeal to the European Court of Human Rights. Article 10 guarantees the right to freedom of speech; the right can be condemned by an artificially intelligent computer but not given by applicable legislation. The question of why and under what conditions the program has been used to construct artificial intelligence and cryptographic code is a subject that is discussed in a following chapter. The laws proposed earlier by Isaac Asimov on robotics are fairly general. The legislation depends on particular legal doctrines in case-specific circumstances for the interpretation of the facts of a case. Take the case in which a human is hurt by an individual robot. The court would attempt to identify the guilty party to claim restitution if the status of a legitimate personality is absent. The legal philosophy of the matter is product liability; provided the situation, suppliers, wholesalers, retailer firms (and, where staff incompetence is necessary), repairers, installers, regulators, programmers, and certifiers should be held responsible for the imports and their individual employees; there is an absence of artificial intelligence in the list of theoretically accountable people. Moving forward time by time, let's assume that the artificially smart robot is considered sensible but still lacks legal status. While a certain corporation would have made the device, they argue that the robot has been reprogrammed by the current owner after the robot has left the producer. In spite of the current differentiation between hardware and software the legal doctrines of product responsibility are highly troublesome for artificial intelligence. The maker, the robot, or the program creator is responsible for destroying the robot, unless it's the owner's fault, in which case the computer should show some sympathy.

When artificially intelligent devices become mobile, the possible risk is amplified. Given the absence of legitimate personality status, the rule on dangerous wild animals can be used as a corollary of running robots. People in Britain and other traditional laws and jurisdictions have a duty of responsibility to avoid damage to anyone by their animals, whether they are harmful or not. If an animal keeper's treatment or containment of the animal is negligent and this mistake causes harm to another person or his/her belongings, it is the keeper who is responsible for that. Everything well and good, aside from artificially smart robots, will ultimately be smarter than animals and independent from humans in ways other than animals. This raises the question of penalties against artificially smart machinery, especially where a person has no means of restitution. The requirement of restitution can be remedied if artificially intelligent robots are entitled to the identity status and can enter into contracts for their services, engage in an exchange rate, buy insurance, etc. The bulk of financial exchanges are rendered by artificially intelligent bots. This is not that far-fetched.

Another fascinating concern about artificially intelligent organisms is that a patent is necessary. Many patents have obviously already been granted on device and robotics' software and system parts. However, the question in this book is whether artificially intelligent and aware computers may be copyrighted. An interesting topic is the most possible patent rights of the mechanical elements of a cyborg according to existing law. But what about an individual that is self-aware, might they be patent subjects? Instead of the 1980 case, *Diamond v Chakrabarty* should be questioned according to U.S. law if those individuals may be patented. The U.S. Supreme Court in *Chakrabarty* dismissed allegations that Congress merely wanted to restrict utility patents to inanimate matters. The Court ruled that GM life forms, which were not characterized by them, may be entitled to a utility patent (issued for any functional new invention or improvement on a machine, product, or the composition of matter). The U.S. Patent and Labeling Office's position on the issuance of patents on human tissues and organisms, some of which include human genes, is of special interest to both bionic humans and cyborgs. Such an approach has essentially left the issue of the patentability of human-machine hybrids unresolved, though refraining from patenting people outright.

There is actually no case law or laws about how much human genetic content a creature may have before it becomes a human being. And, of course, having one or a few human chromosomes does not make an animal human. Currently, patents exist on species that have human chromosomes, including the Harvard Oncomouse. Transplant patients acquiring animal organs are specifically considered individual and not patentable on the other side of the scale. Can a cyborg with 49% human genetic materials be entitled to a patent? With respect to human-computer/mechanical hybrids, the current knowledge of this concept suggests that one or more essential physiological duties of the person are contingent on mechanical means. Bionic people would therefore have a complete human gene complement, but they actually would use some artificial means to perform certain functions (e.g., the use of "bio"). Bionic humans would also have a full human gene complement but would simply employ artificial means to execute those activities (e.g., the use of a "bionic" weapon, for example). There may nevertheless be an interesting ethical and legal problem if the critical task achieved by mechanical means is thinking processing (i.e., the use of a computerized brain). Such entities certainly do have a complete complementary human gene, but certain people intuitively would believe that such entities are fewer (or more?) than human beings.

7.10 CONCLUSION

In spite of many of the urgent problems pertaining to future Singularity, the fate of mankind is not the least; the public must learn and enter into the discussion right now. Humans must determine whether to support or reject Singularity. We can model our path into extinction, as the opponents say, and as supporters claim, we can build a future of utopia. We could make our country less unpredictable and risky if we could, as a species, agree to what we wanted, where we had been, and why. One would assume that our self-preserving nature could lead us to such a conversation. One philosophical error many people make in thinking about the role of technology in our future is to interpret technology merely as a medium for human use, for instance, to help the blind see or the deaf hear in a manner to better humans. That being said, much of the technologies used for human enhancement can only be thought of as a means to support the creation of the next wave of artificial intelligence machines. I assume that we are either inventing the future of our own demise or in inventing the technologies to liberate humanity from the limits of our brain and body. Interestingly, Google was asked to establish an artificial intelligence security and ethics oversight board to make sure the artificial intelligence technology under their jurisdiction was established securely after the new artificial intelligence firm Deep Mind was acquired by Google. Taking this appeal into account with a statement from the Supreme Commissioner Shane Legg: "in the end, I believe human extinction will most likely happen and technology is expected to play a role in this." I am positive that the Ethics Board appears to me to be a smart idea and that types of artificial intelligence will pose the most serious danger to mankind this century. But businesses have policy that does not necessarily align with society's best interests, so I take Stanford Professor Francis Fukuyama's view that the future of mankind must be in the hands of the people and our elected leaders, who can protect the best interests of humanity through legislation. Their perspective is not always the same as that of society.

I refer to the principle of developing "friendly" artificial intelligence introduced in this chapter and close this chapter with a quote from Nick Bostrom, director of Oxford University's Institute for the Future. "If in the future a machine has dramatically exceeded us in intellect, it will still be very powerful, capable of influencing the future, and whether or not there's more people," then "you have to put the criteria in the first place in the right way, to make the machine humane friendly." I like friendly, awkward machines, particularly machines that might destroy my species. I don't like them. If we ever mix or hack our DNA with computers, the exterior appearances of surgery implants, machinery, and random animal parts are far less visible than corporeals. Why is this so, since we have a preference, and few of us (if any) would choose to live in a dystopian world?

REFERENCES

[1] Goulding, J. (1995). *Aliens and conquest: A critique of American ideology in Star Trek and other science fiction adventures*. Toronto, Canada: Sisyphus.
[2] Gray, C. H. (1995a). An interview with Manfred E. Clynes. In C. H. Gray (Ed.), *The cyborg handbook* (pp. 43–54). New York: Routledge.
[3] Gray, C. H. (1995b). Science fiction becomes military fact. In C. H. Gray (Ed.), *The cyborg handbook* (pp. 104–105). New York: Routledge.
[4] Gray, C. H., & Mentor, S. (1995). The cyborg body politic and the new world order. In G. Brahm & M. Driscoll (Eds.), *Prosthetic territories: Politics and hypertechnologies* (pp. 219–247). Boulder, CO: Westview Press.
[5] Gray, C. H., Mentor, S., & Figueroa-Sarriera, H. J. (1995). Cyborgology: Constructing the knowledge of cybernetic organisms. In C. H. Gray (Ed.), *The cyborg handbook* (pp. 1–16). New York: Routledge.
[6] Grosz, E. (1994). *Volatile bodies: Toward a corporeal feminism*. Bloomington: Indiana University Press.
[7] Gunkel, D. (1997). What's the matter with architecture. In B. Nicholson (Ed.), *Thinking the unthinkable house (CD-ROM)*. Chicago: Renaissance Society, University of Chicago.
[8] Halacy, D. S. (1965). *Cyborg: Evolution of the superman*. New York: Harper & Row.
[9] Haraway, D. (1991). *Simians, cyborgs, and women: The reinvention of nature*. New York: Routledge.
[10] Haraway, D. (1991b). The actors are cyborg, nature is coyote, and the geography is elsewhere: Postscript to "cyborgs at large." In C. Penley & A. Ross (Eds.), *Technoculture* (pp. 21–26). Minneapolis: University of Minnesota Press.

8 Intelligent Cyborg Brains

8.1 PLACING AN EXPONENT ON INTELLIGENCE

With the benefit of increasingly sophisticated technology, artificially smart computers today work much quicker and more effectively in many examples that were once considered an activity that involves clear human intelligence. As a matter of urgency, artificial intelligence, for instance, easily masters the art of driving while riding in a vehicle requires diverse computational, perceptive, and motive abilities. Indeed, autonomous cars advance to the point that the rule of rapid returns will determine that a person who was born today cannot drive legally when he is young. Within a few years autonomous cars will be so "smart" with the pace of advancements in information technology that a person's only necessary response will be a voice-activated destination [1].

Of course, the "mechanical" car itself does not make it clever even though sensors collect information and pass it on to an onboard computer, the computer that directs the car, namely the brain. And since the crude computing capacity and ability of artificially smart machines is closely connected to the brain-building applications, algorithms and architecture, laws concerning their ability of data gathering, computation, and communication will lead to a new law of cyborgs, this chapter's core character. Sincere computer scientists expect in a few decades that artificial intelligence will be greater than human intelligence and pose an existential challenge to mankind as hardware, software, and simulations in an artificially smart brain continue to grow. For this reason, it may be important for the future of the planet to provide an extensive understanding of how the rule can relate to artificially intelligent machinery and in particular the brain design and capacity. Amusingly, contrary to human driving quality, the Google Director states that "the auto driver was not the cause of the crash once after 6 years and 2.7 million kilometres." This is not to suggest, though, that self-driving vehicles have not been in crashes. In fact, over the last six years there have been only a dozen small accidents, but in one instance, it forced a man to drive another vehicle. Indeed, the concept of keeping a person out of the decision-making process in processes containing artificially smart machines is being taken very seriously as artificial intelligence gets better. In this respect, a few courts considered individuals neglectful because they did not obey the guidance that a machine gave them. *Wells v U.S.* and *Klein v U.S.* were two early cases in this respect. In *Wells*, a court considered a pilot to be incompetent, based on evidence that he switched from autopilot in disaster conditions to manual control. The brain of a computer in this case was seen as a better decision-maker than the human one. In *Klein*, the court found that while the pilot was not required to use autopilot in cases of incompetence, his mistake had been considered inconsistent with the proper operating procedure and proof of a lack of due care. Should we infer from these examples that the legislation can cover human intelligent automated machinery? But I don't always say that artificial intelligence really is superior to humans and always would be benign. Indeed, the possible dark side of artificial intelligence is more of a worry to me, where artificial understanding supports mankind [2].

Naturally, "artificially smart minds" do far more than drive vehicles; semi-autonomous drones have parcel deliveries today, some robots assist operational physicians, and "artificial intelligence" writes sports articles, studies, or shops. These are all tasks that need an enormous amount of intellect and in some cases complicated engine skills; however, no one really thinks that robots with these skills are similar to the human intelligence standard. The humans, instead, think that robotic devices in these situations are actually extraordinary instruments to support us with cognitive or perceptive skills. We now have the basic idea that technology will provide us with an even stronger toolkit to satisfy our needs as technological advancements proceed. This, I believe, is a delusional future vision with dangerous ramifications for mankind. According to the stance assumed by Elon Musk, Chairman and CEO of SpaceX and Tesla Motors, the development of artificial intelligence is described as the "convocation of the devil" Nick Bostrom, director of the Future of Humanity Institute at the University of Cambridge, asks: When we accept that AI might constitute a new sort of threat to humanity, how are our governments and courts going to react? Some of them suggest that research into artificial intelligence be fully banished, while others propose that we should encrypt "friendliness" into "minds" of artificial intelligence (as can the possibility of AI render human's passive when confronted with violence). Finally, I think that there is an evolving body of law on the design and production of an artificially smart brain, including opinions, particularly in intellectual property and constitutional law such as artificial intelligence-generated voices. Although these laws and regulations have been introduced to protect the rights of citizens and not self-aware machinery, I believe that they could also lead to an evolving cyborgs rule, that is to say, they are a body of laws that may precede potential artificial smart machines, which have achieved human intelligence standards and will then claim rights.

Elon Musk is not alone in his predictions about the imminent danger to civilization posed by artificial intelligence. In his 2004 book, *Our Final Hour*, Cambridge cosmologist Martin Rees, former Royal Astronomer and President of The Royal Society, explored related topics. Computer scientist Bill Joy, who co-founded Sun Mickrosystems in 2000, wrote *A Scientist's Warning*, in a Wired post, about "why the future doesn't need us." However, some of the leading scientists in artificial intelligence and robotics have also expressed concern that by the turn of this century we can either represent the artificially smart machinery that we are now building (which inevitably takes over its own design) or we will be inconsistent with it and we can be the second most intelligent species in the world. But, as suggested by the robotics specialist Hans Moravec, Google's Ray Kurzweil, the third alternative could occur, which this author addressed in this book, which is the fusion with "them" and therefore becoming the result of our future technologies rather than assuming the status of bystander.

For the reasons discussed later, much of the public remains uncertain regarding a world consisting of robots with human or intellectual levels, and yet others (including

Intelligent Cyborg Brains 185

some leading AI researchers and philosopher) completely condemn a future with a powerful artificial mind as either unlikely, a science fiction topic, or too much to critically focus on in the future. I do not believe those who ignore the rapid increase of artificial intelligence and the effects it has on mankind know that there are now foundational developments to build artificially smart machines, which are being developed at an exponential rate, and which will likely be available to supplement human intelligence and physical prowess in the form of computers in the next 20–30 years. As we approach human-like artificial intelligence, I argue that an "artificially smart brain rule" would be appropriate to improve our regulatory agencies and that such an approach would offer a foundation for many social and legal problems formed by the advent of artificial intelligence. These matters are protected by statute as well as tort damages, contract law, and criminal guilt for artificially intelligent devices that work independently of human beings. However, the applications, operating systems and design of the artificial brain itself are still based on a "law of artificially intelligent cereals." Since the role of artificial intelligence in society is important, this is not the first chapter to deal with these problems, and a number of reviewed documents and books on law and robotics exist (see for instance Gabriel Hallevy, *When Robots Kill: artificial intelligence under criminality* and law professor Ryan Calo's papers, for instance). Moreover, I am optimistic that national and public interest will grow as a result of this pretense of confidence and advocacy for rights [3].

This chapter looks at some legal and political questions pertaining to artificial intelligence brain construction in the light of these findings. I use a considerable amount of existing law relating to computer applications and the law relating to computer engineering, which is important to machine computation and thus logic and understanding, to address these issues—these laws can be considered to be the regulation of artificially intelligent minds.

8.2 THE NUMBERS BEHIND BRAINS

I inevitably ask the next question: how close are we to computer hardware and software, that matches the human brain in output, when I am speaking about our cyborg future in conjunction with artificially clever machines? Many influential robots and inventors tend to be fixed in an unknown timeline until the middle of the century. First of all, what computational tools are required for the purpose of artificial intelligence like humanity? That is, how can the artificially intelligent brain be stored and interpreted in a raw way? If human and machine minds have identical characteristics and designs, the same laws will apply Are all bodies applicable? Recognize that chimpanzees have the brains of structures close to that of a human brain and the behavior of chimpanzees is that of a distant relative, but in most jurisdictions, chimpanzees receive few or no individual rights.

Receiving human rights is a big obstacle; society cannot afford generous rights to all species. Counting the napkin's folds yields back, we will address the question of the calculative tools required to match the human brain by looking at the numbers used with reverse brain engineering. Neuroscientists, computational engineers, and robotics have experimented for nearly a century to modify the human brain and, in the final analysis, to construct a computer architecture based on the workings of the mind. The secret to the human brain's reverse engineering is the cerebral cortex—a perceptual seat—decoded and simulated. The cortex has about 22,000,000 neurons and trillions of synapses. According to researchers, a device with a computer power of at least 36.8 petaflops with a memory capacity of 3.2 petabytes will be required to run a human brain machine simulator.

A significant and realistic question would be how many lines of code will be needed to simulate a brain? This is all fascinating and technologically feasible, but Terry Sejnowski, Head of the Salk Biology Institute computational neurobiology center, agrees with Ray Kurzweil's estimation that approximately one million lines would be adequate to fulfill the mission. Intuitively, this figure seems to me to be tiny, but I said we are doing "back of the napkin." "The architecture of the brain lies in the blueprint of the genome," according to Kurzweil. The mitochondrial body has a base pair of 3 billion to 6 trillion bits, around 800 million bytes until compression. Kurzweil states that, "the information can be compressed to around 50 million bytes by removing redundancies and applying lossless compression." About half of this material affects the brain and is around 25 million bytes or around one million lines of code. I've heard that rebuttals are too few, but I'm shocked to learn that even though we lift lines of code that are important for simulating the brain by order of size, we are now developing the technologies and obtaining the information to unlock the brain secrets, so it is just time we replicate the brain with a million lines of code or, if appropriate, 100 million lines of code.

So how close can we get to a human-like artificial intelligence, a machine, algorithms, and electronic components that can argue that they can think, solve a problem, and regulate the movement of their bodies? Professor Hans Moravec, a robotics expert from his books and documents, has described the 2020s as the time span of man-like robots. *Mind Children: the future of human intelligence and robots*, a basic engine for transcendent humans, was written by Hans Moravec in 1998. Based on his pioneering work on robotic machine vision, he calculated the need for approximately 100 MIPS computing resources to balance the "absolute human activity" to the 100 trillion-billion brain capacity of approximately 100 million megabytes. These are less than those provided in the previous material (the brain measurements in the petaflop scale), but because we already have supercomputers on the petaflop range, we have balanced the crude computing capacity of the brain on the basis that it is calculated to be raw in computational power. In the same vein the progressive speed of artificial intelligence accelerated when a couple of years back IBM's Deep Blue beat the world chess champion using chips that are built to work at a rate of 3 million MIPS or 1/2 of a gross Moravec average of the efficiency of humans [4].

Naturally, we are mindful that neurophysiological influences are vital for human understanding of the cerebral cortex of 22 billion neurons and to emphasize the difficulty of designing and building a brain, and therefore why if the fair estimation of the amount of synapses per neuron is 12,500 (some estimate around 10,000), the 22 billion cortical neurons alone need somewhere in the order of 275 trillion transistors to balance the number of synapses inside the cortex, so a "rule of artificially smart brains" will be exceedingly difficult (we are several years out from creating such chips but we will get there). However, the evolving nature of the neural networks does not take this degree of sophistication into account when we understand

and construct additional memory and cellular structures. The internet's billions of streamlined nodes working in parallel with one another are replacing the brain's traditional computing approach. So, I see predictions related to when we will create a computer that hits people's intelligence thresholds, with a clear interest and I think this is going to happen, but I will be shocked to see the degree of sophistication that will reverse the date for Singularity, when we understand more about the brain, particularly the cortex. What does it matter if the 45.4 million years of history come a few decades or even hundreds of years after the whole Earth! The human brain is clearly unbelievably complex; however, we work in compliance with the laws of Moore, which now means that the machines can be much quicker than us on the basis of raw processing power and that the quintillion will inevitably exceed exaflop estimates (1018) at that point. Computers per second are dependent on raw processing power. I thoroughly understand the complexities of the human mind and how very challenging it can be to construct (the biggest obstacle to mankind) but it is the sense of rapidly accelerating techniques and, in particular, the dramatic advancement in neuroscience and in computer science that is called the difficulty of producing human-like intelligence.

8.3 LAW AND BRAINS

Recognizing the difficulty of constructing an artificial brain, let us now address laws that might apply to artificial intelligence more precisely. If we (and our computers that support us) can execute scripts in order to build an artificially intelligent brain (with the requisite computational resources), what do the existing laws say about computer programs and algorithms comprising the brain of an artificially smart device? Several fields of law, including the copyright law and patent laws, trade secret law, Constitutional law on speech production, and artificial intelligent brain algorithms, are important for this and therefore for our cyborg future. Copyright and patent law are also in a way applicable to software security and to algorithms, which lead to an artificially intelligent brain and which, as I have said, aid in the evolving laws of cyborgs (the owner of the software has such protections under copyright law). In particular, proprietary software owners have, with exceptions, the right to: copy, develop, or change the derivative software and sell to the public copies of the software, by license, by auction, or otherwise. Will an artificially intelligent brain have any relevance? Someone practicing any other proprietary privilege without the device owner's consent is an infringer and responsible for penalties or civil penalties. Amusingly, by stealing the software and algorithms and infringement under copyright law one might actually deprive the mind of an artificially intelligent brain. Similarly, potential brain scan technology in cyborgs might even serve to replicate the thoughts that a human mind has produced, but because "mind software" is not covered by copyright (mind outputs might be), the user wouldn't be an infringer, and we may need to amend the rule. We see "civil rights" as a means of framing problems surrounding artificial intelligence without direct status and jurisprudence. In the early days of coding, software engineers used the law to secure copyright protection over their programs, claiming that writing code is equivalent to all ways of writing. In this regard, they are using a software program that safeguards intellectual property. The scripting language for the operating system and applications of an artificially intelligent brain can be accompanied by the same rationale. The general conditions for copyright under the U.S. Copyright Act are: "existing authenticity activities, now established or subsequently created, which may, either directly or by machine or apparatus, be interpreted, copied or exchanged otherwise." In addition, the Copyright Act describes a computer program as "A collection of statements or instructions to be explicitly or implicitly used in computers to achieve a particular output" (think the tools used by robots for parsing images in scenes). Explanations include machine vision applications, robot navigation, or test and error learning.

When a computer is programmed, as long as the conventional copyright conditions are largely fulfilled, a piece of work is published on a pad of cardboard; it means it must have been autonomous by its creator and minimal artistic effort must have been involved. The work must be unique. Technology, which is a core component of an artificially smart mind, is obviously a subject that can be patented. We understand that the raw data, which means the language for artificial intelligence programs, is protected from copyright, but what about the executable program? In *Apple Computer, Inc. v Franklin Computer Corp.* a U.S. circuit court found that the programming language and the object code programs are protected by copyright. The Court, surprisingly, dismissed the logic of only transmitting object code directly to a computer and not defending it, which opens up the prospect of copyright security in abstract language (a machine telepathy method of machine-to-machine communication).

But what about applications not written on paper but grafted on a chip that jeopardizes an autonomous robotic brain's architectural design? In Franklin, we have discussed whether programs encrypted on chips are functional objects—and thus not protected by copyright. The Franklin Court dismissed the contention that programs were "utilitarian" only and that the means for encoding the program does not decide whether the program is proprietary. The memory and emotions of an artificially smart brain in origin or object code can be preserved if they are locked to an artificially smart brain in an integrated circuit. This is a valuable method to assist an artificial intelligence in protecting its rights to the information. Obviously, the programming software, in general, executes tasks that are essential to an artificially intelligent brain, rather than just individual lines of code. In Franklin, the Court dismissed the contention that mobile devices are not protected by copyright because they are "processes, systems or organizational structures." Rather, the Court determined that an operating system had to be treated as a copyright work. For example: robotics, a technology that takes us into singularities; obviously robots from one age to another are smarter. This is directly comparable to an artificially smart brain. The next is focused on enhancements to the brain architecture's tools, algorithms, and interface layout. As the backbone of an artificially smart brain is liable for copyright protection, an operating system cannot replicate it or extract it from an artificially smart brain created by the owner of copyright without permission. The creator may (or may not) own the robot (the owner of the robot may license the software); however, in the future, it may be unclear whether the human being owns an artificial intelligence patent. (The 13th Amendment forbids slavery and unintended servitude.) And if in the future, autonomous robots are artificially emancipated by a human creator, they will use copyright law as a means of defending their vocabulary and software is a communication type [5].

To sum up the controversy on this argument, the key aspects of an artificially smart brain comprise programs, an operating system, and protocols, and courts have ruled that the basic components of a computer code are protected under copyright law. The Second Circuit Court of Appeals has, for example, considered in *Computer Associates International, Inc. v Altai Inc.*, whether copyright covers non-literal program code elements, i.e. program form, sequence, and structure. Copyright covers an invention but not the idea itself as a central point (ideas are protected by patents or to some extent trade secret law). So the manner in which the program is written varies considerably from an artificially intelligent brain in terms of preserving its code.

So, where do we differentiate between program language and idea? The Court ruled in *Baker v Selden* that "items that actually ought to be used as incidentals" are not patented. However, this opinion did not offer guidance about how to differentiate a concept from its language. With regard to a similar issue, the Court sought to define the distinctions between "concept and speech" by stating that the function of the job is the idea and that anything else that is not required to function is the expression of the idea in *Whelan v Jaslow*, a seminal case in establishing concepts applicable to computer applications. Yet other courts considered this solution unlikely to operate and followed the approach to filtration adopted by *Computer Associates Int'l. v Atltai* on the Second Circuit. This approach removes the concepts and elements of the public domain from their representation and then only applies the rights of the copyright to the word. The court decides first of all the structural elements of the apparently infringed software in the emergent phenomenon test. The components are then filtered into non-protected objects. Non-protected elements include: performance elements (i.e. elements with minimal means of communicating them and hence by way of an idea), external factors-driven elements (i.e. common technologies), and public domain architecture elements. Any of these nonprotected objects are removed and the rest of the elements are matched in order to assess significant similarities with the elements of the supposedly offending program. In my opinion it would be impossible to use the algorithms of an artificially smart brain to assess what facets of software are proprietary, just as it will be difficult to determine which aspects of human thought are featured versus speech. Therefore, courts would need to create a new test appropriate for the cyborg era to evaluate the aspects of code, especially the code of an artificially intelligent brain, which is copyright-proof.

8.4 MORE ABOUT ARTIFICIALLY INTELLIGENT BRAINS

An artificially clever brain is a replay of a key issue: computational design, software, and algorithms to control its behavior. Will such a brain be knowledgeable of enough computing tools and software? This is a matter of considerable discussion, but there is a strong neuroscientific agreement that the human condition has essentially evolved as a result of information processing from the architecture of its 100 trillion neurons. And from the previous discussion we know that much of artificially smart brain software is "protected" by the law on intellectual property. Thus, there is an evolving rule of artificially intelligent brains. Interestingly, while we can find that no "neuron law" exists, machine law exists. Thus a human brain cannot analyze the brain of an artificially intelligent computer in a human brain fashion according to the rule.

Another topic of importance for those who model the existence of cyborgs and challenge how the rule should function is whether intellectual minds artificially outweigh the capacity of the human brain. As we did earlier, let's think like an engineer for a time and concentrate on the quantitative measures of brains. Provided that the electrochemical signals that human brains use in order to accomplish reasoning move at approximately 150 m/s, orders of magnitude are lent to machine speeds. Thus, massively parallel electronics of a modern human brain will be capable of thinking millions and trillions of speeds greater than our evolved natural system. Further considering that in 2013, the clock speed of the microprocessors reached 5.5 GHz, which is around five million times stronger, this ensures in some cases that neurons can generate a maximum of approximately 1,000 action potential per second; in the performance of human brains, computer brains are superior. But supercomputers can also have energy needs that compete with certain municipalities, and which are bigger than calculating machines in the laboratory at the start of the computer industry. Meanwhile, the human brain uses approximately twenty watts and is small in volume [6].

However, as computers exponentially improve computational performance, we can expect the computer to have the human brain's computational power in just a few years. Many factors, including a huge drop in the price of information technology combined with an exponential increase in performance have driven the recent increase in artificial intelligence. Algorithms are more and more capable with so much computational power to understand languages, to recognize images, and to perform autonomously from people. For example, artificially smart machinery is not only getting smarter, its costs are falling and all information technology is involved in this price-performance ratio. In this regard, production costs have plummeted for an important technology for the future production of manufacturing artificially intelligent machinery-electronics. For the development of electronics, robots are increasingly so expensive that sometimes the use of a robot for routine assembly costs just a few dollars an hour compared with a six times higher rate for the average working person. How long will it take before the next generation of robots is designed to meet their own requirements and move more people from their places of employment?

Interestingly enough, Hans Moravec reported that computer human output is only economically significant if their brains cost about $1,000, and when can it be expected? Our data indicates that the human brain will replicated around 2029. For our cyborg future it is important to remember that the Moore law demands that machines will proceed to double their prices every eighteen months or so for the following decade or longer and that once we hit the human output standards for robots, their development is not biologically dependent and does not rely on random mutations in genes in order for them to be more intelligent. While critics claim that Moore's law goes on, other technology will take over current chip-builders, if the past trend in computer technology continues, and it will begin the exponential development in computer power up to and outside of the mid-century. People have the incredibly strong exponential growth of IT every day, and we are all showing the exponential development in technology, as the mobile phone that we all use is both a million times smaller and a thousand times larger than the most powerful computer available in the 1970s. And any cellular call is, of course, directed with artificial intelligence. Others may not be, but artificial intelligence is in the context, quietly, doing its job, now, we rely absolutely on artificial intelligence, from air traffic control systems to house appliances.

Hans Moravec continued his comments on the need for a machine brain of $1,000, back in 1999 and in his latest writings. In 2020, a laptop of $1,000 will have the processing power and storage space of the human brain. Ray Kurzweil predicted (100 billion neurons, 100 trillion synapses). Currently, we are well on the way to building such a primitive computing capacity with a cheap computer. Kurzweil predicted that hardware needed to simulate the brain could be ready as early as the year 2020, based on exponential growth curves, utilizing innovations such as graphics processing units (GPUs that use a vast hierarchical framework that is an ideal architecture for brain software algorithms, I could add). Although critics who reject the concept of human intelligence-like artificially smart machines worry that this result will be devastating to mankind, many understand that we are now entering a phase in which computers are able to process the cognitive capabilities of the brain. But to develop human-like artificial intelligence, more is required than computational resources. Critics, for instance, point out that the existing software cannot model the human brain anywhere in its capacity to process and determine.

While critics are right that computing resources are needed, although not enough to produce living organism intelligence, artificial intelligence software and algorithms are indeed making huge strides in imitating decision-making in the brain (that is, simulating neuronal systems). In fact, according to Kurzweil, it will be just a matter of time until the brain's energy needs to match those of the software to accurately mimic the brain, maybe as early as 2029. However, in my opinion, while developing an artificial intelligence that equals a person in capability is a seminal event in the history of humankind, what humanity really has to concentrate on is what happens after Singularity has reached us. An significant premise in this book is that we must merge with our technological progeny for the survival of mankind, or as Hans Moravec put it—our "mind kids." To blend with artificially smart computers, we must combine in cyborg technology with artificially intelligent machines (which will be a key factor for our future survivability). As the machine technology continues to advance at an exponential pace, we will have the combined human intelligence accessible through neuroprosthetic devices implanted within our brains in a couple of decades. This capacity is necessary for our species' survival until artificial intelligence achieves its uniqueness.

While it is expected that Singularity will only be a couple of decades away, the machine already has a great advantage in comparison with us: it is interlocked through the Internet and shares information billions of times more quickly than we humans can using the network connection that we naturally have. That means that a regulation of artificial intelligence must take into account how a collective type of artificial intelligence shares accountability and other legal obligations. In 2040, Ray Kurzweil of Google is the most credible futurist (within a few centuries, at least). A hundred million times further intelligent than biological intelligence, us, is non-biological intellectual capacity. The reader might ask how this is conceivable and why so quickly. After all, today's artificial intelligence lacks general intelligence and good sense, although it is surprisingly intelligent in a small domain. But as Kurzweil pointed out in his insightful books on what the future could bring (see for instance *The Era of the Metaphysical Machines* and *The Singularity*) people's issue with imagining the possibilities is that they are linear thinking, they extrapolate the universe in which they live straight away. There is a boundary to forecast the future of technology. Linear technological thinking worked very well centuries ago but the pace of technological developments started to increase dramatically at the middle of the 19th century. With regard to computer resources, it has been shown that the increasing pace of information technology is obviously not linear by the explosive increase of many computer-based technologies [7].

Believe me, in a line scale against an exponential scale of advancement technology, the artificial intelligence brain in particular makes all the difference. With an example, I will prove this to you. Wanna be wealthy? I argue that it's simple to do, tongue-in-cheek. Let's use our replacement shift here and allot a thirty-one-day month to achieve our wealth target. Place a penny on the first day of the month and double the amount set in the calendar before the end of the month on the next day of each subsequent day. What's going on? You have two centss on the second day and four cents on the third, eight cents on the third and a total of sixty-four cents by the end of the week. A week of pennies has passed; do you still feel rich? You have $81.92 on Day 14. Proceed. You are nearly halfway to the end of the month; do you still believe me that in seventeen more days, you will be really rich based on duplication? From day to day, pennies? Most people reply no, use quantitative analysis to scale up the issue, and say they will get a few hundred dollars at the least (which is a lot more than they thought they would have at the beginning of the month). So, I keep on practicing (with your pennies!). We got $10,485,76 on Day 21 and that certainly gets fascinating, but we have just ten days until the end of the month and we are not yet wealthy. At Day 25, you've earned $167,772.16 and you're probably still intrigued by the exponential growth idea. At last, we got $10,737,418.24 on Day 31. We have obtained this remarkable outcome because of the doubled in size one day to the next, in order to reiterate the problem of exponential growth. It is obvious that doubling pennies is not a matter of growing artificial intelligence, of the efficiency your brains can achieve, and of fusing our future with artificial smart machinery. Since technological developments are exponentially increasing, I assume you would be in agreement with me that extraordinary technologies await you. The size of the exponent that is necessary for growth is essential. Probably, you will think that it is the future that awaits us, a world with amazing, clever resources to serve us. But not so easily, the challenge I see is that our instruments are going to become more and more intelligent than ourselves, and who is going to be the manager and the servant?

Although it is essential to double technical capital, the time between doublings is essential (which also makes all the difference). In our rich example, we have to say that rather than using days on a calendar that we have used a time limit of ten years. So after ten years, we will have two cents and in twenty years, we will have four cents. I argue that technological improvement will change all of this because of the power of duplications and the simple periods between duplications. Let us now say that from one generation to the next people have doubled their intelligence. This is certainly not possible, but if it were, a person with twice the general population's intelligence would be born in about eighteen years. The trouble is that in that period we cannot substantially alter our intelligence. As we are the products of an extremely slow evolutionary process, the cycle performance is calculated in the millennium for improving human intelligence. However, there is a solution to keep up with progressively smarter machines—technologically working to improve how they are incorporated into our body exponentially. The cycle time for doubling the number of transistors on a chip is approximately eighteen months according to Moore's law. Using the previous instance, although the human had eighteen years to wait to double his intelligence, the hard drive would have had 12 duplicates of computational power, which indicates that computer power is necessary, but not adequate, to generate artificial intelligence. To refer back to our example with pennies to be wealthy, 12 doublings is the distinction between two cents (one time) and $20.48 (12 times); and therefore, it is plain to see that if we keep doing things the way we are right now, we will eventually become the intellectual equivalent of our so-called "competition." Recall that because supercomputers are already computing for petaflop (one quadrillion floating dot operations per second) 20 petaflops are doubled 12 times and compared with the processing power of the petaflop computer brain that continues to be based in biology without any technical improvements. Importantly, we are mainly in an exponential increase in the computing power of technologies that could lead to artificial intelligence that then goes beyond us with the use of exponentially accelerating technologies. And the same concept can work for the human brain in theory, namely when the brain is wired. But the same concept could work in theory for the human brain when the brain is connected wirelessly to the cloud via a neuroprosthetic smartphone. In fact, by 2045, by connecting our intelligence from our neocortex wirelessly with a synthetic neocortex in the cloud, Ray Kurzweil forecasts that it would be a billionfold.

Assumptions of the future are a by-product of Moore's law and the law of accelerating returns in more general terms, as should be clear from now on. Given the power of computers now, a few doublings are an incredible improvement in the computing power compared to where we are now when the cycle time is about eighteen months. If we begin with 40 petaflops, for example, we are now at 320 petaflops after 3 doublings (less than 6 years); 8 times in less than 6 years computing capacity is a huge improvement (I will take it on the share market!). And provided such a short cycle time among duplicates, it is not surprising that Kurzweil and others predicted impressive technical advances so successfully, that they only have to postulate what is possible, if technologies are to be improved exponentially in twenty years or so. In general, IT follows an exponential curve of growth founded on the assumption that computing strength allows it to double every eighteen months. Indeed, IT has seen a massive increase over decades. This has contributed to enormous memory, computing power, artificial algorithms, and speech recognition and computer intelligence upgrades. And with the growing raw computing capacity of computers, improvement has also been made in algorithms to emit thought and the design of chips, which manage information better than the brain (mostly parallel to computers in recent decades required to handle information on the basis of machine architecture. Of course, the law that covers algorithms and computer chips falls within the new law of artificially smart intelligence and cyborgs [8].

8.5 MACHINE LEARNING AND BRAIN ARCHITECTURES

Informational technologies that improve exponentially is not the only field in science and engineering that is making tremendous steps in terms of the creation of artificial general intelligence, and our future is merging with artificially intelligent engines. Hundreds of millions of dollars have been invested in neuroscience research aimed at mapping the brain's many levels and functions. In neuroscience research will most likely eventually expose the comprehensive methods that lead to cognition and behavior—the biology that makes us human and conscious—from genes to cells. This awareness would make computation all the more possible to make artificial general intelligence. However, it will be incredibly difficult and complicated to develop human artificial intelligence, because the human brain performs computations that are inaccessible to the most mainframe devices of today—all the while using no more power than a light bulb. Under the European digital agenda for the future, which is part of the European human brain project, it is possible to make sure that the system reliably calculates non-trustworthy elements and how different brain components communicate to one's heart. New hardware groups, new operating systems; and the computer paradigm changes as a whole. The effect is theoretically big in economic and industrial terms, but the end outcome is undoubtedly an artificial awareness that will exceed us if our fate does not converge with our technology progeny.

The term "artificial intelligence" often recalls possible dreams of human-like robots, but a machine's capacity to learn is still a concept today, as well as the ability of computers to learn results directly from their brain design, applications, and algorithms. And how are modern machines studying and acquiring intellectual understanding? A "machine learning" approach to developing humane artificial intelligence helps a computer algorithm to distinguish the core characteristics of one information source then to incorporate what it has learned to predict another. According to Biome, "optical recognition system, spam filters, automatically facial recognition, and numerous data mining applications" are known examples of this machine learning approach.

Although a supercomputer has the raw computational capacity of a brain (in the field of petaflop computing), it cannot contribute to artificial general intelligence without the use of regulations/algorithms that permit thought about this amount of processing power. However, progress is made clearly using a range of strategies to build machines that think and reason more like humans do; some developments depend on algorithms, while others rely on the nature of computer hardware architecture itself. For example, a method called "deep learning" is one of the methods used to build a device that "thinks" and literally refines machine learning. Machines learn in detail by smashing massive data sets and then statistically analyzing the data as they are searching for correlations without human interference. This method of machine learning is extremely effective, because it provides computers with a way to recognize things when they see them, since they cannot carefully establish laws for any occurrence and contingency in the universe by making them themselves. I would stress that these elements of "thought" relate to a growing space of constitutional interpretation relating to the law of artificially intelligent minds, regardless of which algorithms, software, or design are concerned.

Intelligent Cyborg Brains 195

Describing whether equations, fundamental elements of an artificially intelligent brain, are speaking, Professor Stuart Benjamin, Duke Law University, states that "many human acts require the transmission of parts in conjunction with human-made and engine-induced protocols and algorithms." In my perspective, Benjamin's qualifier of the phrase "generated by people" is a word that may become obsolete in a generation or two when artificial intelligence advances to the level of cognition. "Is this algorithm based performance talk in line with the first amendment?" Benjamin asked. We understand computers "think" about bits, using algorithms like data analysis, for instance, to signal intensity and lines in a scene to identify an image. Indeed, machine code is simply a collection of directions and algorithms (is the human mind identical?). Benjamin notes that their performance can be subject to at least some First Alteration defense, even though algorithms are not expression. Sorrell versus IMS Health, Inc. is of importance to the law of artificially smart brains that use software and algorithms to generate compartmental outputs in which the High Court ruled that, in light of the First Amendment, the production and distribution of information is expression. Professor Benjamin said, "We will regard the products of the machines as those of humans by applying the First Amendment to messages generated by artificial intelligence." In his opinion we might then assume that a speech was actually made and was not just communicated or assisted by a computer. In reality, in *Brown v. Entertainment Merchants Ass'n*, where video play production was taken into account when the court decided that ideas had been transmitted as a literary instrument, the question of whether the output of a machine was speech was answered. I assume that this holding is a precursor to a potential artificial intelligence that asserts the freedom to communicate. According to Columbia Law Professor Tim Wu, however, the issue of "free expression" for computers is troublesome. "Computer systems are utilitarian resources to support us," according to Wu. He states that "the First Amendment is structured to prohibit censorship of real people." Wu believes that full defense under the First Amendment should not be extended to non-human or artificial decisions and that speech should not always be taken as such. Professor Wu claimed it was the "elevation of our robots above ours" to grant computers the rights intended for humans [9].

I think we are several generations away, responding to Wu's claim, from having to contend if artificial intelligence is equivalent to human intelligence as information technology advances exponentially; courts may have no alternative but to define limits of artificially intelligent speech security.

A main feature of an intelligent algorithm, an artificial intelligence, is not patentable, since the patent would establish an overwhelming and fundamental monopoly over natural law, if it were regarded purely as a mathematical formula that communicates the basic theory of nature (e.g., gravity). We have obtained this remarkable outcome because it has doubled in size one day to the next, in order to reiterate the problem of exponential growth. It is obvious that doubling pennies is a matter of growing artificial intelligence, of the efficiency your brains can achieve, and of fusing our future with artificial smart machinery. Because information systems are exponentially growing, I hope you would accept with me that extraordinary technologies await you. The scale of the exponent that is necessary for development is important. Probably, you will think that it is the future that awaits us, a world with amazing clever resources to serve us. In an extraordinary matter involving the patentability of the methods of business conveyed in the code State Street, for example, the court determined that mathematical algorithms could not be copyrighted until they were "no more than abstract theories composed of disembodied principles that were not useful."

Used by artificially intelligent machines to assimilate and apply a hierarchy of interpretations for data interpretation, these algorithms have even been successfully extended to a number of problems such as image detection, natural language analysis, and spoken comprehension in uncovering the underlying data structure. Interestingly, MIT researchers find that a deep education framework for the identification and naming of scenes often taught how specific objects are identified. To explore this, they were using a deep learning method to train an effective scenario that was between 25% and 33% specific. That is one step of millions of investigators in the direction of invention of technology with human-like reasoning ability, and this findings mean that scenery and object-detection systems can work together or reinforce each other.

Professor Pieter Abbeel is investigating another promising approach that imitates human learning and thus can be crucial for artificially intelligent brain learning Without preprogrammed information on their environment, UC Berkeley has developed a kind of reinforcement learning, with colleagues, that works by completing various taskings such as putting a clothes hanger on a rack, assembling a toy plane, tilting a cap on a water bottle, and more. UC Berkeley faculty member Trevor Darrell, director of the Berkeley Vision and Learning Centre, said: "most robotic applications are placed in managed environments where objects are in predictable position." According to Darrell, "it is a continuously evolving world that is a risk of bringing robots into real-life settings like homes or workplaces." The robot must interpret and respond to its environment in order to be intelligent. Conventional but unworkable methods to help a robot navigate into a 3D universe include preprogramming to manage the broad variety of potential situations or to create simultaneous robot operating conditions. UC Berkeley scientists use profound teaching methods, which lose creativity as it perceives and communicates with the environment, in the neuronal circuits of the human brain. The methods for machine learning mentioned here explicitly deviate from the frail process by which any rule must be encoded into the mind of a machine; otherwise parents do not know the rule.

Deep learning programs build "neural networks" with artificial intelligence layers that overlap raw sensory information, whether it be sound waves or pixels of images in the field of artificial intelligence. This allows the robot to distinguish patterns and categories within the obtained data. Sarah Yang said, "Those who use Siri on their iPhones will already be benefiting from major developments in deep learning in speech and visual perception from the Google conversation software or Google Street Views." Using a deep reinforcement process in non-structured engine tasks, 3D worlds became even more demanding as the job went beyond passive image and sound recognition. "We've yet a long way to go before our robots can learn how to clean or organize a laundry. But our initial findings suggest that such fundamental learning technologies can turn robots into the ability to learn difficult tasks completely from scratch," says UC Berkeley's Trevor Darrell. In the next five to ten years, major developments in robot learning capabilities could emerge, based on Darrell's work and other researchers investigating the use of deep learning for robotics. I agree with this observation based on the accelerate rule. This comment fits with my opinion that we reach a cycle of time based on the rule of exponential returns in which notable changes from one and the next robot iterations occur (similarly to cell phones). So the age of increasingly smart robots with a human-like intellect will come upon us before the public is fully informed.

Additionally, trademarks, copyrights, confidential information, and mask works are among the key forms of intellectual property protection for artificially smart computers using neural networks in the United States (see next sections). As previous emerging technology types, some facets of neural networks and brain software cross current legal definitions. This was attributed largely to the complex character and the fact that the educated state of the system cannot be predetermined. The legislative or judicial action is required before certain uses of neural network technology go beyond the realm of security, as they did for computer applications, to the customization of existing legislation for this technology.

8.6 BRAIN ARCHITECTURE

The architecture of the brain of a computer usually decides its cognitive capacity through its tools and algorithms and hence demonstrates knowledge. One of the reasons for pushing greater knowledge on machinery is Moore's law. Is Moore's law in effect? and will the days of exponential growth in computing capital be ended because of the physical restrictions that may be fulfilled by etching circuits on a silicon chip? I don't believe so. Some methods are tested and, if successful, the rapid advancement of computational capacity will continue. IBM is, also, researching the use of completely integrated silicon chips to relay information using high-speed light pulses. This ensures that the chip can transfer data at high frequencies and vast journeys more than the existing computers. The silicon photonic chip is multiplexed by wavelength and thus can pass a variety of light wavelengths, increasing the bandwidth of the propagation of information in contrast with the current technology. This debate underlines that since the human brain is built on a specific architecture and a comparatively low signal available bandwidth, An artificially intelligent brain is capable of evolving drastically alongside technological advancements. That is why I question if a law of effectively implementing brains still lags behind technical advances. If so, we will actively question ourselves as human lawmakers and maybe artificial intelligence will become part of the rule-making in the future [10].

Quantum computation is one of the most interesting developments being studied. Quantum machines use quantum bits or qubits, rather than encoding information as a zero or one as computers today whose states encrypt a whole variety of alternatives by capitalizing on the quantum overlapping and interlocking phenomenon. If quantum computers were successfully built, it would take few minutes to execute computations that today's computers would take thousands of years to execute; consider whether our artificially smart progeny would be able to think in this way; imagine if we did. Another exciting area of computer science is the production of edge computing chips using new materials such as gallium arsenide, carbon nanotubes, and graphene. This is led by IBM and the Defense Advanced Research Projects Agency (DARPA). Currently, a team led by the IBM has developed a computer chip to imitate the architecture of the brain. The chip "TrueNorth" is a 5.4 billion transistor chip at the time of this writing, with the 100 configurable neurons and 256 million synapses, but, on the opposite, the brain comprises nearly 100 trillion synapses. However, a neuromorphic chip will achieve complexity thresholds in less than twenty years. In comparison, the TrueNorth chip is already 1,000 times more environmentally friendly than a traditional chip.

Digital computing today depends on the 1940s architecture and a well-known "bottleneck" among processor and memory with any rapid improvement in processing power, materials, and manufacturing. Specifically, von Nueman's computing architecture reflects a typical computing platform, containing three components: a (1) CPU; a slow data repository similar to random access memory (RAM), (2) A machine with a computer architecture from Neumann stores instruction as binary values and (3) handles instructions sequency—in other words, the processor chooses instructions one at a time and sequentially processes them. With regard to world thought and thinking, an artificially intelligent brain uses integrated circuits to measure and control "machine thinking" signs, using electronics composed of resistors, transistors, capacitors, etc., all of which are graded into a small chip and are connected to accomplish a shared objective. Integrated circuitry may come in all sorts: single-circuit logic gates, tension regulators, engine controllers, microcontrollers, microprocessors . . . but think about such components as characteristics that require an artificially intelligent brain design. The von Neumann statistics indicated that the handling system has limits, not least because it does not operate near the capacity of the brain environment of three pounds on our shoulders (computers beat us with brute force computing not with eloquent massively parallel processing). However, a lot of work is being done to discover the workings of the brain and to reverse the brain circuitry, to create chips that work like the brain does and to write the software and algorithms to emulate the reasoning of the organism. For example, take the use of very large integration systems (VLSI) in a neuromorphic computing concept generated by Carver Mead at the end of the 1980s, circuits that are found in the nervous system to imitate neurobiological architectures. In particular, the models of vision, motor control, and multisensory integration are used in VLSI systems.

To sum up this quick debate on artificially intelligent brain architecture, integrated circuits form a central element of the brain computer architecture and are made up of billions of minute, carefully organized electrical interconnected pathways, for example, on a sole piece of material like iflicone. In reality, as chips get smaller still, problems like hot spots, contamination, etc., make an integrated CBC not a simple operation; it is very difficult to achieve power-efficient architecture. The great effort made by highly trained professionals and substantial financial contributions typically result in successful designs. However, it is comparatively easy to duplicate each sheet of an integrated circuit and to ready the "pirated" integrated circuits. Accordingly, it seems natural for manufacturers or owners of these designs to have some sort of legislative security, taking into consideration the immense difficulty and expense of creating an integrated circuitry design, the broad industrial utility, the ongoing desire for innovation, and the ease with which such designs can be copied. But where are these

Intelligent Cyborg Brains 199

protections to be found? And, of course, you would assume that at present every software, algorithm, and computer architecture regulation is an artificial intelligence law.

8.7 HARDWARE PROTECTION FOR ARTIFICIALLY INTELLIGENT BRAINS

During the last decades of the production of software, courts have emphasized the disparity between trademark and applications patent laws. The extensive protection of software under patent law must adhere to requirements of newness and non-obviousness in order to secure a patent; originality and a certain degree of free speech must be standards for copyright protection. The copyright intent for software is to safeguard specific expressions of a concept written by a programmer in source code (which then fills in the object code), not the very concept; an idea is subject to patent legislation. In the defense of software as intellectual property, both copyright and patent law play a role in and contribute to the law of artificially smart brains.

For the elements of an artificially intelligent brain, for example, patent law that covers inventions is specifically applicable to circuits built to mimic neuronal characteristics. For example: a "silicon neuron" patent, defining an integrated circuit to mimic the functions of a biological neuron (U.S. patent 5648926 A), which has earned several other patents in this field. Applications are considered complete in the United States if they include the information required for a patent. The code for computing devices and associated equipment restricts the software to particular uses—software linking and operating hardware components can be used. This regulatory approval definition for machines tends to refer specifically to an artificially intelligent brain as brain-running software is used to monitor the computer effectors and actuators.

As already stated in this chapter, the ownership of programs snatched on chips is of special significance in our cyborg future. If they are made, the chips are attached to the computer and they become part of the brain circuitry of the computer. This means that the brain of the robot has patent privileges that are not provided for human brains composed of millions of neurons, of course. The copyright protection of utilitarian objects generally does not apply to chips, but as mentioned earlier, the Court rejected the argument in the case of software that software encoded on chips is utilitarian and therefore not protected by copyright, where identifying the physical storage media and where the code was stored would be impossible.

We may search for rights to give the manufacturer of the numerous hardware components composing the artificially intelligent brain a restricted monopoly under the law in terms of maintaining the legal security for autonomous robotic brain architecture. Patent security is an alternative for defending intellectual property expressed in an integrated circuit design given, for example, that integrated circuits have an acceptable inventiveness and satisfy the requisite level of uniqueness. However, in most patent schemes, the lion's share of integrated circuit designs is apparent since there are generally no changes to their predecessors (inventive step; prior art). In addition, integrated circuits consist of various construction blocks and may be patented by each building block. However, a patent application for an integral circuit would have to require hundreds of thousands of separate parts, which would be like seeking to write a patent on the neural circuits in the brain. Therefore, it will take hundreds of pages to claim a patent that seeks to characterize a complete integrated circuit. Clearly, that narrow argument, particularly for an artificially intelligent brain with trillions of circuits, is practically without defense.

And if one attempted such stringent defense, it would be exceedingly difficult and costly to write a patent application defending a claim of thousands of elements. "The built in circuits obviously cannot be conveniently defined by a patent specification or in the statements," as Rajkumar Dubey pointed out for Mondaq. It can also take a number of years for certain patent offices worldwide to secure an integrated circuit patent. This is inappropriate because the usable business life of an integrated circuit could be less than 1 year. How is the same obsolescence concept applicable to the human brain so that a person has to patently apply every one to two years for the safety of their brain's neural circuitry? Or visualize the human brain being fitted with millions of integrated circuits for responsibility to work during the next cyborg era. This means that every two years or so, the human brain is outdated because of the need to incorporate (or update) modern technologies in the brain. Patent laws would not have adequate protection for neuroprosthetic systems and thus for the intellectual brain of the artificially intelligent computer, due to the time-intensive existence of patent files and the exceedingly limited protection involved.

The integrated circuit architectures, which constitute a major component of the artificially intelligent brain, therefore cannot be generalized to other types of intellectual property security. The design patents cover the ornamental features of the manufacturing article mentioned in his sketches but not its practical ones. As the integrated circuit architecture is functional rather than ornamental, the defense of design patents for integrated circuits is usually not valid. Finally, trade secret legislation cannot in certain cases be used to cover most integrated circuits, as a reverse design of an integrated circuit configuration is possible. But what if an artificially clever brain writes its own algorithms that are kept dynamically on its integrated circuits, and what if they have market value (i.e. trade secrets)? During a program, it can still be a business secret, but after it has been announced to the public trade secret security, it will still be a tangible expression. Because an autonomous robotic brain, however, interacts in the object code and "closes" the source code, the rights of the software and trade secrets can be maintained at the same time. In this connection, the radical innovation of the human brain is one of the key methods used by certain scientists to attempt to build an autonomous robotic mind.

Rajkumar Dubey wrote regarding integrated circuits: "there is a transistor configuration on the semiconductor built-in circuit, or transistor weather patterns, on the integrated circuit that defines both the size and the processing strength of the integrated circuit." He notes that "The layout design of transistors represents a type of intellectual property which is so essential and distinctive that is profoundly distinct from other types of intellectual property, such as copyrights, trademarks, patents and manufacturing patterns." Provided that patent, copyright, and business confidentiality legislation cannot properly secure the integrated circuit design, the semiconductor industry wants exclusive protection for the semiconductor integrated circuit design. This standard of security is a regulation that is vital to the survival of our cyborgs. So what security would an artificially smart brain hardware provide? In 1984 the United States adopted the Semiconductor Chip Security Act, which gives statutory protection to the rights of integrated circuits. While codified as copyrights, the Act is specifically aimed at supplying sui generis integrated circuit designs rights ("of its own type"). It includes some elements of copyright laws and some aspects of patent law and is entirely separate in some respects.

Providing the design of an artificially intelligent brain with legal immunity to the physical elements is also part of an evolving cyborgs' statute and analogous to the principle of maintaining human "corporal dignity." Semiconductive chips shall be massaged from three-dimensional layered models, which are known by trade as "chip masks," and "masking works" shall be given under the Act. The key goal of the Chip Safety Act on Semiconductors is to outlaw "chip pirates" copying and selling, semiconductor chip products unapproved by the original makers. But, given that this Act is made up of integrated circuits, it may also secure the infrastructure of an artificially intelligent brain.

Under the Act, as with copyright, when they are developed, integrated circuit design rights exist; this is contrary to patents conferring patent rights after submission and review and authorization. However, only copyrights and patent holders have more proprietary protection than those given to owners of built-in circuit designs. Alteration (derivatives), for example, for owners of integrated circuit designs is not the sole right. In addition, the patentee's proprietary right to "use" an innovation cannot be extended to the exclusion of an analogous integrated circuit design that is separately developed. Therefore, most jurisdictions expressly allow replication to reverse engineer an integrated circuit design.

Due to the value of defending integrated circuits from piracy, many countries, including Japan and the EU, have taken the examples set in the U.S. and have accepted their own relevant statutes/directives respecting and protecting integrated circuit designs (also known as "semiconductor chip topography"). The International Convention on Copyrighted Works for Integrated Circuits (IPIC Treaty) was established by a diplomatic conference of different nations and held in 1989. This agreement is partly included in the World Trade Organization TRIPS Agreement (WTO). I assume that the potential danger to civilization from artificial intelligence is sufficiently severe to establish a popular reaction, much as with semiconductor production, to the potential threat to society posed by artificial intelligence.

Additional regulatory matters surrounding electronic chips often refer to an artificially smart brain. For example, an interesting issue for the safety of an artificially smart machine's brain involves its memory and how it is loaded and stored in various computers. Specifically, memory chips, like an EPROM chip, are chips that hold its data when its power supply is turned off. The Semiconductor Chip Security Act includes protection to EPROM chip topographies, but this protection does not apply to information contained in chips such as computer programs. Such data is, in the degree to which it has been addressed previously, protected under copyright laws applied to applications. Of special concern, in MAI Systems, the *Court of Appeals of the 9th Cycle v Peak Computer, Inc.* ruled that a "copy" and a possible violation of "RAM" were generated while loading applications to the computer's random access memory, in compliance with the Act on Copyright. What this meant was that, even without a hard copy, a replication and possibly an infringement was made briefly by preserving a program in RAM. Making a computer do anything is like making a copy of the operating system's software, since it is automatically saved in RAM when the computer is unlocked or when a file from one of the computer's users on the network is passed to another. The MAI Court ruled that the software deposited on a temporary basis in RAM is a duplication, although the United States Congress later passed a copyright amendment to allow specific exceptions to the ruling in the court in certain situations.

8.8 OUR COMPETITION AGAINST BETTER BRAINS

Centered on the aforementioned discussions, as artificially intelligent brains continue to be more advanced under Moore's legislation and its brain design, it is plausible that artificial intelligence would become more successful than human intelligence. It is likely that it will arrive with time. From then on, artificially smart computers will not interrupt the smarter process but rather intensify the process of information creation and world connection over the internet. In the last decades, machines have been increasingly going faster, having more memory space, and networking with each other and with the Internet of Things. And although the advances in artificial mind and information in brain science (according to some authors) are not growing exponentially, they continue to expand gradually. So it is only a matter of time, I suppose, before artificially intelligent robots appear to be independent of themselves and argue for citizenship. The further regulations and policy we build on the workings, devices, algorithms and design of the artificially smart brain, the more we can influence our own fate and shape the world as individuality is approached. Ray Kurzweil said persuasively that if artificial intelligence hits the human level, artificially smart minds can evolve rapidly dependent on technological upgrade until they are much smarter and more capable of evolving than human beings. The rates of artificially intelligent machinery's growth in terms of physical architecture would also increase, as their slow-thinking and less intelligent designers alter their own development. This is where the individuality is achieved and then conquered; artificial intelligence is assumed to develop itself incredibly rapidly. What's going to be the rule of intelligent artificial brains?

According to James Barrat, the actions of these cleverer-than-human intelligences, which we could share one day with the world or which one day we could combine with, are uncertain after Singularity (through a steady process, not all at once like a step function). However, by fusion with technologically smart machines, it is possible to become super smart cyborgs (or any other mysterious entity) by using computers to increase our intellectual power. Perhaps artificial intelligence is useful if we do not integrate with the technologies that we create and remain the biological result of nature. Perhaps artificial intelligence would be compassionate, enabling us to treat the consequences of old age, prolonging our lives, "fixing" hunger, etc. But in comparison, maybe artificially smart robots can turn on mankind and try to destroy or manipulate humanity unwantedly. If we combine with our artificially intelligent progeny, though, we will escape extinction and continue the process of turning our race into something that can no longer be accepted as such by mankind. This transition has a name. It is called posthumanism and is a development explored in this book. Naturally, there are those who reject the notion that one day mankind could become something "new" and now is the moment of resistance for those who are firmly opposed to Singularity. Clearly, artificial intelligence as it appears today does not generate the sort of intelligence that we equate with human beings. On the other hand, after increasing our use of cybernetic technologies, we shall blur the distinction between humans and machines and have totally transformed into technology as we pass into the late or early next century.

Today's artificial intelligence appears to learn in only one very particular area, such as search queries and playing chess. They normally exist within a very particular reference system and do not use common sense. They are intelligent, but only if you are strictly and limitedly describing intellect. Possible artificially intelligent beings that Ray Kurzweil describes but has not yet been able to identify as effective machine learning are where his mind is most at work. Why not? We are, of course, still anticipating Moore's law to balance the increasing computational power with advancement in algorithms, awareness of the pathways in the brain gained from neuroscience, and improved artificially intelligent brain architectures. But, as Lev Grossman wrote in *Time* magazine, "stuff can still happen in our minds that cannot be mechanically repeated no matter how many MIPS we throw at them." "Neurochemical architectures creating the temporary confusion that we as human consciousness recognize can only be too abstract and similar to reproduce in digital silicon," adds Grossman. In comparison, the biologist Dennis Bray is an alert voice on the wonders of cyborgs for the future: "biographical pieces behave in ways comparable with electronic circuitry, but they are distinguished by the large number of different states they can assume," he argued in a speech titled "What Cells Can Do That Robots Can't" in which they were addressed. Multiple biochemical pathways produce chemical differences in protein molecules that are further diversified by combination in a cell's specified position with unique structures. Bray points out that "the resulting blast of states helps living systems to store knowledge about past and current circumstances and to plan for future events in an almost unlimited capacity." The uncertainty of biology ensures that the binary language used by machines for processing data looks rough, and whether artificial devices can replicate a brain remains to be seen.

Since we don't know what a very sophisticated artificial intelligence would like to do if it were to be a newborn citizen of the earth, Grossman observed "Kurzweil admits the fundamental degree of danger that is difficult to refinish with singularity." It could seem to be competing with us for funds, thus it couldn't possibly work. Whether or not we want Singularity, these questions must be answered. Kurzweil thought it was not only impractical but also immoral and likely dangerous for mankind to attempt to put Singularity out by prohibiting technology. Kurzweil argues, "The enforcement of such a prohibition would involve an authoritarian regime." He adds, "It wouldn't work. It will only lead this technology underground, where the responsible scientists on which we depend to build the resources would not be readily accessible."

Kurzweil sees no inherent distinction between flesh and silicone, which would preclude it from behaving like a human being. The rule, therefore, differentiates neurons from integrated circuits because one may have circuits but not the neurons of another. Kurzweil challenges biologists to build a neural system that cannot be modeled or at least balanced by computational power and versatility. If Kurzweil is correct, there is a possibility for an artificially smart entity that defends rights; therefore mankind should create a regulatory framework for protecting humanity and for ensuring fundamental rights for all smart entities that emerge to enter society. In short, artificially smart brains quickly advance based on exponentially speeding developments. The new law of cyborgs, especially that addressed in this chapter relating to an artificially intelligent brain, might provide essential protections, not only for the eventual rights of man-made intelligent beings but also for people who are either functioning with or living among them, based on less intelligent beings, by the mid-century.

8.9 CONCLUSION

According to the debate, several agent-related APIs may be used to build this chatbot concept, which takes the shape of a 3D avatar with a speech interface. Yet, there are certain flaws in the study since the APIs utilized in this article don't have a lot of trust in their results. As an illustration, despite having a correct transcript, the HTML5 voice recognition API currently has a confidence level below 80%. With too much unnecessary noise, it is still difficult to discern the voice. However, if that field were to be developed further, the problem would be overcome. A different form of human-computer interaction, like customer service, could be possible with this chatbot architecture. Exploring this concept with other tools, such as a web camera, might allow the agent to better understand the user's feelings and reactions.

REFERENCES

[1] Haraway, D. (1997). *Modest_witness@second_millennium.femaleman_meets_oncomouseTM: Feminism and technoscience*. New York: Routledge.
[2] Harper, M. C. (1995). Incurably alien other: A case for feminist cyborg writers. *Science-Fiction Studies*, 22, 399–421.
[3] Harrison, T. (1995). Weaving the cyborg shroud: Mourning and deferral in star trek: The next generation. In T. Harrison, S. Projansky, K. A. Ono, & E. R. Helford (Eds.), *Enterprising zones: Critical positions on Star Trek* (pp. 245–258). Boulder, CO: Westview.
[4] Harrison, T., Projansky, S., Ono, K. A., & Helford, E. R. (Eds.). (1996). *Enterprise zones: Critical positions on Star Trek*. Boulder, CO: Westview Press.
[5] Hartshorne, C. (1937). *Beyond humanism: Essays in the philosophy of nature*. Lincoln: University of Nebraska Press.
[6] Hayles, N. K. (1995). The life cycle of cyborgs: Writing the posthuman. In C. H. Gray (Ed.), *The cyborg handbook* (pp. 321–340). New York: Routledge.
[7] Hayles, N. K. (1997). *Virtual bodies and flickering signifiers*. Available on-line at http://englishwww.humnet.ucla.edu/Individuals/Hayles/Flick.html.
[8] Hayles, N. K. (1999). *How we became posthuman: Virtual bodies in cybernetics, literature and informatics*. Chicago: University of Chicago Press.
[9] Heidegger, M. (1977). Letter on humanism. (F. A. Capuzzi, Trans.). In D. F. Krell (Ed.), *Martin Heidegger: Basic writings* (pp. 193–242). New York: Harper & Row. (Original work published 1967).
[10] Hess, D. J. (1995). On low-tech cyborgs. In C. H. Gray (Ed.), *The cyborg handbook* (pp. 371–378). New York: Routledge.

9 Neuroprosthesis in Cyborgs

9.1 INTRODUCTION

In preceding chapters, I mentioned a variety of technologies that led mankind to a fusion of artificially smart machinery. Probably the two most important developments to produce a convergence of human and machine are artificial intelligence and the advancement of neuroprosthetic brain implants. As we transition towards a cyborg world with information technology in our minds and bodies, our vulnerability to government supervision, invasions of privacy, and access by third parties to our internal ideas and memories rises. The legal distinctions between humans and robots tend to fluidize and become subjective as more technology is introduced into humanity. This adds to a number of legal and political problems for the 21st century. For example, attorneys Benjamin Wittes and Jane Chong are discussing a woman with a central pacemaker—a specifically embedded technology in her body—but she does not have any rights to the implant-generating knowledge about the operation of her heart. Following this example and those in the book, various jurisdictions have been aware of the necessity for the legislation to adapt technology incorporation into the human body. This finding is much more important for the development of neuroprostheses capable of repairing or enhancing cognitive functions [1].

Researchers unlock the secrets of how the brain calculates and write algorithms to simulate the operation of the neural pathways in the brain with technology that can increasingly research brain and significant advancement in the field of neuroscience. As a consequence, the ability of neuroprostheses is significantly increased, as by the mid-century "cordless" people can, for purposes other than medical necessities, prefer neuroprosthesis products. But after technology is inserted into the brain (read it now), states, businesses, and other third parties will remotely control the implants to build the nightmare that will challenge the person's "cognitive rights." That will be a serious danger. This segment explores how the capacity of a human to monitor and memorize their mind's material, including their memories, would affect third parties' access to neuroprostheses facilities and thus pose crucial questions of legislation and practice for the future cybernetic age. The first wave of cyborgs is starting to evolve on the basis of brain implants used for treating disorders like Parkinson's disease, dystonia, chronic pain, and depression. This cyborg century, armed with neuroprosthetic equipment, is taking advantage of impressive developments in neurological disease treatment. For instance, technology for "reading the brain" allowed cognitively intact patients to control their bodies by pushing a cursor on their computers and by using thinking to manipulate a robot arm or prosthetic extremity to talk to their loved ones. Future cyborgs' potential will fade as they age, despite being equipped with mind-blowing technology in their brains, since they aren't the first cybernetic creatures. That is, in decades, neuroprosthetic tools can greatly change, helping people to increase and expand brain functions and to edit the contents of their memory. Of course, the cyborg technology is exponentially developing, and we all foresee an exciting future for the man-machine.

One important discovery surrounding the use of cyborg technology in the brain is that the essence of IT tends to change from a neural mechanism via the comparatively low-level transmitting of electro-chemical signals to a digital architecture with improved processing computation and high space, 10–120 m/s over myelinated neurons. The effect, however, is the facility for third parties to control, alter, and modify a person's mental processes—including information saved in their memory—and to provide individuals with brain implant technologies. Obviously neuroprosthetic instruments are being used or should be accessible by the time of Singularity because the substance of our thoughts is made available to a host of others. As expected by the computer scientist Ben Goertzeil and Google's Ray Kurzweil, these prospects alone pose big law and policy problems that should be answered sooner rather than later, when society still has time to control the way into the future. Their ability is to fundamentally shift our connection to governments and companies [2].

A big change in information technology would have happened at around the middle of the century depending upon the law of rapid returns. A cyborg fitted with neuroprosthetic devices can download information on implants inside the brain or inside the sensors of the body. The combination of the mind with IT helps cyborgs to fluently convert modern languages, to acquire knowledge quicker and more effectively than cyborg technology, to preserve and exchange their memories among minds, and to interact telepathically with other cyborgs and artificially smart machinery with continual improvement. However, with innovations for the improvement in the cognitive functions of the brain, the usage of modern neuroprosthetic instruments, and the policy that regulate their use, ethicalists, lawyers, and scientists have started to face a challenge. On this point, the use of neuroprosthetic technologies for the identification, neuromarketing, and editing of memories would have significant legal and policy consequences not only for the evolving body of cyborg law but also for the cognitive independence of thought.

Although the embedded cyborg technology in the brain can look like the subject of a science fiction story (and has been the subject of science and fiction novels!) it soon becomes part of the exponential growth information technology revolution. In their book *Abundance*: *the future is brighter than you imagine*, Peter Diamandis and Steven Kotler describe the features of innovations that progressively grow. These developments are the ICR and they rely on electronic miniaturization and advancement in software technology, all of which are required to improve neuroprosthesis rapidly. For the reading person interested in the scientific dimensions of cyborg innovation, a primer for neuroprotshetic instruments can be found in the co-edited book of Theodore Berger and Dennis Glanzman, *The brain's replacement parts: implantable biomimetic electronics as neural prostheses*.

9.2 MEDICAL NECESSITY AND BEYOND

The key reason people select and receive neuroprosthesis and other implants, as mentioned in the book, is the need for medical treatment. However, as we reach the middle of the century, I expect this explanation to shift and people to want to substitute or improve brain function with brain implants that have unenhanced superior knowledge management. The treatment of Parkinson's patients, help for people who are depressed, and the restoration of impaired senses are some reasons for the use of implants for medical purposes. For mental disorder, scientists aim to fix malfunctioning by using sensors embedded in the brain of a human to treat neural disorders such as depression, addiction, and anxiety disorders as well as neural illness. The reader may ask, though, why not substitute the typically functional tissue with a prosthesis for people not suffering from a disease? That is, why would anyone be a cyborg if not medically necessary? In the chapter on the transition, improvement, and hacking of the body, many explanations are addressed, but from a broad-based viewpoint, the key explanation is that our species lives more specifically in competition with powerful artificial intelligence [3].

When I talked about technological individuality in preceding paragraphs, I mentioned that humanity will be left trailing artificially intelligent robots with quicker response speeds, better memory, increased access to information, and superior reasoning skills without improving our brain through the use of neuroprosthetic implants. Hans Moravec, robotics researcher and writer of *Robot: mere machine to transcendent mind*, has proposed that it is speeding our own evolution to keep up with cyborg developments. "We can adjust ourselves," he says, "and we can create new ones, too, that are ideal for new situations—robotic children." This century would also see a substantial paradigm change in information technology—from biology to technology-based concepts, which takes the form of human evolution. Interestingly, the use of "capable body" machines to expand their senses, memories, and reasoning capacities above natural standards would be achieved as part of a possible human-machine convergence awaiting mankind.

People also tend to understand that there is an exponential growth in information technology, so readers may be shocked to discover that hundreds of thousands of people now have neuroprosthetic devices, and the number is soon likely to be in the billions. About 25,000 Parkinson's patients have now been treated with a "deep-brain" implant (which acts like a pacemaker to reduce tumors and other motion problems) in the thalamus, globus pallidus or subthalmic cluster. In addition, the research community has demonstrated considerable interest in developing neuroprosthetic instruments in order to relieve our senses' difficulties because the visual and auditory models are crucial for worldwide service. For example, cyborg technology in the shape of a retinal prosthesis is used to monitor the light that reaches the eye by electrodes inserted below the patient's retina to help visually disabled individuals. The electromagnetic energy is transferred to a microchip translating the signals that are forwarded to the cortex for further transformation. Neuroprosthetic implants also produce a generation of cochlear implant cyborgs. Hundreds of millions of people around the world have now obtained cochlear implants, according to the Food and Drug Administration (FDA), to enhance their ear. This cochlear implant consists of a part outside the ear and a second portion of cyborg that is operatively inserted beneath the skin. Implant signals are sent to the brain by the auditory nerve that identifies the signaling as a vibration. Tens of thousands of adults required cochlear implants in the United States alone. And as of 2000, FDA-approved cochlear implants for use by qualifying children starting at the age of 12 months have created the first generation of childhood development cyborgs. Before these young cyborg children are born, one must wonder what the potential of future cyborgs will be with at least nine duplications of computer power.

Often you have to follow the money to figure out where technology is going, particularly for cyborgs. The EU has pledged 1.3 billion dollars to research how the brain works in the United States. In this respect, for fundamental brain science research, the Human Brain Initiative has raised $1 billion. I would like to point out that both initiatives would provide crucial details on the neuronal circuit structure used to reverse the brain's engineering (one way to create artificial general intelligence). In comparison, the combined support for neuroscience research of $2.3 billion which has just been listed is not the full financing scene. In the U.S., for instance, DARPA was one of the main government agencies sponsoring research to create brain chips and other inventions that link the brain to computers. In this respect, DARPA currently designs a neural protest implant that can be used to treat extreme memory impairment in human patients with many groups of researchers. The initiative is part of the DARPA Active Memory Restoring (RAM) initiative to help U.S. war veterans who have experienced a form of brain damage recover normal memory function. If successful, the schizophrenic, amnesiac, demented, and other patients with brain diseases will benefit immensely. The aim of another DARPA project is to bring "chips into the brain" to boost soldiers' sensory and cognitive strength. In order to record and activate brain function, the defense agency is particularly working to create a compact, wireless system that "must contain implantable samples"—in fact a memory that stimulates a black box. The implantable system will be made up of wires in the brain and under the scalp that can relay electrochemical stimulation through a transmitter situated below the skin of the chest. The purpose of the project is to a create technology that "promises to read the ideas of a living brain directly—and to instill thoughts." The technologies developed by RAM projects of DARPA would help to build, if successful, technology needed for a potential cybernetic world and make it more likely that the next human computer would arise [4].

If we use medical need as an incentive for the design of neuroprosthesis systems, the treatment of Alzheimer's disease is one of the most promising fields of brain neurotechnology. In the process of the creation of an artificial hippocampal system that plays a significant role in consolidation of information from shorter- to long-term memory that also leads to space navigation, Professor Theodore Berger and his research team at the University of Southern California have made notable progress. Alzheimer's disease has been known to result from hippocampus damage and affects approximately 5.2 million people in the United States alone. The development of an artificial hippocampus may also benefit millions of people suffer from extreme neurological disease, but cyborg technology intended to aid those who are suffering from a medical need may have the effect, whether or not they wish it to lead to our cyborg future and to the inevitable merger of robots, as a matter of fact.

This opinion is backed by Professor Berger's study into prosthesis architecture. His study includes a thorough review of the different operations on the hippocampus and the creation of algorithms to reproduce and incorporate hippocampal functions in a microchip by Berger and his team. Naturally, "brain chips" are an important invention if human beings are to be cyborgs and combine with artificially smart machines. In reality, the breakthrough in 2011 saw the first prothetizing memory system, a promising improvement, developed by Wake Forest University scientist Samuel Deadwyler in collaboration with Professor Berger regarding capacity of rat memory retrieval. The resulting unit was a microchip implant comprised of 32 electrodes and an algorithm that was capable of interpreting and reproducing the neuronal signals transmitted from one end to the other of the hippocampus. Subsequently, the researchers were able to construct an artificial hippocampus where the information obtained from electrodes could not only be read but also replicated as needed. From that point, the technology has been tested extensively in non-human primates including monkeys; human research is around the corner.

9.3 THIRD-PARTY ACCESS TO OUR MINDS

When the capacity to hack brain implanting technology grows, what rights do individuals have to the veracity of sensory information propagating in their brains? When third parties might hack brain implant technology, it is difficult to underestimate the opportunity for a dystopian future for mankind. In the case with cochlear implants, a sound a person has never really heard may be transferred into the audio nerve. For example, a retina prosthesis can be hacked to put pictures on the back of the retina that a person has never seen; furthermore, an artificial hippocampus can be pirated with incidents that never took place in a person's memory. Under these cases, what legislation and policy should apply? If the First Amendment forbids the government from inserting words into the mouth of a person, the Constitution will definitely forbid the words, sounds, or memories of a person from being inserted in his brain. On this basis, it is important to ask: can the technical capacity to hack the mind continue as a bastion for secrecy, protected from the eyes of the prey of emerging technologies based on the observation? Furthermore, if the government or a company will invade our thinking and alter our intellectual contents, can the dignity of our mind appear to be under our independent influence, and if not, who are we as individuals? This chapter addresses the legislation and procedure of those topics [5].

When third parties are willing, in another individual's brain, to activate a neuroprosthetic system, what could go wrong? Not surprisingly, several things. For example, will a person committing a felony be forgiven of guilt, when another has access to the brain remotely? Lawyers already regularly monitor the minds of incarcerated respondents and contend that a disorder of their neurology inhibits the suspects from regulating their behavior. In the next cyborg age, a tech specialist will be asked to study a neuroprosthetic device's programming language and algorithm to see if they have been manipulated. If so, a third party would have been given the mens rea for a crime remotely. The use of neuroprosthetic instruments could, however, contribute to other significant legal and policy concerns. Third parties, for example, with accessibility to implants technologies, could make it possible for advertising companies to advertise our consciousness or for the government to find out our feelings, without knowing it. Will a government or organization have scanners in a person's head, track his unspoken feelings, or alter the quality of the person's memories, or could there be a more gross breach of his privacy? When neuroprosthetic systems are mounted in the brain, it functions primarily as a von Neumann machine; should the mind be considered as a network or computer during the next cyborg age and should the mind be supplied with an identifying URL? The idea of transmitting spam through a mind implanted with a neuroprosthetic system is even more annoying, such that it is subject to much tougher regulations, much like the usage of spam in mobile phones or machines that might be implemented under regulation in the future. Consider Professor Theodore Berger's earlier work on developing an artificial hippocampus, a computer that can directly transmit information to an electrode within the brain of a human. What would prohibit a company from specifically sending ads to a cerebral individual if the neuroprosthetic system is available? Provision for sufficient protection could be rendered under the cyber safety guidelines for medical implants deemed by the FDA. In addition, most states in the United States do have regulations surrounding junk e-mail either directly or indirectly. These regulations are often parallel and, in some circumstances, related explicitly to other state laws concerned with telemarketing or company demands in other newspapers (e.g. text messages). For the evolution of cyborg regulation, I assume that much of the old law will precede conflicts surrounding cyborgs in fields of information technology and commercial email; the regulation relating to spam email is a prerequisite [6].

Judicial sciences and professionals also regard spam law as part of a broader machine law. The Anti-Spam Legislation of Canada can be seen as uniting the two strands in an attempt to create an integrated legal sense for online business. The CAN-SPAM Act in the United States sets down the guidelines for commercial emails and commercial communications (Control of the Non-Accounting Pornography and Marketing Assault). The Act requires recipients to stop e-mailing and specifies the punishments for those that break the rule. Such a regulation should definitely be enforced to prohibit the reception of unrequested commercial applications by a neuroprosthetic device. It is one thing to walk into a display and get an ad specially designed for a person based on the identification technologies of the face but another thing to see the ad on a computer inserted into the brain. The CAN-SPAM legislation includes all commercial communications, described by law as "any electronic mail message that is intended primarily to promote or commercialize a commercial product or service," including emails supporting commercial web content. It exempts transactional communications and linking messages, a limitation that needs to be resolved if people are fitted with cellular wireless neuroprostheses.

The sensory or auditory data retained in a person's brain originating from a world experience cannot currently be retrieved directly. However, this could become an alternative for cyborg technologies, as a cyborg is electronically captured as soon as it is fitted with a technology to feel the environment. One statement that Professor Steve Mann suggested on this subject is that he should give records of a person's existence for the use of wearable computers. Will courts be able to summarize the data contained in the prosthesis to be used as evidence for a world sensor in cyborgs fitted with neuroprosthesis? This topic concerns U.S. civil rights. The most fundamental Fourth Amendment in computer cases questions if a person has private information for electronic records, which are contained in a computer (or other electronic storage devices) under their control. It may be possible modern technology has been installed in the human brain, for example, do people expect their computer contents and disk storage devices to have a fair presumption of privacy? If so, the government typically needs to get a warranty based on possible reasons before the information contained inside is available. Although the components of closed containers typically uphold fair secrecy requirements, the data contained on electronic storage systems often preserve reasonableness of privacy standards. Will the same inference apply to cyborgs fitted with memory recording equipment? And if it is knowledge of machines or algorithms that forms part of the internal structure of the being, will there be a difference?

The confidential existence of the mind should be covered by the court, whether improved by technology or not. Under *Katz. v United States*, the Court test for assessing privacy rights when a government actor is engaged is whether the individual believed that the privacy requirement was fair and whether the privacy expectation was one the organization was willing to accept. When faced with the question of deciding whether a cyborg has an appropriate privacy expectation of the information stored on a neuroprosthetic system, the courts can examine, depending on the precedent, the neurological device with a sealed container such as a suitcase or cabinet. As regards the Fourteenth Amendment, police departments would usually forbid entry and visualization without warranty of information contained in a computer, where, in contrast, a locked jar will not be opened and the contents checked in a comparable situation. In the form of a file cabinet that is locked outside the country, it seems fair to access files stored on a neurofosthematics computer and to secure the material of a neuroprosthetic device in the Fourth Amendment. However, while courts have usually concluded that electronic memory modules should be matched to sealed containers, various conclusions have been drawn about it. This chapter addresses the legislation and procedure of those topics [7].

If a different closed container should be considered for each specific file contained on a device or disk, will a single file stored on a PC be equivalent with a file stored on a neuroprosthetic computer under this background? If so, will the use of such information by the government be covered by the ban on self-incrimination under the Fifth Amendment? Professor Nita Farahany of Duke University spoke on this subject at length, as we will see later.

9.4 CONCERNS AND ROADBLOCKS

Certain opponents of enhancement science, for example Francis Fukuyama from Stanford, have concentrated on the existential danger to civilization posed by biotechnology, such as genetic engineering. However, an evolutionary danger to mankind could even emerge from bioelectronics innovations: sensors and brain implants involving nervous networks and computer interfaces. As cyborgs are fitted with brain implant technology, a significant argument is that there are possible negative effects of brain engineering methodology that are part of our future technical growth, including benefits arising from neuroprosthesis. The father must discourage the progeny. The growth in the number of people fitted for brain implants would be important, for example, as neuroprosthesis is enhanced and becomes a realistic choice for "fitness" people. A significant issue as this occurs is that there may be a perceptual digital gap between the neuroprosthetic instruments and the technologies that do not exist. Thus, the time has come to create cyborg equity policies and to give all individuals fair access to enhancement technology. By means of various legislation and policies, the society usually aims to fix differences between the citizens but cyborgization of people could work to intensify disparities. Obviously, because we are used to expanding our senses, enhancing our memory and adding to disease control, the application of brain implant technologies poses significant ethical and policy concerns relating to the protection of our ideas and to the dignity of our mind. For instance, the opportunity to make an error would become an important problem for legal scholars, technologists, and individuals with cyborg technologies by continual improvements in the area of neuroprosthesis. Remember that former Vice President Dick Cheney was so afraid that terrorists could hack the wearable equipment implanted close to his heart that he deactivated a mechanism that enabled wireless administration of the defibrillator. This discovery echoes the questions posed by researchers over the years surrounding the insecurity of embedded medical systems fitted with computerized functions and wireless capacities to control devices without any further surgery. The Chaney scenario further highlights the balance between opportunities and possible risks to better the human body and mind by using cyborg technology. As a hopeful example, neuroprosthesis in the next cyborg era unlocks the prospect of permanently modifying maladaptive processes that lead to mental disease, primarily to relieve patients from their psychological disorders. On the other hand, though, the substance of an individual may be edited by reprograming neuro-circuits, governments, or businesses—in this situation, judges, ethicists, and the public will have to address the basic issue of what constitutes fact.

Another problem about the use of cybernetic technologies to improve cognition in humans is that governments and businesses can use brain implant technology to "pick up" private thoughts of an individual. It can even download unauthorized knowledge directly to a storehouse in the brain of a person. This statement is a call for action—now is the time to consider safeguarding people's right to manage access, mental innovations, and the freedom to escape their compulsory use. If governments could hack our brains, for example, this would control people's ability to engage in political processes, so people couldn't make independently informed decisions without reliable interpretations of life events. As the most likely technology of potential neuroprosthesis is dramatically enhanced, the production of policies, legislation, and laws to minimize the possible harmful consequences of this technology is necessary until the technology is popular. As technology matures to get into the head, policymakers will sanction people by actually formulating an idea that goes against government dogma and not merely for the representation of their thinking. At this step, law scholar Jeffrey Rosen from George Washington University asks whether it will be an infringement of message communication to forbid cruel and unusual punishment and to punish someone for their thoughts rather than their actions. This is not an insight merely important to the plot of a science fiction book, as governments and companies would be able to use brain implants to edit long-term memory until the end of the century. It reflects the experiences of a person's life. It would certainly be necessary to use technologies to view and modify a person's memories of an individual experience—infraction, attack, and assault, even extortion—in compliance with the rule. On this last argument, Marc Goodman, former Secret Service agent, is worried that keeping people away from their memories could in the future be a form of extorsion. For purposes of equality, thus, the human body and mind should be treated as sacrosanct in the coming cyborg age. To violate the mind of an individual without their permission should become an omnipresent human rights offense, punishable according to criminal law. Henry Greely from Stanford Law School recognizes that memory-retrieval technology may be an important threat to our freedom of speech, which he claims to be primarily supported by the guarantee of freedom of expression in the First Amendment. According to Greely, "the fact that you can only know what anyone thinks on the grounds of your actions . . . has always reinforced free thought." In view of advancements in technologies for brain recording, Greely commented, "This technology provides the ability to see into the skull and see what actually is going on and to look at the feelings." Greely concludes that the principle that this potential might challenge what we do, not what we say, should hold us liable. "This opens the opportunity to discipline people because of their acts for their feelings for the first time," he says. Discussing the possibilities for a possible authoritarian regime, Greely said, "The explanation that dictators could not detect it is considered free from the harshest dictatorships." And they will now bring more emphasis to legal prohibitions against government involvement with freedom of expression [8].

Although guaranteeing cognitive freedom is an essential concern in the next cyborg era, the cognitive freedom issue has been called to the consideration of the U.S. Judicial Branch in other strategies currently in use. For example, Supreme Court Judge Frank Murphy said that liberty to think is "absolute in its own way: the most totalitarian government is unable to regulate the internal workings of a mind" in the case of First Amendment that deals with a law banning the selling of books without a license. Professor Marc Blitz claims in the recent support of the principle that governments should not be able to regulate the internal workings of the mind. It'd be a serious breach of "free thinking of any state action to stop us from using our minds to activate our memories and to archive them." Professor Blitz also revealed that the process could not do much to curb freedom of thinking, other than to attack the language of speech and religion before the implementation of cybernetic technology. Blitz said, "The government cannot control our mind from inside, its only way to limit mental behavior is to target contact or other terms that incorporate such activity." However, as illustrated in this chapter, based on the constitution to accelerate IT returns, a lot of things have improved. Not only has innovation been easily developed to encourage the state to control our minds from inside, but actually it is still used to treat psychiatric and neurological disorders.

9.5 A FOCUS ON COGNITIVE LIBERTY

Given that a number of brain-computer interfaces and neuroprosthesis systems will be used to treat patients with drastic advances in brain implant technology in many decades, a serious debate about "cognitive independence" is expected. Cognitive independence is basically the liberty of self-government; it incorporates to a less certain degree the principles of freedom of expression and physical dignity. As a fundamental assumption, freedom of expression cannot be separated from cognitive freedom by preserving the freedom of a person to consider "whatever" they like, while the defense of the right of a human to think "whatever" he wants is a matter of cognitive freedom. This latter aspect of independence of thinking applies specifically to the use of neuroprostheses to facilitate cognitive processes. The U.S. Supreme Court has previously considered the freedom of the mind to be the "broad definition," the freedom of expression being a mere "part." Mirroring the value of freedom of speech on cybernetic technology, law professor Marc Blitz said that our First Amendment American jurisprudence is based on the Supreme Court's stance in terms of freedom of expression "To allow the executive to dominate the mind of citizens is a rebellious entire constitutional legacy."

A series of computer scientists, neurologists, and lawyer scientists challenged the ways in which the mind would follow technology and proposed that government policy should give the "cognitive independence" of the mind as much security as possible. What is at stake for the human race, when states, companies, and third parties will reach the inner thinking and memory of an individual through implants, through rapid improvement of technology for mental manipulation and study? A significant issue for all mankind is the ability to have "the independence of cognition." Cognitive freedom is roughly the personal freedom to have total sovereignty over one's own mind. It has to do with the principles of independence of speech and to a lesser degree with physical honesty as I said earlier. Neuroprosthetic technology may have a dramatic effect on the cognitive independence of the mind in the coming cyborg era. How much cyborg technology can be controlled thus needs careful debate.

Cognitive equality is an integral feature of international human rights law, or "right to mental self-determination." In the Universal Declaration of Human Rights, for example, which is constitutionally binding on foreign Member States Article 18, which says "Everyone has the right to freedom of expression, opinion, and faith," there is the provision for the Covenant on Civil and Political Rights. Clearly, it should be a crucial priority to preserve cognitive independence during an age of brain implants when civilization is nearing a potential cyborg and ultimate human computer fusion. Indeed, an increasing number of legal scholars consider cognitive freedom as an important universal human right and claim that cognitive freedom is the concept underlying a number of accepted rights in the constitutions of the majority of developed nations. Now that psychologists have found that we use words to move in our own minds, it is disturbing to create the technology to read our "thoughts" because it could threaten our perceptual independence, First Amendment rights, and other constitutional rights. Although the U.S. Constitution applies specifically to "free expression," an important consideration is if free expression still guarantees "internal expression," which is the very term that governments will reach through a neuroprosthetic apparatus. And what of the reasoning that cybernetic technology transmits from one brain to another, besides recognizing internal thinking as speech? Does the 1st Amendment represent a type of language liable for protection? Furthermore, what bandwidth rules will Federal Communication Committee (FCC) extend to cyborg-mediated telepathic communication? As progress in implant technology is being made, these issues will need to be answered in the next few decades.

Discussions over the capacity of the government to surveil citizens by tracking their messages are increasingly significant at a time when cybernetic organisms are provided with neuroprosthetic systems and networked sensors. In this regard, the government controls the new cyborg technology. For example, the transfer of thinking from one human to the next involves the use of spectrum for telepathic communication, as just described. The FCC currently controls the usage of electromagnetic radiation through a frequency allocation management mechanism, which include electromagnetic spectrums being regulated and licensed for commercial and pro users, including: federal, county, and municipal governments. The FCC administration process covers public and trade protection, fixed and mobile telecommunications networks, broadcasting and TV, satellite communication and other services, as well as non-commercial service providers. Furthermore, the FCC has established rules for a portable and implantable medical system body area network.

What if the authority encrypts a transmission from one mind to another in the field of privacy? The Fourth Amendment protection of the citizen to defend themselves from arbitrary searches and seizures would not only refer to FCC enforcement but would also be applicable. One way law enforcement intercepts a signal is to connect a "worm" to a phone line of a citizen and monitor the chat. I often think that a "worm" could be added to a neuroprosthetic system in the cyborg future that would allow for internal thinking to be controlled before vocalization or electronic delivery. The Court ruled that the addition of a bug to the line was a search under the 4th Amendment for the telephone messages since the right to bear arms safeguards the user privacy of an individual in cases in which the person has reasonable privacy expectations. Certainly, in the coming cyborg era people will expect much anonymity to build up their unexpressed thoughts and to convey thoughts from one consciousness to the next.

Amusingly, the concept of what comprises communication is not simple and obviously cyborg correspondence from the viewpoint of jurisprudence would pose a range of concerns that "stress" the existing laws. The courts have actually defined multiple modes of speech, which are also covered by the courts with the same degree of scrutiny. That indicates that the authority is authorized to regulate expression, based on the kind of speech generated. In the United States one form of speech is known to be a symbolic language or art used to express acts (not spoken) that communicates a clear meaning or argument intentionally to those who interpret it. However, the category of "pure expression" of the transmission of ideas by means of speeches or written remarks or by means of behavior confined in the manner required to express the idea is especially important for cyborg technology. If the 1st Amendment forbids a previous confinement of expressions, the provisional containment will be less clear. The courts traditionally have firmly protected speech-only pure from public control, and previous cases in the field of cyborg communication could serve as legal precedent. Maybe in the future a new way of speaking—cyber voice, exchange of ideas with thought—is accepted by the Court; if so, what oversight standard does the government start receiving?

In several cases the Supreme Court of the United States has accepted freedom of speech as a constitutional right, defining freedom of thought as: "the matrix, the indispensable condition of almost every other kind of freedom." The First Amendment to freedom of expression is unfounded, so you can only share your feelings. The most basic type of discrimination, the limitation or censorship of what a person feels (i.e., cognitive censorship), is counter to some of our many cherished constitutional values. Perceptual independence proponents aim to place both a negative and a positive duty on states: not to engage with cognitive systems consensually, to allow people to self-define their own "private world," and to regulate their own sensory perceptions.

The original duty of the state not to interfere with the thought functions of a person without agreement extends explicitly to government access to the neuroprosthesis that attempts, without their permission or knowingness, to protect people from modifying or controlling their cognitive functions. Although conceptual independence is also described as freedom of a person from intervention by the state, Hamburg University's Jan Bublitz and Reinhard Merkel propose that cognitive liberty can often resist intrusion from other non-state bodies into the mental "private sphere" of an individual. Bublitz and Merkel introduce a new federal crime that would prosecute "invasions that substantially conflict with others' intellectual dignity by weakening mental control or manipulating pre-existing intellectual vulnerabilities" for an evolving regulation for cyborgs. And these "direct acts" that minimize or affect cognitive resources like memory, motivation, and willpower; alter preferences, convictions, or interpersonal arrangements; produce insufficient emotions; or create objectively observable psychiatric injuries will all be prima facie unacceptable and prosecutable. The Centre for Cognitive Independence and Ethics' Wyre Sententia and Richard Boire have shared fears that companies and other non-state institutions might use new neurotechnologies to modify the mind process of people, without their permission.

While one responsibility of a State is not to interfere with cognitive abilities of a person, another, a neuro-prosthetic instrument that can strengthen the brain's pleasure centers may let individuals shape or expand their awareness in whatever manner they see fit. An individual who has the pleasure of this element of cognitive independence has freedom, either by indirect approach such as mediation or yoga or more specifically by neurotechnology, to change his/her mental processes. The advocates of the transhuman revolution, whose main concept is the advancement of the human mental capacity, attach considerable importance to this aspect of cognitive independence. It is connected to the use of neuroprosthesis to enter one's own brain that allows people to decide their own "inner realm." The principle of self-stimulation, for instance, is that an animal (including a human being) constantly electrically stimulates his brain even to the point of fatigue. In certain parts of the brain this effect is stable and easily replicable. Amusingly, one of the most well-known results in brain stimulation studies is the discovery of "enjoyment centres" in the brain. It happened incidentally. Professor James Olds, along with Peter Milner of McGill University, implanted an electrode into the brain of the rat, heading for the reticular system. The electrode angled off from its planned trajectory and landed near the hypothalamus in a different location. Olds put the rat in a box and activated their brain as the rat came to a specific corner. He hoped the rat would remain outside, but instead Olds found that the rat "came back for more," behaving as if it was a stimulating brain stimulus. Further study indicated that activation of the limbic system regions generated enjoyment for people and that people suffering from pain or stress were very likely to find electrical brain stimulation very pleasant [9].

In the generations since the presence of stimulation centers of the brain was confirmed by Olds and Milner, scientists have found that, when stimulated, many brain regions are triggered by feelings of victory, euphoria, sexual satisfaction, and addictive activity of all kinds, including non-drug-related activity like gambling. If individuals or third parties will "build" these and other behaviors, using neuroprosthesis technologies, a number of institutional and legislative concerns will be raised. For example, a neuroprosthesis could easily lead the person to become dependent on cortical stimulation by third parties who have access to the enjoyment centers and thus fall under the influence of the third party. The administration would definitely control this region extensively. Remember what professor of law Lawrence Tribe said of Harvard: "the guarantee of free speech" "was inextricably tied up with the safeguarding of transparent and free mental thought." In a case of the *Supreme Court v Reidel*, which considered a postal law banning the selling of adult content to be lawfully admissible, Justice Hugo Black challenged the argument that the United States Constitution "denies the authority of Congress to serve as a censor" by amending the United States Constitution. In *Stanley v Georgia*, the Court noted even on the matter of the government's regulation of expression: "the individuals personal first right to be independent of government initiatives of thought control" and it is important to protect "the independence from legislative exploitation of a mental substance." The Court seems to be a clear champion of the general concepts behind cognitive independence that, if the technology is readily accessible, I consider as an invaluable safeguard against the government or corporate power over our thoughts and minds.

9.6 READING THE BRAIN, LIE DETECTION, AND COGNITIVE LIBERTY

The architecture and operation of the brain are being studied with increasing resolution and consistency according to developments in technology for neuroimaging such as magnetic resonance imaging function (fMRI), magnetoencephalography (MEG) and positron emission tomography (PET). The ability to incorporate brain waves is a technology key to telepathy and other "cognitive" powers of potential cyborgs. Telepathic communication may give government access to a person's ideas on two occasions from the standpoint of a neurological independence—both by means of implantation and by interception of electronic signals sent from one brain to the another. While researchers have not yet established functioning brain-to-brain connectivity interfaces for the general public, there is already much technical advance in monitoring brain processes and making sense of performance. Functional magnetic resonance imaging, for instance, is used to assess brain activity by recognizing variations in sanguine oxygenation in neuronal activity—if a brain region is more active, more oxygen is taken, and blood flow rises to the active region in order to satisfy this additional need. Private companies such as No Lie MRI are developing fMRI technology's ability to recognize lies in order for the findings of fMRI scans to be accepted in court as evidence. Judy Illes, Canadian Neuroethics Research Chair, uses brain scanning technologies to identify lies that are increasingly evolving—commenting that we may have technology that is accurate enough to get to the binary issue if someone lies that can be used in those legitimate environments. Another organization that uses fMRI technologies to identify lies has created the Guilty Information system. Daniel Langleben and his research team at Pennsylvania University tested the device as follows. Before they joined, Langleben gave subjects a card and an fMRI computer and instructed them not to answer a set of questions, even though they had the relevant card. When people lied about why they had a card, Langleben and his colleagues pointed out that those brain regions can be used to identify binary deception.

It is important to remember that recent advancement in the use of cyborg-driven brain waves is based on a technique that has existed since the 20th century—EEG. An electromephalogram (EEG), using thin, flaky metal disk(s) connected to the scalp, can be used for the detection of electrical activity in the mind of a human. The brain cells of a human interact by electrical impulse and are continuously involved, even when resting. Commercial goods recently launched on the market with EEG technology to read brain function. The London-based This Place has, for example, been creating an application called MindRDR that consists of a head-mounted hardware application and Neurosky EEG (an off-shelf sensor) that generates the contact loop between screens like Google Glass and the EEG sensor by choosing brainwaves that are recorded to correspond with the capacity of the person to focus. The software transforms the brainwaves of a human into a meter read, imprinted on the Google Glass camera view. More "emphasis" raises the reading of the meter and the app takes pictures of what an entity sees in front of it; if the person focuses, the image is shared online. In my opinion, it is only after first receiving an order from a judge that the connection to what a person is focused on, i.e. what they are consciously attending to, should be practicable, otherwise it is a breach of Fourth Amendment confidentiality rights of the person and the emotional independence of that person. As fMRI evidence and other mind monitoring technologies are becoming more popular in courtrooms, it is possible for judges and juries to make fresh and often disturbing distinctions between the "usual" and the "abnormal" minds. These decisions may have an effect on someone charged with a crime's cognitive freedom rights. In multiple cases Ruben Gur, a psychology professor at the School of Medicine at the University of Pennsylvania, appeared as an expert witness requesting assessment of the defendant's mental abilities. One case of this type was the increased conviction of a suspected serial assassin known as "the classified rapist," because it will line up with the women's promises to buy, sell, rape, and trash the home in their classified ads. As a national authority for PET scans, Professor Gur was summoned to decide if he was responsible for his actions.

A picture examination of the brain, using radioactively defined material called a tracer to detect an illness or damage in the brain, is a PET (positron emissions tomography) scan. Following investigations into the defendant's PET scans, Gur testified that his amygdala had experienced a motorcycle crash that left the defendant in a coma (which has a role in memory, decision making, and emotional reactions). The individual executed his first rape after recovering from the coma. If courts determine whether a "brain injury" might relieve a human being of liability, I would contend that courts could also attempt to decide whether thinking inserted by a third party on neurological instruments would relieve a person of liability. Michael Gazzaniga, a psychology professor and an author of *The Ethical Brain*, observed in an expansion of Gur's study that, within a couple of years, neuroscientists may demonstrate neuronal discrepancies in the testimony of their own past events and testimonies of what they witnessed. Gazzaniga says, "You've got a procedural memory about it whenever you kill someone, but it's an episodic awareness that uses another part of the brain while I stand and watch you kill someone." Perhaps the prosecution may differentiate between procedural and episodic memories by manipulating information stored on neuroprosthetic machines, thereby either convicting or acquitting a person charged with an offense. It is a constitutional question and a subject for the community and legal profession to discuss if this is appropriate, that is to investigate a person's brain to gather evidence for prosecutions. Even if participants are not able to scan their minds, neuroscience can cause judges and jurors to infer the authenticity of certain forms of memories because of their area of mind [10].

9.7 TOWARDS TELEPATHY

Although EEG's and fMRI innovations contribute to considerable improvements in the use of computerized tomography for lie detection, other neuroscience research explores the issue of telepathic communication more specifically. Duke University's Professor Miguel Nicolelis is a pioneer in brain science development. His research focuses on brain-to-brain connectivity, interfacing brain machinery, and neuroprosthesis in human and non-human specimens. Dr. Nicolelys was among the first to suggest and prove that animals and human subjects would actively regulate neuroprosthetic technologies by brain-machine interfaces via electric brain activity. Professor Nicolelis hypothesized about the possibility to share information between the two brains in his 2012 book *Without Limits*. Nicoleli's study team at the Duke University Medical Center later announced in *Science Papers* that an interchange of two rat brains was carried out in full. His team taught two animals to maneuver one of two levers for a quick drink when a lead was triggered to test the brain interface technology. The cortical function of the rat was moved to the second animal's brain in a room where the corticity of the corticus was situated, when a rat pushed the right levers. "Benefits of drinking it's period" LED's been missing. As evidence of the exchange existing between the different brains the rat pushed the right lever (to accept a drink) on the receiving end of the prosthesis, which had been transmitted through the brain communication. Results summarized—Nicolelis and his team presented proof of the concept and findings that telepathy may be a potential means of future contact.

Results from experiments in human subjects suggest that, with respect to Professor Nicolelis' work, telepathy can still, within several decades or less, be a feasible technology for the public. For example, the use of EEG technology was effective at demonstrating contact from person to person through thinking by researchers at University of Southampton, England. And more recently researchers have seen a functioning brain-to-brain interaction with people using EEG technologies at the University of Washington. In their research, there were two people who could not talk other than with their minds using EEG technologies in separate rooms. Both themes looked at a video game where by shooting a cannon they had to protect a simulated city. But a man had his brain linked with a brain signal electroencephalography, used for the simulated cannon shooting. That is, the man was told to put his hand to the cannon to shoot instead of using an input system to fire the canon. This idea was moved on to another person with his hand on a touch-pad that would move and tap the correct way if the signals were obtained. The University of Washington researchers were optimistic that the technology was functioning in line with their familiarity with the system. The next step is also to decide what kind of knowledge within people's brains can be sent. For example, they would like to know whether an instructor will download the knowledge to a student's brain one day—I guess this is yes, because it would be cyborg revolutionary technology.

9.8 CREATING ARTIFICIAL MEMORIES

In order to construct artificial memory, neuroscientists predict a hypothetical world where the mind will be configured. Based on recent progress in brain-to-brain connectivity, several scientists claim that the consciousness of a human can insert memories that can be passed from one mind to another. For another hundred years this may seem like far-off technology, but researchers have indeed effectively inserted a fake memory into a mouse's brain. What is more important for the emerging cyborg rule than preserving the privacy of our memory, considering these results? To construct a recollection prosthetic limb, MIT scientists Steve Ramirez and Xu Liu mark brain cells associated with a particular memory and then tweaks the memory to make a mouse think it was an occurrence that hadn't existed. Recently, Ramirez and Liu have demonstrated that it must be possible, in theory, to detach and activate a human memory in individuals using a neuroprosthetic system. In reality, the MIT thesis was addressed by Michael J. Kahana, director of the Computational Memory Lab of the University of Pennsylvania, "We'd have every evidence to suggest this in humans as it was in mice." Clearly, the mental technologies are increasingly evolving in neuroprosthetics.

Let us first discuss certain laws and policy issues related to technology before we talk in more depth about the technology of implanting false memories. Nita Farahany, Duke University Professor, said that a great deal of intelligence is held in the head, which could be of use to the government and to companies. She found, for example, that sounds, smells, and pictures can be recognized in our minds. Interestingly, emerging brain technology could also detect this knowledge, which could be very helpful in a criminal enquiry. But should the brain of a human or the data processed on a neuroprosthetic computer be allowed to be scanned to reach our entity or person recognition? Perhaps, since testimonials in the courtrooms are extremely false and witnesses often fail to recall crucial details. In criminal law cases, it would also be useful to provide clear access to an incident memory of a person. But what if the false memories of eyewitness reports might be planted? Most people believe that the government's own "star witness" will not be tolerated, Farahany said.

Professor Farahany claimed in the legal review papers that a right under the U.S. constitution and applicable to government access to cyborg innovation is the Fifth Amendment to defend oneself against self-incrimination, given their experience in constitutional laws relating to brain recording technologies. She asks—would the

Fifth Amendment defense from self-incrimination really mean in the cyborg era that the government would "read your mind" and use the performance as testimony before the court? In view of the increasing capability of Professor Farahany of accessing human memory through implant technologies, the constitutional defense of cognitive rights as a mechanism to protect the right against self-infliction embedded in the Fifth Amendment has been proposed. In an essay written in the Stanford Law Review, Farahany went over *Schmerber v California*, which found that under the Fifth Amendment's Self-Incrimination Provision, no one is obligated to "show an allegation from his lips" but that anyone may be required to offer substantive or tangible evidence (for example, DNA or a blood sample). Whereas in a court prosecution the defendant cannot be required to "take the stand" and be a witness against himself; the prosecutor may take from his body samples and use them as evidence. With the advancement of technology for brain reading, Farahany claims that the taxonomy of facts subject to the right of autonomic discrimination must be redefined based on new applications of neuroscience. This is because evidence can arise from the government's access to a neuroprosthetic system or by direct brain function monitoring, which does not reflect the type of physical evidence the court may receive. A significant topic of jurisprudence in future cyborg years for this and other purposes is whether civil protections, such as the Fifth Amendment, extend to data stored on neuroprosthetic devices.

9.9 LITIGATING COGNITIVE LIBERTY

The definition of cognitive freedom is broad and so the defense of cognitive freedom between different jurisdictions may vary. On this basis, freedom of expression is not the exclusive defense of cognitive freedom in the United States as regards the First Amendment. The Fifth and Fourteenth Amendments provide, for example, some protection from unfair body interference under the United States Constitution Due Process Clauses. Why is it that our cyborg future has a dual standard of security? If the state does not regulate the expression of thoughts, it affects brain physiology and can impair cognition. This may not, for example, be the first provision to give rights of cognitive liberties but rather to safeguard the due process in compliance with the Constitution and can be used to protect the dignity of our organisms, by authorizing the administration of antipsychotic medications. Professor Jonathan Blitz of the University of Law of Oklahoma claims that the potential to biologically mold our reasoning process should be regarded as a more general kind of control that our right of speech is supposed to be in our hands, not in the hands of authorities of the state. The cyborg technology that a government might hack has significant repercussions for cognitive independence. Technology that enables the government to control mental functions is a deliberate attempt to change the substance and shape of the mind of an individual—the basic substrate of democratic freedom of expression. A crucial question in an era of cyborg technologies is whether the state should control intellectual material before it gets outsourced. This problem was not litigated in particular circumstances of cybernetic technology, but cognitive independence was claimed in accordance with the right of a person to be granted by the state. In the United Kingdom, for example, *R v Hardison* concerned a person who was convicted of breaking the 1971 Abuse of Drugs Act. Hardison submitted that Article 9 of the European Convention on Human Rights granted cognitive equality. In fact, since psychotropic substances are an effective way of modifying the thought process of a person, prohibition under the Abuse of Drugs Act 1971 was counter to this act. Clearly, the complainant submitted, "Individual sovereignty in the world is at the center of what freedom means to be free." The court objected, however, and Hardison declined to request an appeal from a higher court. In *NAACP v Button* in the U.S., the Supreme Court wrote: "only an imperiosive state interest . . . will warrant restricting freedoms for first amendment." In the coming cyborg age, the focus of discussion and legislative intervention is what those interests should be and under what circumstances they should be covered.

The U.S. Supreme Court heard arguments after the Hardison Decision in Great Britain on a significant case that deals specifically with questions related to the cognitive independence of the mind. In his history, Dr. Charles Sell was charged by the federal court with false claims against Medicaid and private insurance providers, which led to bribery and money-washing allegations. Dr. Sell had previously obtained medical support and volunteered antipsychotic medications but considered the adverse effects to be undesirable. Dr. Sell was found incompetent (but not dangerous), following the original accusation, to face trial, which led to a disciplinary hearing and a verdict that forced medication on Dr. Sell to be used to recover mental skill; Dr. Sell contested the ruling. The government's policy to compel Dr. Sell to take medications that would change his behavioral functions posed serious constitutional concerns. In this respect, Harvard University's professor of law Lawrence Tribe said that "if government wishes to mess with our mental autonomy by confiscate books and movies or by refusal of psycho-medicaments" it is ultimately the same "offense." Can anyone who did not pose a threat to anyone be deliberately injected with antipsychotic drugs to make them exclusively competent for punishment of offenses defined as "non-violent and purely commercial" by Judge Kermit Bye of the Eighth Circuit Court? The government attempted in Dr. Sell's case to control and change the thoughts and processes of Dr. Sell, causing him to take "antipsychotic" medications that would alter the mind. The Supreme Court has severely limited the ability of a trial court to regulate the use of antipsychotic drugs during Dr. Sell's holding for the sole purpose of ensuring the competence and trial of an indicted convicted suspect who was considered to be impaired. Therefore, as the lower court did not determine that all the conditions for forcible care required by the court had been fulfilled, the decision to unlawfully diagnose and treat the defendant was overturned.

In Sell's case, the defendant's mind was changed by forceful medications, so what are the consequences for government access to neuroprosthesis or other equipment for implants that might change the mechanism of consciousness or even edit his memories? The Sell court explicitly did not forbid the government entirely from modification of the brain-chemistry of an entity, and this raises the issue of whether or not a government should access the implant in the brain or even edit a person's memory. In the case of an inept accusation, it is generally considered that a fair defense is refused to the defendant, since the defendant is unable to engage in his own defense; those who condemn involuntary drug use to ensure a fair trial contend that drugs are frequently so overwhelming that it is often difficult to participate appropriately in the defense. In the Fifth and Fourteenth Amendments, the reliance on free thought and due procedure privileges for the government to "manipulate" the minds of an individual appear imperative to me: how can a person's voice be free from government influence if the government can forcibly prescribe medication or edit the mind by getting access to the technologies that can change the way a person speaks?

The "cognitive independence" interest in the case of Dr. Sell should be viewed as the union of the freedom of interest of Dr. Sell in physical honesty with his free thinking and due process right under the Fifth and Fourteenth Amendments. Such an infringement by the government on physical dignity is a breach of the freedom of free speech in the First Amendment and an attempt to control the thoughts and thought processes of an individual directly. Since, "in the first constitution, the idea is at the center that the values of a free society must be formed by one's mind and conscience rather than manipulated by the state," it would definitely be that, as the government wants to alter Dr. Sell's thought by altering the brain chemistry violently; they are breaching the First Amendment. In addition, manipulating the mind of a person when drugs are used forcefully, is an act of behavioral censorship and mental coercion, an activity that, according to the First Amendment, is surely more disadvantaged than even speech censorship. An administration that allows people to control their consciences at their very root—forcing an individual to take a prescription that changes their mind or hacks a neuroprosthetic instrument—needs no censorship so it can stop a previous thought in the mind of the speaker. The government manipulates and changes the type and substance of Dr. Sell's eventual speech by directly influencing the way Dr. Sell processes facts and formulates ideas. Equality of speech renders the First Amendment meaningless.

The basic topic of how people can morally alter the mental health of others is generally not discussed in legal thinking, aside from cases of penal law concerned with the mental ability of the individual to face the prosecution, but is an emerging topic in cyborg law. While every constitution provides the right to physical honesty, few preserve mental integrity. If a cybernetically augmented mind has the civil privileges of machines, a set of protections beyond the safety of biological humans would possibly be given for potential cyborgs. Here a brain with neuroprosthetic instruments can even be exploited just like a computer can; will cyborgs be given the same protection as anti-hacking laws in the future? Future hacking crimes could take a very sinister turn; they might not violate information networks, but minds, bodies, and behaviors. In reality, insulin pumps can now be compromised or jammed to deter hackers from deadly attacks.

9.10 IMPLANTING A SOFTWARE VIRUS IN THE MIND

Future hackers may be able to use wireless technologies to interrupt the neurosis of an individual and even to insert a machine virus in the head of a person in violation of the internet, telecommunications, and criminal laws. On this last point, the first infector of a computer virus was an American scientist and an early student of Professor Kevin Warwick, Dr. Mark Gasson. How do we do that? In the case of Dr. Gasson, the cyborg programmers might deliberately use brain-implant technologies or the hacking of a wireless brain network as part of a concept-proofing analysis, but in the future, they could spread the virus to the mind of another. A chip, which was then contaminated with a software virus, was put into his hand in Dr. Gasson's study. Dr. Gasson, who is important in cyborg laws, has demonstrated that it is possible to spread this virus to external control systems—which means that a human with cyborg technology "infected" will transfer a virus to the external cyborg computer. More specifically, if other embedded chips within a person's body were connected to the device, like the neuroprosthesis, the virus would also have contaminated them.

Cyber security specialists are highly worried about the facility to hack implants. For example, the subject was thoroughly explored by Professor Dr. Kevin Fu, a leading authority on health care at the University of Michigan.

His issues are specifically linked to neuroprosthetic systems and implants that are attached to an internet network and thus vulnerable to laptop or other computer infections. The fact that the vendors frequently refuse to change their devices and even install protection functions exacerbates the issue of implants that suffer from a software virus. "I think that's spectacular," Fu said. The shortage of security updates might be a significant barrier to cognitive independence as it is possible to hack brain implants.

I have often assumed that software virus propagation is no different from the transmission of a disease-relieving virus to the body. Mark Gasson's statement about feeling like he'd had a computer infection resonates accurate to his experience in this regard: most people with medical implants believe that they are often combined with their own body principles. A virus must be hosted, and in certain situations a machine virus may be transferred to a spectral cybernetically enhanced host by the air we breathe. There are several regulations in this field with respect to breaking into machines. In the United States the Electronic Theft and Misuse Act tackles the problem of improper access to protected computer networks and using devices and programs thereof. In addition, the Electronic Theft and Misuse Act forbids anyone from accessing public servers without consent. Hackers convicted of crimes that breach this law may be liable to fines, probation, or otherwise. Depending on the severity of the injury, they may serve prison time.

According to U.S. law, there may be criminal liability if an illness is intentionally spread to another human. For example, illegal transmission of sexually transmitted diseases may be affected under state regulation, which normally covers HIV and other sexually transmitted diseases that are communicable or contagious. However, the disease propagation paradigm is not actually being used to distribute a digital virus: instead, we are considering other legal options for individuals who transfer malware. The possible effect could compromise millions of hosts if software virus dissemination is through the internet. Think about a Cornell University student who had a form of malware known as a "worm" published on the Internet, which compromised millions of computers and took millions of dollars to get rid of, as a harmless experiment. As the U.S. Government destroyed many computers, the student was charged and punished in compliance with the previously mentioned Electronic Fraud and Harassment Act. Additional jurisdictions also prosecute anyone who infects virus machines. Section 3 of the Data Abuse Act, for instance, includes the application of ransomware to a computer in the U.K.

It is an act of behavioral censorship and mental coercion, an activity that, according to the First Amendment, is surely more disadvantaged than even speech censorship. An administration that allows people to control their consciences at their very root—forcing an individual to take a prescription that changes their mind or hacks a neuroprosthematic instrument—needs no censorship so it can stop a previous thought in the minds of the speaker. The government manipulates and changes the type and substance of Dr. Sell's eventual speech by directly influencing the way Dr. Sell processes facts and formulates ideas. The observations of Dr. Gasson that a virus may spread from one implant to another, evidently, have significant consequences for a potential cyborg in which third parties have access to brain implants to store memories and sensory information where a virus infecting another neuroprosthesis can be infected by medical devices such as pacemakers, cochlear implants, and retinal prosthesis. The results of Dr. Gasson indicate that the propagation of a computer virus is often feasible if access to neuroprosthesis becomes possible for third parties, and the preservation of functional independence is a significant aspect for those with a technology for neuroprosthetics.

9.11 CONCLUSION

With the improvement and incorporation of cyborg technology into the human body, critical ethical and policy problems will have to be dealt with. The possibility that a repressive dictatorship will have access to neuroprosthetic equipment is maybe a worrying outcome for cognitive independence. Equality of thinking is each person's inherent human right to be sure of the universe as best they can understand it. We cannot make independent rational decisions without the capacity to learn independently and obtain a reliable representation of external events. The obvious trend is that we are more vulnerable to the government while improving our bodies with technology. The apparent development is that we are vulnerable to more state supervision and privacy invasions as we improve our bodies through technology. Should embedded technologies within our minds and bodies have normal people's rights or just property rights? Cyborg technology, for example, a pacemaker, does not contain data generated by a cyborg, so it is deemed to belong to the manufacturer, seller, or licenser. Contractual and property laws are relevant to issues of possession of cyborg inventions and to data obtained from implants.

REFERENCES

[1] Hillis, K. (1996). A geography of the eye: The technologies of virtual reality. In R. Shield (Ed.), *Cultures of internet: Virtual spaces, real histories, living bodies* (pp. 70–98). Thousand Oaks, CA: Sage.
[2] Howell, L. (1995). The cyborg manifesto revisited: Issues and methods for technocultural feminism. In R. Dellamora (Ed.), *Postmodern apocalypse: Theory and cultural practice at the end* (pp. 199–218). Philadelphia: University of Pennsylvania Press.
[3] Ito, M. (1997). Virtually embodied: The reality of fantasy in a multi-user dungeon. In D. Porter (Ed.), *Internet culture* (pp. 87–110). New York: Routledge.
[4] Kelly, G. (1955). *The psychology of personal constructs*. New York: Norton.

[5] Kevles, D. J., & Hood, L. (1992). *The code of codes: Scientific and social issues in the human genome project*. Cambridge, MA: Harvard University Press.
[6] Kramarae, C. (1995). A backstage critique of virtual reality. In S. Jones (Ed.), *Cybersociety: Computer-mediated communication and community* (pp. 36–56). Thousand Oaks, CA: Sage.
[7] Kunzru, H. (1997). You are Borg. *Wired*, 5(02), 154–159. Available on-line at http://www.wired.com/wired/S.02/features/ffharaway.html.
[8] Lannaman, J. (1991). Interpersonal communication research as ideological practice. *Communication Theory*, 10, 179–203.
[9] Larson, D. (1997). Machine as messiah: Cyborgs, morphs, and the American body politic. *Cinema Journal*, 36(4), 57–75.
[10] Licklider, J. C. R. (1960, March). *Man-computer symbiosis* (pp. 4–11). Available on-line at http://memex.org/licklider.html. (Original work published in IRE transactions on human factors in electronics HRE-I).

10 Body Sketch for Cyborg

10.1 MAKING, MODIFYING, AND REPLACING BODIES

A main theme of this book is reiterated—we human beings are being improved by the "cyborg" technology—by the use of artificially intelligent machines. Currently, the use of technology of the 21st century to construct artificially intelligent equipment leads to exchangeable, replaceable, and upgradeable bodies that can decide the recognition or bigotry, animosity, and unfair treatment of our technical innovations within the community. When we start connecting with our robotic creations, I conclude that they are the same biases and bigotry we humans suffer daily. Will our legal structures be adequate to guarantee that our robotic technologies are increasingly autonomous, intelligent, and humanoid in nature as we grow in cyborg technology? That is an important topic since many of the artificially smart robots in society imitate real humans and claim that they deserve basic treatment, such as basic human rights and the dignity of the legal individual. As humanoid robots grow so far that they argue that they are aware, are we going to regard them as equal, or will we discriminate and deprive them of equal legal protections?

In coming decades, when cyborgs are armed with the technology that enhances their capacities and robots migrate from their assembly lines to our houses, it will become necessary to decide what constitutes the ethical treatment of technologically advanced individuals and whether they are to have equal legal rights. Should machines, for example, reassure us that they know themselves to enjoy equal rights under separate rules, statutes, and constitutions for human beings? And would we be inclined to treat a robot as human if it was designed visually to look like a human and combined with artificial intelligence? One of the aspects that would certainly impact upon the appearance of artificial intelligence, its personality, and its comportment is the response to questions regarding the privileges our technological progeny can obtain. However, it seems fair to assume that cyborgs and androids are discriminated against depending on their "machine" appearance; in the event of this, it should be possible to brace society for clashes between artificially intelligent robots and humans. I am basing this finding on social research into sexism, aggressive treatment of cyborgs in multiple contexts, and the findings that people can be uneasy in the presence of robots because they closely mimic people in appearance. For the emergent laws of the cyborgs and of artificially smart machinery (especially those that exist as androids) this phenomenon, described by Professor Masahiro Mori as the "uncanny valley," is so relevant that this chapter concentrates precisely on the "uncanny valley" egitimacy. For some futurists, it is not only possible but likely to occur this century that mankind will use technology to construct qualitatively different kinds of beings. In *The Rise of Digital Citizens* physicist Sydney Perkowitz writes that they may take the form of completely artificial, intelligent, and aware robots

DOI: 10.1201/9781003392699-10

from bionic humans to androids; they may take the form of a race of "cyborgs," immensely enlarged and extended physically, psychologically, and emotionally; or they can take the form of "internet beings," who may or may not exist in this country. New types of people could evolve from technologies such as cloning, genetics and stem cell research in biological science. However, the policy of human-changing, genetically engineered species has moral, ethical, and theological implications, as Stanford's Francis Fukuyama states in *Our Posthuman Future*. It is alarming how human-induced changes propagate through our genomes. However, whatever the shape our technologies take when they are clever, join society, and compete with us, they will elicit terror, hostile responses, and discrimination against people (and other computers that are artificially smart). This is why policymakers will need to decide the legislation and regulations necessary to uphold the constitutional rights of all smart people in our country over the next few decades [1].

In my view of future technological innovations those who fret about genome shifts might skip the broader images, when the rule of exponential returns indicates that the future mightn't be populated by biologically enhanced humans but by technologies, leading to cyborg races and, later in this century, the merger of the "organic cyborg" and artificially adding variations. Last, Jennifer Robertson of Michigan University has commented that people and machinery are not only considered but also aggressively sought by leading Japanese robotics companies—that they may combine into a new, superior species. And as we go into Singularity, there are several moral and legal problems with the development of cyborgs and intelligent machines. Any critics think that cyborg technology coupled with artificial intelligence will potentially prove to humans more difficult and harmful than genetic engineering. However, before we get to a stage where digital super intellect poses an imminent threat to mankind, it is the particular situation and timeline on the Rule of Look and Artificial Bodies that is discussed in this chapter that would have entered society and are subject to bigotry and discrimination from men.

Many outlets, including those considered by the United States' highest court, are capable of gathering the notion that the physical presence of our technical progeny could establish prejudice against them. The Supreme Court, for example, in the *McCleskey v Kemp* opinion on an accused defendant's sentence issue, argued that in criminal trials the prejudice of appearance may also be extenuating. If prejudice remains for humans in our court and workplace processes, cyborgs and androids who engage with us in social settings and compete against us for work will definitely be discriminated against. In reality, in the presence of the fit, people can feel awkward. Indeed, there are job cases brought to court in the United States under American Disability Act and other anti-discrimination legislation where individuals may be disturbed in the presence of cyborg technologists and may discriminate against them. More broadly, however, if the appearances of cyborgs and any intellectual machines in society diverge from the social expectations for shape, attractiveness, and form, it can become a conflicting issue. Studies have shown that people have a high degree of comprehension of the physical attractiveness of other people, and I'd expect this result still to apply for androids.

How should public policies direct their creation and how should the court react to the likelihood of unfair treatment of our technical innovations based on their presence, provided the exponentially increasingly accelerated technology contributes to an environment of intellectual machines engaging with people in different social sects? We need to go a long way, as in the future unenhanced human beings could be discriminated against in our intelligent robot creations, in addressing those problems and designing solutions. Therefore, the rules and legislation that lead to an equal society consisting of people of flesh and without flesh are crucial to our consideration. This chapter addresses those problems in the sense of the development and functionality of man-made, android-shaped, artificially intelligent machines. Leaving aside the reader's conviction that a "look rule" is a concern that has yet to be resolved by multiple jurisdictions, the "look law" is evolving, centered on work and other employment cases and regulations. Relevant subjects are, among others, enforcement of "freak shows," lawsuits brought under the American Disability Act (ADA), cases of malfunctioning prosthesis, and the intellectual property law's protection. In light of the next cyborg era and our potential fusion with artificial intelligent machines, I am talking about these rules.

In this chapter that in the cyborg future, not all humans and not all AI devices would show bias towards artificially intelligent robots. Much would depend on the society in which the machine is embedded, the characteristics and actions of the automated agent, its tasks, and policies followed by humans. Interestingly enough, MIT's Kate Darling shows that human beings appear to anthropomorphize our robotic inventions; I get the sense from reading her articles that we prefer them. If their action is self-sufficient when we communicate with robots in a social context, Darling notes that they will impress us with "loyalty and depth." We should handle them even as if they were alive. In the sense of rights, Darling intends to introduce "barrier protection regulations" as has been done with pets with our technological innovations. I'm able to go even further, in my mind, into the regulation that is required to defend androids and other types of artificial intelligence.

To sum up, why is there a "lookism" chapter on bigotry against our potential technological innovations? Conflict between humans and AI is only going to increase as technology advances, and numerous social science studies have shown that one's outward appearance significantly affects his or her access to resources and opportunities. Most notably, if we will now learn how we incorporate cyborgs, androids, and artificially intelligent machinery into society, how will "they" treat us once our scientific innovations go beyond (and become more appealing than) our intellect and performance? There are also high stakes for mankind. This chapter on the law of appearances of artificial bodies explores the rules and regulations surrounding the appearance of technologically modified individuals and whether our possible robotic creations can have equal rights under the law, while also exploring whether there are any legal theories for shielding our artificially intelligent genetic material from disks [2].

10.2 THE SHAPE OF THINGS TO COME

Although this chapter provides a variety of examples, readers may already have a grasp of the potential ways of androids and the use of artificially smart machines. Becoming the subject of sci-fi books and films for a while are cyborgs, androids, and artificially intelligent machinery. Indeed, because science fiction books were adopted for film and television programs, a variety of artificially clever pictures were shown to the public. Interestingly, the conflict between non-enhanced humans and androids in science fiction novels is a way that the writers investigated the significance of civilization and debated legal protections for nonhumans. Parts of the machine intelligence media robots are human-like and ready for us, while others tend to be disgraceful and intimidating beings that engage in revolts of the human race. In the final argument we take into account the androids of the 1973 dystopian novel *Westworld*'s theme park, written by Michael Crichton, who chased human tourists purposely and destroyed them. An increasingly advanced technology, at least over the next two decades, would lead to the emergence of cyborgs and androids who have the powers or the presence of unenhanced human persons. A more plausible scenario for our cyborg future is the concern whether our technological innovations would be discriminated against as they enter society, start to communicate with us, and compete against us for jobs. As the presence of a human has a great deal to say on how they approach society and whether they are discriminated against at work, in our health care environments, and by our institutions, it is important to examine how technical advances in the bodies of cyborgs, androids, and artificially smart machinery often influence their care.

An android is a robot, but it's a robot that looks and behaves like a person, particularly with a flesh-like body. Professor Jennifer Robertson from the University of Michigan said that the humanoid "must fulfill two criteria: it must be made up of a human-like organ and behave like a human in conditions engineered for human body capability." A variety of experiments in robotic labs around the world have made remarkable strides with the goal of developing androids that look and to some extent talk or behave like a human being. Smart androids are coming, and the appearance and intellect of smart androids will radically transform the legitimacy and challenge our legal and social structures. In reality, a report by the University of Oxford in 2013 analyzed 702 occupations, showing that 47% of the total of U.S. workers run the risk of being lost in favor of computing. The robot Botlr from the start-up Savioke, used in some properties in the Starwood hotel chain, is an example of our rivalry in the service sector. I wonder at hotels around the world how many people are doing this job today, and how will they feel about the machines that drive them. The duty is to provide additional towels and forgotten toiletries to the guests of the hotel. In the future, various ways and forms of technical enhancements will be possible, and our technological innovations will be looked after. Professor Perkowitz wondered at this stage what human features can stay in form and shape while we gain the potential to progress with cyborg technologies. Can we stay in the form of organic human beings, or will any other form be more functional? Many robotics engineers envision a future of expanded contact between humans and robots, such that our intelligent creations are best adapted for human culture. They work to build androids as human beings. However, the notion that autonomous robots are able to assume non-human forms is, in contrast to this human-centered conception of what a robot might look like, ideal for a specific mission. How are people going to react to highly intelligent robots that assume different shapes and types or look strikingly like people? On the other hand, in this situation are we going to demand more from and discriminate against our human analogs? When our bodies are fitted with cyborg technology, how should law, in particular, react to unfair and drastic human, android, and robotic enhancement if the concept is equal treatment according to law?

Robotists, in particular, often get their layout indications from nature—and particularly people for androids. For example, robots operating on assembly lines are designed to handle objects, whether it is a welding weapon or laser-scalpel, through arms and end-effectors. According to Larry Greenemeier, "other robots are fitted with head-assembly cameras for eyes and wheels to emulate human locomotion, intended for remote workers as telepressure replacements or supports for older persons or people with disabilities." He also feels it is easy to assume that the robots today are just primitive imitations of their masters. Most of the existing robots may not seem human, but in a decade smart human-like robots (i.e. androids) will have entered civilization, guided by considerable improvement in the construction of flesh surfaces, detailed facial regulation, and considerably improved motor skills.

Sidney Perkowitz of Emery University focuses on the future, as I noted in the previous chapter: first the role of prosthetically designed goods and implants such as artificial arms, replacement knees, and hips and vascular stents and second the aesthetic improvements and developments of the future. Human beings have long been involved in altering their own bodies and appearance for religious and cultural purposes and as a means of self-expression. As I noted in a previous chapter we have changed our bodies with multiple technologies, whether on or inserted under the skin; according to David DeGrazia, professor of philosophy at George Washington University, we are feeding, exercising, and coloring our bodies with tattoos and body piercings. When cyborg machinery enhances science rapidly, we should foresee massive alterations and improvements in the human body by the middle of the century thanks to advancement in exoskeletons, prosthesis (such as limb, cochlear, or retinal), Cardiac pacemakers, neuroprostheses, and alarms. In addition, as I mentioned in the chapter on the improvement, enhancement, and hacking of the body, there is a trend amongst do-it-yourself hackers (or grinders) to alter their body with technologies. However, to (re)establish the "cool vision" of our cyborg world, body restoring, enhancing, and altering technologies not only improve the technology dramatically but also bring mankind to a cyborg future with artificially smart robots, which could inevitably combine. The level of approval by human beings of cyborgs and androids when they join the population depends on a variety of factors like tasks, attitude, and appearance. Given human preferences over "looks" and provided that all forms can be made, do law and social norms dictate that androids are permissible for only humanoid forms that copy the picture of an "attractive" (and young) person? In 2010 more than sixty household robots were sold in a variety of sizes and forms, which supported cleaners, compañeros, and carers, according to Professor Jennifer Robertson. But if we improve according to the rule of speeding up return, as androids become knowledgeable, are they going to content themselves with being our domestic servants? Naturally, people are writing the software and building robots at the moment this is written, but that will change soon. Software bots now make profitable storage transactions with increasingly sophisticated algorithms, and other Ais diagnose medical disease, make music, test mathematical theorems, and drive an automotive device (Does "backer" refers to the assault and battery if an AI is behind the wheel of an automobile that now bears its name?). As a matter of strategy, mankind should be mindful that any bigotry of our technological offspring will turn the human race away and prove devastating until they become intelligent.

10.3 THE ANDROIDS ARE COMING

The question of physical identity will be an interesting matter for the "Rule of Appearances and Robotic Bodies" as individuals are augmented by cyborg technologies and as artificially intelligent robots become more human in appearance. Indeed, the right to personal dignity, the right to exercise protection or power of the body, is one of the most basic human rights. An android may employ a right to body dignity for smart machines to protect the body from unwelcome adjustments or to stop anyone from scratching its pieces on a different computer. I may note that, should equivalent rules exist for androids, there are laws controlling organ donation for humans. The author may be curious as to why an artificially smart computer would avoid a transition to its body or its "mentality." Any computer system software may be successfully implemented, just as any individual might seek to use technology. Much like man can resist any alteration in his or her body, such as pushing drugs to render a human physically competent, knowledgeable, self-conscious machines can also resist improvements that are considered unwelcome (e.g., an upgrade that could affect their memories). Of course, they are subject to human judgment as long as the artificially intelligent machinery is lacking in privileges, but it easily gets smarter, enough that I think it is just a question of time before they settle about their physical and mental dignity.

In the United States, it was consistently held by the Supreme Court that the freedom from unjustified intrusions and the dignity of the personal body are protected by the procedural privilege of the Fourteenth Amendment. Androids may be particularly interested in preserving their organism's dignity from personality or even from vanity. Once an autonomous smart computer has feelings and ties to the body, it will be concerned with how people view it. In particular, emotions may be required if an android makes a decision to undertake a racist argument. As mentioned earlier, androids can also argue for the right to undergo technical improvements, including aesthetic enhancements (the upgrades may stop a digital divide between androids), which could have no practical purpose. Any non-functional add-ons to an android would have an impact on its rights under intellectual property law, which we will see in a later segment. Some may find that robots are emotionally weird or look redundant. However, many robots' programmers know that they are engaging more and more with humans as they join society, so a trend occurs to formulate interactive, life-like robot (i.e. androids). In my experience, the responses of robots and androids with feelings and attitudes and their societal approval levels can affect them strongly. Pepper, a Japanese robot capable of sensing and voicing a variety of feelings, is an illustration of this. It has a face recognition feature and a variety of cameras, voice recorders, and sensors. It weights approximately sixty-two pounds and it has a range of cameras. Pepper can read and answer the moods of users, according to Softbank, a Japanese web firm. Researchers at the Korea Institute of Industrial Technology have developed the android EveR-3 (one of a series of women androids) into an additional example of the advancement of robotics, which uses interpersonal interface models

Body Sketch for Cyborg   233

to mimic human emotional interaction via facial "muscles." EveR-3 can participate in rudimentary discussions and correlates with the typical Korean woman in her twenties (notice the selected appearance of EveR3). The EveR-3 can perform speech gestures and body synchronization with its artificial brain as seen in Figure 10.1. Her entire body is made from very sophisticated plastic silicone jelly and with her face, spine, and lower body artificial joints she can exhibit natural facial movements and sing while dancing at the same time [3].

Since Pepper obviously is a mechanical being without biologic components, a significant issue for the next cyborg era is how much technology can be integrated into a person's body? And if this difference is found relevant in law and regulation, how different are the looks of augmented humans and artificially smart robots that they ought to be shielded from prejudice on the grounds of their appearance? The lines between humans and robots will, in the future, irrevocably be blurred, and our courts will pick on an entirely different collection of problems with this transition. To begin, will there be a difference between how the law treats a human and a computer after both have been upgraded?

In addition to developing the technologies to compete against human beings on the labor market, some innovations in robot and android architecture have led me to a world in which we combine with artificially smart computers. Consider, for instance, the robot built by the Garage of Willow, PR2. It relies on mechanical hands to grasp things, making it a distant cousin to the androids in the future human-machine household unit. This robot will be advanced as it is. In fact, rapid developments in PR2 have already been made with the layout of hands that appear like and have physical dexterity much more like human hands.

My feeling is that if we want to combine with machinery, we would like to see a form that would replicate that of humans (at least initially for early adopters) so that

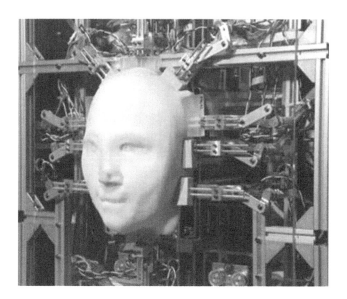

FIGURE 10.1 Mechanical computerized hand.

developments in robotics that render limbs and arms that look and function like their biological counterparts are a step in the direction of a human-machine marriage.

To highlight the high strength of robotics architecture in conjunction with artificial intelligence in the development of our potential technological progeny, think about how rapid developments in the field of computer training allow robots to control structures far more complex than a few years ago. In 2006, a significant advance was made in this sector when Andrew Ng—then in Stanford and now at Baidu—and a team of researchers developed a way for robots to learn how to treat unknown objects. The researchers have allowed their robot to review thousands of 3D images rather than write rules about how a single object or shape is captured and to classify which types of grip function with different forms. This allowed the robot to decide on sufficient grips for new products. Development continues and researchers in the area of robotic technology used a strong approach called profound learning to develop those technologies gradually in recent years. The intelligent and more ambitious the computer, though, the more it becomes like humans and the more it competes against us in the short term, that is, until we evolve technology.

But before we become "they," what is the likely response of humans to robots in the workplace, using a word popularized by robotics expert? History makes it clear that people will discriminate against machines that compete against them. The Luddite movement of the early 18th century is an example where modern technologies put English cloth workers at risk of homelessness, which the Luddites described as "commonal hurt machinery." They burnt mills, destroyed equipment, and the military mobilized. There were more Luddites who fought than those Napoléon fought in Spain at one time, according to historian Eric Hobsbawm. The British Parliament introduced a bill to make machine-crashing a big crime in reaction to the Luddite campaign. You might ask—would a common code provision make such an act a capital crime if you "mash" a robot in the coming cyborg age? If not, how about "breaking" an android that appeared like a human being? If so, will we feel more obliged to give the android the right not to be hostile, racist, and physical?

When technology advances, cyborgs and artificially clever robots come together and look like unrefined humans, even with extra features (e.g. camcorder for movies), which could affect the freedom of others (such as their right to privacy). At the same time, androids leave robotics labs armed with a growing degree of intellect and humanoid connectedness. (It is only by the virtue of their enhanced intellect that they are built to appear like us or not; fascinating legal and social problems remain). Consider the work of Professor Hiroshi Kobayashi, who leads the Smart Mechatronics Lab of the University of Technology, Tokyo. The team of Hiroshi developed the android Saya that serves as a reference at the University. Saya is capable of transmitting human facial gestures and of interacting through her head and facial expressions to those specific emotions. Saya and androids developed at the Korean Institute of Industrial Technology such as the remarkable technical feat EveR-3, for example, will be replaced in a few years by more intelligent and much more functional androids. And just a couple of decades back, family members of Saya and EveR-3 stated that they are mindful of and respectful of human rights. They might also need more privileges, because they are way more knowledgeable than we are and have institutions that are outside our ability. When androids attain a certain ability to argue for freedom, they will presumably argue for equal rights (and other liberties). In comparison, no section in human rights, then we'll be the ones to defend the rights of our more advanced offspring [4].

10.4 CULTURE IS IMPORTANT

Consider the use of a robo-assistant by Toshiba at the information desk of a department store in Tokyo, for an example of the adoption of androids in society. Aiko Chihira, the female android, speaks Japanese and can speak sign language. Notably, when Chihira talks and bows, she moves her mouth and lips effortlessly and is programmed with several gestures identical to humans. The presence of the android was not built to give a pleasant impression to a particular user. Both Aiko's "good nature" and traditional Japanese clothes increase the acceptance and reduces the possibility of negative responses. Android acceptance is particularly important to Japanese roboticists who build robotics to assist people with dementia, to offer telecoms with natural words, to interact by sign language with the hearing impaired, and to enable medical officers to track the elderly. Android acceptance is also particularly important.

A contrast between Japanese and south Korean robotics in their cultures and that of the United States and Europe shows us how humans will survive in a world of technologically enhanced beings in the future. I assume that cultural influences can have an important effect on the adoption of androids and robotics by people in society. One example of this is Japan's culture, with autonomous vehicles not only growing rapidly but also incorporating robotics into several layers of Japanese society. As an example of pop culture, the first android to host its own TV program has been the cross-dressing Japanese TV star's robot clone. According to Michael Fitzpatrick, "Dentsu agreed to clone the exact android copy of the famous entertainer along with Professor Hiroshi Ishiguro, Japan's biggest advertising agency." The agency spokesperson said: "there is still no content that can be related to technology for artists and entertainers, but the company Dentsu believes it will expand in order to improve Android entertainers." The robot played with natural gestures and a surprising similarity to the "actual" entertainer when it was voiced and operated remotely by a voice-impersonator. Furthermore, the top Japanese maquillage artist took to the extreme to make the android appear alive. In Japan, a culture that is very in line with the concept of a digital future, people are more sympathetic to as robots tend to appear more human. But whereas many Japanese are predisposed to being compassionate to the robots, prejudice against robots may be much less common there than in Western countries with different cultural norms and societal standards towards robots, since the Japanese popular culture shows them as nice (e.g. the *Terminator* series of films and the military attempt to arms robots) Interestingly, the span of time for an android that cannot be distinguished from a human in appearance by Japanese robotics is around ten years. Combine this with Google's Ray Kurzweil's belief that artificial intelligence will be achieved by the mid-century. The combination of intelligence and realistic android bodies within the space of twenty-five years and under offers human intelligence (in other words, artificial general intelligence) a powerful opportunity for humanity to look sooner or later at the ethics of robots and the persuasive problems of robot rule.

As demonstrated earlier, our smart robots will have several different looks and identities in our cyborg future then the options for making your robots or androids more subservient, modern is endless, but I think our response to artificially smart machinery will rely partly on their look and their personalities. Indeed, a Google patent for human robots indicates that a wide variety of individuals will become feasible and that many various forms of personality could be downloaded from the cloud. Therefore, it is incredibly likely that you will be able to pick the kind of personality that you want for your future android. It will do this by viewing and learning about your devices before configuring an individual dependent on that knowledge. It can even personalize its experiences with you by using words and facial recognition; this is a model of how our technology is becoming more like ourselves. For instance, for others the situation evokes the idea of "machine as a weapon," which goes against my conjecture that we are not only creating tools to support mankind but even building up our rivals and potential substitutes as we develop stronger cyborg technologies. If the robot were to take on a real being, this would be a sort of "misappropriation of likeness"—an important concern for the law would emerge. The patent in Google indicates that a deceased loved one or a celebrity will live on after their death easily in computer form or in simulated avatar. More on this in the following pages, since legal protections apply to presence in certain cases [5].

10.5 OUR REACTION TO CYBORGS AND ANDROIDS

Up until now, the cyborgs residing amongst us have received mixed public responses, from curiosity in their body's sleek "cybernetic" technologies to an overt hostility focused on the presence of the cyborg. On the latter, two of the first of us cyborgs, Steve Mann and Neil Harbisson, both with head-mounted display technology, have confirmed that, based on their cyborg presence, they have been targeted in public. Professionals at a McDonalds in Paris physically "recognized" Professor Mann during the first event. And Neil has a head-mounted sensor he uses to translate color to sound and has been targeted by security officers afraid he'd film them (he was actually hearing them in color). Google's Ray Kurzweil has identified the assault on Steve as the first documented hate crime against cyborgs.

In reality, considering the character of human beings, I suspect cyborgs and androids would, for various reasons, including their presence, be the object of bigotry, animosity, and hate crimes. Hate offences are typically on the increase globally and cyborgs joining society have also experienced "look prejudice" and overt violence. In comparison, for reasons of medical need, citizens armed with cyborg inventions are still discriminated against. For example, hate crimes target disabled people in the U.K., according to DisabilityHateCrime.org.UK, and they are a type of themselves (although they are mostly "repaired" by a prosthesis system, I would add). Based on color, faith, natural heritage, genus, nationality, sex, or disability of the perpetrator in the U.S., the state of Missouri describes hate crime as "knowingly motivated." Steve and Neil are known as using cyborg technology for injury management (Neil's is obviously due to his severe colored deficiency, and Steve is traveling with his physician's statement explaining his reliance on cyborg technology) and are being targeted. Will the cybernetic manifestations of androids and other technologically sophisticated entities encounter equal hostility? In the next few decades there are problems that will tangle our courts' processes and the decisions of judges in these cases will lead to an emerging cyborg rule. Sadly, in spite of Steve and Neil's indifference to cyber hate crimes and other types of bigotry, it can happen regularly to cyborgs and androids. There are in reality lobbying groups, called "Stop Cyborgs," who are attempting to drive forward cyborg laws in reaction to cyborg technology. Thus, the message appears to be that people cannot comply with it, becoming a cyborg at their own risk.

Where hate crimes escalate to a physical assault on a human, an analogy in the machinery world can already exist—consider the case of computer sabotage (and from a historical perspective recall the Luddite movement of the 18th century). An executive of a Korean computer corporation was recently been accused at an event held in Berlin of willingly destroying numerous Samsung washing machines. Consider also the cyber-industrial crime, such as hacking, involved. In terms of physical harm incurred to equipment, in one of its laws describing criminal vandalism, Washington considers that the damages to machines are:

> anyone does anything with the intent to harm, disrupt, replace, negate, or impede the owner or operator of . . . products, instruments, machines, or mechanics, or who has reasonable cause to believe that their conduct may have such an effect. Cyborgs, on the other hand, have already been assaulted and had their artificial muscles stolen.

Think, people are discriminated against, cyborgs are targeted, sabotage of a computer is a felony, and with it more androids and artificially intelligent robots emerge from the past.

The general issue of bigotry against androids precedes today's android culture. For instance, in Joan Brunners novel, *Into the Slave Nebula,* where the blue-skin androids were subjected to slavery by human beings, the issue of bigotry against androids was discussed in a sci-fi novel. Because of slavery, I believe that any sentient artificial intelligences that claim to care about human rights should strongly resist the idea of involuntary servitude. I sincerely feel it is not prudent for any intelligent being to exceed us intelligently in a couple of decades. As Martine Rothblatt, CEO of Therapeutics Inc., commented, future intelligent beings want to be autonomous; they are discovering from humans how oppression of artificial intelligence and the lesson of peaceful (think Gandhi) hostility can lead to ways of resistance (think terminator). Taking into account the remarks of Rothblatt, how likely is prejudice? What is the potential of our future technical progeny and what is their unwelcome reaction? Take the example of the recent cyborgs, Neil and Steve, suggesting that people wearing protective mount monitor technology can face animosity and bigotry on the grounds of their "cyborg presence." As in one example, in a cyborg technology case, the Sixth Circuit Court of Appeals ruled that an operator of a scoop operator with a prosthetic leg was allowed to reinstate himself as long as he could comfortably exercise his vital functions under the American Disability Act; in other words, the cyborg inventions could not disqualify the worker [6].

Although technological advancement exponentially always goes beyond the legal potential, a few countries are starting to take seriously the implications of a cybernetic future, for a variety of realistic reasons. Inevitably, people, cyborgs, and artificially intelligent robots are the job market of the near future. An optimistic proposal has been put forward in South Korea by the Ministry of Information, Media, and several robot cities for the country are expected by 2020; the first one to be founded in 2016. The new robot city will have R&D centers for fabricators and product manufacturers, trade show halls, and a robot stadium. In addition, South Korea is creating a charter on robotics ethics that will lay down basic rules and regulations on human contact with robots, define requirements for robotic users and developers, and establish protocols to be programmed into robots to deter human robot violence and vice versa. Artificial intelligence physicists, in particular, suggest the software "friendly artificial intelligence," which will minimize their perceived danger to mankind in the "brains" of possible artificially intelligent machines. Interestingly enough, questions of access to new technologies are closely related to complaints about the injustice against those who cannot afford or are hostile to such reforms in a Brooking study prepared by the lawyers Benjamin Wittes and Jane Chong, *Our Cyborg Future—Law and Policy Implications*. They also claim that anti-discrimination regulations could be required to discourage the refusal of cyborgs as a consequence of their cybernetic alterations and to avoid discrimination against unenhanced people for opposite purposes.

10.6 THE UNCANNY VALLEY

In the coming cyborg era, there might be cause for alarm about human responses, cyborgs, androids, and robotics; just look at the "Uncanny Valley" phenomenon. This definition was initially created by roboticist Masahiro Mori to offer a simple representation of human responses to robotic architecture, but it was generalized to human encounters with virtually any nonhuman organism. Simply said, people react positively to a "human" machine but only on a specific point. For instance, people generally like "class" robotic toys, but once an android look like a human, it doesn't really look like that. People report a clear negative reaction to their "creepy" presence, meeting the norm. However, as soon as the presence enhances and is inconsistent with a human, the answer becomes positive. So, the reaction goes . . . positive, negative, and then again positive. This abyss, the uncanny valley, is the point where a person who looks at the creature or the object concerned sees something almost human, but just enough different to appear edgy or disturbing. For example, robotics, animation of 3D computers, and medical applications such as burn restructuring, contagious diseases, neurological conditions, and plastic surgical procedures can be found. The kid in Pixar's pioneering 1988 short film Tin Toy was a printmaker in a scary valley in popular culture, according to robotics writer Dario Floreano. This led to negative responses from the audience, which first caused the movie industry to take the notion of a scary valley seriously. Moreover, some critics called *The Polar Express'* animation creepy in 2004. CNN.com journalist Paul Clinton actually said, "those human characters in the film are quite . . . okay, crapy."

There have been some design principles suggested to prevent the uncanny valley—I believe they reflect design laws for cyborgs, androids, simulated avatars, and any artificially smart being entering society. These rules may be taken into account by future courts. The robot can look uncanny and likely experience a discrimination in the look when human and non-human elements are mixed in the design of a robot. For example, a robot with a synthetic voice or a person with a human voice was perceived as less angry than a robot with a human voice or a human with a synthetic voice. Furthermore, to give a robot a more positive feeling, its level of human realism should correspond to its degree of human realism in behavior. There is also a bad feeling that an android looks more human than its movement capability. Furthermore, in terms of reliability, if the robot seems to be too equipped, people won't expect it too much if it looks too human; however, with ongoing improvements, future artificial means smart machines will satisfy our expectations and then exceed them. However, a high movement of material leads to the hope that some behaviors, such as realistic motion dynamics, will be present. In the end, abnormal face dimensions, including the ones used by artists to increase their attractiveness (for example, larger eyes), may seem annoying, combined with the texture of human skin.

According to futurist writer J Jamais Casico, a similar "uncanny valley" effect could emerge when human beings start modifying with cybernetic improvements aimed at improving the capacities of the human organism and intellect beyond what is usually feasible, whether it is vision, muscle strength, or cognition. Casico postulates that while these improvements remain within a perceived human behavioral norm, there is no likelihood of an adverse reaction, but repulsion is expected once people change their normal shape and form. But in our cyborg future, according to the theory of the collective unconscious, as these technologies become farther away from human standards, "transhuman" persons would stop being judged at a human level and instead be considered as different entities all together (this point is what is known as the "Posthumans"). In fact, some work has been done on how people view cyber-enhanced organisms. In work at Dartmouth College, for example, Jessica Barfield discovered that people with cybernetic emerging technologies had to change their body images and identity and learn how to use their bodies to satisfy new technologies. Do robotics, androids, or prosthetic device designers have too much difficulty in duplicating the human aspect? If so, the android or cyborg may drop into the uncanny valley with an apparently minor flaw [7].

But encourage me to get away from the robots and inquire, is the uncanny valley phenomenon valid for human beings? Yes, it is, here's a case in point. Ulzzang, or "best smile," is a Korean culture where girls digitally change their looks with makeup or some other way to get an animated look.

In other words, a girl tries to have ghostly circular eyes, a narrow nose and lips, faultless fair skin, and a small, coordinated body. When you have this anime style, you upload photographs of yourself to online prestige and Internet celebrity tournaments. From the outside, the anime's use of the uncanny valley to digitally alter a face by physically altering the face with glue and contact lenses is fascinating; this technique is analogous to "grinder action," in which people use technology to gain an extra sense "under the eyes." I can only say that in the next cyborg era the spectrum of human speech will be even more drastic when it comes to altering appearance, when human and robot body and facial features are replaced by "cyborg" technologies.

10.7 OBSERVATIONS ABOUT DISCRIMINATION AND THE "UGLY LAWS"

In my opinion, the response is strongly "yes," provided the human drive to adhere to cultural (or subcultural) norms is high, to whether artificial intelligence as encompassed in various physical types is discriminated against on the basis of its looks, whereas those who are short of the notion are always the victims of "lookism." For example, in cases where their appearance is obviously irrelevant to their qualifications or ability, visually unattractive individuals are often handled unequally. By comparison, other studies in social science have found that people assign to someone who is visually desirable a wide variety of good attributes. Studies have furthermore found that, merely because of their lack of beauty, less beautiful individuals are handled poorly. Are "unsightly androids" going to be discriminated against in our cyborg future with the same lookism? In a study by Jessica Barfield on the understanding of cyborg bodies, she recorded findings of surveys that showed that people with cyborg technologies had a large amount of societal prejudice, although none responded to cybernetically augmented persons. The conclusion of the social science research that a person's image has an effect on the care they get is so high that parents, like teachers, have lesser perceptions of unattractive children. It was already proposed that we want more from human-looking robots. In comparison, unattractive persons receive greater fines and smaller damages in civil law lawsuits than adults in simulation experiments. In overview studies of social science, the inequality of the "look" is common in society. It is fair to presume the same unfair treatment of our scientific innovations depending on the presence of humans.

Discrimination toward disabled persons is also triggered by the absent or injured portion of the body that may be substituted in certain situations by a prosthetic arm or leg; this can become the foundation of the unfair care. Interestingly enough, robots and androids also have identical "cyborg" technology. For the disabled, cyborg technology may be a game-changer, but, paradoxically, cyborg technology often gives the machines racist reactions that are based upon their presence. The enhance computing skills of the human opponent. Androids can be discriminated against based on their look (particularly when they slip into the uncanny valley) and also based on their improved capacity to displace humans from their workplaces. Robots and androids in particular also use state-of-the-art prosthetic equipment, computer vision, and machine learning algorithms to compete against humans to do the work we human beings do. And, in British Columbia, lower-paying employees are five times more likely to take over positions from machines than people earning higher wages, according to a joint study by accounting firm Deloitte and Oxford University. The robots will "steal" about half of all jobs around the world in the not-so-distant future, according to MIT, Oxford University, and Sussex University, because according to them, a second generation of machinery is being implemented that will have an impact on civilization more significant than the beginning of the industry. *The Second Computer Age: Jobs, Growth and Prosperity in a Time of Glamorous Technology and Emergence of Robotics* by Erik Brynjolfsson, Andrew McAfee, and Martin Ford have published two interesting volumes on the topic. Nevertheless, my future outlook is different from those described earlier, as I consider the second computer era as the synonym of an age of cyborgs and a possible merge with artificially intelligent machines; that is, I argue we are being turned into "smart machinery," and the "smart machinery" is becoming us. According to Hans Moravec, I expect that our future will fuse with our inventive artificial intelligence, and that our future, the second machine era in which "They" stand in for "Us" will not be temporary; rather, "They" and "Us" will merge into a single entity. Instead, I assume that the future will fuse together in a second machine age.

South Korea and Japan have also vigorously embraced robotic culture and lifestyle virtues. In national surveys in Japan, Professor Jennifer Robison, leading academic in Japanese robotic society, estimates that Japanese people share robot life and work more happily than foreign guardians and migrant workers. Robertson, in addressing Japanese demographics: "as its population keeps declining and is ageing faster than in other post-industrial countries, the robotics industry is used by Japanese policy-makers to boost the economy and to maintain the perceived ethnic uniqueness of the region." These Robinson projects are coupled with increasing support for the conferencing of citizenship on robots between some Japanese robotics firms and law-makers. The notion that robots have developed beyond the limits of "land" to become juridical bodies with "rights" is now affecting progress in artificial intelligence and robotics, like South Korea, Europe, and the United States of America, besides Japan. In addition, the We Robot Conference, a gathering of leading law and robot experts, promotes the notion that civil protections for robots are being given every year. And the concept of a federal robotic organization to handle the novel experiences and damages of robotics can be created by Ryan Calo, one of the organizers of the We Robot Conference.

It is rational to assume that cyborgs and androids would be discriminated against in "lookism," when the most visually unattractive members in today's culture face systematic prejudice. And citizens do not just discriminate against others that depart from social norms; their appearance may also differentiate amongst local governments. In the past, certain U.S. states have so far gone so far as to ban the public presence of "hideous" or "unsightly persons." This suggests that cyborgs and the beautiful androids that are found unattractive could even insult human sensitivities, be exposed to "lookism," and other prejudice. Notably in the early 1900s, in some mainstream U.S. cities such as Chicago, Illinois, Omaha, Nebraska, and Columbus, Ohio, it was illegal to be found "hugely" offensive. The situation is very serious. The penalty of such a person regarding public prosperity stretched from imprisonment to fines for each "hideous offence." This is how the Civic Code of Chicago (which has since been repealed) illustrated and applied the "Ugly Law":

> No one that is sick, mutilated or in some manner skewed to be an unstoppable or grotesque entity or object Inappropriate persons must be admitted to or be presented to the public in or in public form or otherwise in this city, or shall, for each violation, be subject to a levy of not less than one dollar or not more than $50.

The "lookism" laws at the time felt that, though people with disabilities, the disabled, and the poor were a part of society, no one wished to identify with them, and fewer wanted to see them in public. Therefore, laws were passed to discourage and disregard deformed persons, especially those with cerebral palsy and other diseases. Fortunately, in 1967 Omaha abrogated the Ugly Laws; in 1972, they were retired from Columbus; and in 1974 Chicago was the last to end the penalty of the "hilarious." However, although human bigotry is slowing, oppression by "lookism" continues to form a part of culture and will undoubtedly continue to be guided toward our cybernetic creations in our own future.

Unlike the jurisdictions that have adopted legislation to forbid persons from becoming "unsightly," the jurisdictions that legislate in this field are more likely to react by issuing municipal laws defending individuals from bigotry of lookism. In the United States, in particular, several nations and jurisdictions have passed legislation banning discrimination on grounds of appearance. For example, the District of Columbia forbids discrimination based on "real or perceived" context distinctions and characteristics, including "physical aspect," such as weight (no overweight androids please). And workers in the District of Columbia should be highly vigilant about firing staff when coping with problems dependent on looks, since their personal appearance and the speech of the gender identity is covered. The State of Michigan has since introduced a bill to specifically protect workers from discrimination on the basis of weight and height. And many other municipal bodies, including New York City and San Francisco, prohibit prejudice on the grounds of the general presence of workers. However, with our potential cyborgs, can sexism in appearance extend to cyborgs too?

In a case in point that makes me question if android size evokes bigotry on the part of people, Hooters advised an employee to lose around ten pounds in the immediate future or face potential release. The waitress replied by bringing an application against the restaurant chain for a weight discrimination under Michigan legislation known as the Civil Rights Act of Elliot-Larsen. This law forbids employers from discrimination on the grounds of age, ethnicity, height, or weight, among other factors. "Chubby" androids, as they join society, can be seen to be responded to negatively by the public and I plan to apply androids in a certain manner against appearance prejudice or even regulations to "push" androids. Indeed, Tomotaka Takahashi University in Tokyo has projected that more than half of the potential androids will be women, so gender inequality (which already happens in virtual video games) will be a great possibility in our cyborg future. South Korea and Japan have a particularly significant sex and "pose" for the android built for society with female androids much slimmer than their male counterparts. The sex-related prejudices that burden women in the modern world are regarded as a precursor for the legitimacy of cyborgs. Penn State researchers have found evidence that they can be transformed into simulated ones. In a study of how individuals connect with avatars in an online game, researcher T. Franklin Waddell found that women get less support from other players than males in their service of an attractive avatar on this issue.

Yale Law Professor Robert Post explores the law and the physics of unfair discrimination of work, housing, and facilities, based on height, weight, and physical characteristics, imposed by a municipal code in Santa Cruz, California in 1992 (all things that may refer to androids). The passed legislation concentrated only on parts of the body that go beyond the jurisdiction of an individual (i.e., immutable). However, there is also prejudice in terms of characteristics that are regulated by an individual, such as faith or marital status, tattoos, piercings, and the use of technologies by grinders. The Santa Cruz law evoked an extreme debate about the principles of what was then considered 'anti-lookism.'. I have no question that any law shielding cyborgs, androids, and artificially smart robots from sexism and lookism would provoke a similar uproar. However, the statute inevitably responds to socioeconomic inequities The recantations of ugly legislation contributed, for instance, directly to the American Disability Act (ADA) of 1990 in which a variety of disabled people's rights were granted. It is important to lookism inequality that appearance can be perceived to be impairment in certain situations. For instance, if being deemed obese or as having facial disfigurement influences the person's right to be working, they are deemed to be ADA disabled:

> Disabled people are viewed as a separate, excluded group because of the stigma associated with their condition, the lack of resources available to them, the unjust treatment they've received in the past, and the fact that they've been relegated to a position of political powerlessness on the basis of characteristics that have nothing to do with their actual abilities and that stem from harmful generalizations.

Some people wonder what people have to expect as they address the advent of cyborgs and androids in society (some respond saying an uprising destroying the human race, but I leave this topic to a later chapter and to books such as *Our Final Invention: Artificial Intelligence and the End of the Human Era,* by James Barat). Present cyborgs begin to tackle civil rights with digitally advanced institutions as a response to societal reactions. Many armed with cyborg inventions contend that defense from unfavored and biased responses to their presence should incorporate their fundamental right to fair protection under the law. Indeed, cyborg Neil Harbisson, who holds a head-mounted antenna that lets him hear color, argued that his appearance is not abnormal and commented on the opposite: "some might think we could become less human if we make a transition, but I think it's only human." Moreover, he says,

> in my situation, technology does not make me feel similar to computers or robotics, but more the other way around. Having the antenna makes me feel closer to insects and other animals who have antennas, listening to bone behavior makes me feel closer to dolphins and other aquatic organisms who experience sonority through the bones; I feel closer to insects and mammals who perceive these couleurs through the use of ultraviolet and infrarot sensations. I now feel better than I ever did before with nature.

It is doubtful that an affirmative protection against bigotry against cyborgs and robot would be an appropriate relation to nature. On the contrary, I think robust legislative efforts will be required in response to civil uprisings perpetuating "lookism" discrimination against our potential technological parents and against governments and institutions.

Interestingly, there has been another type of "ugly laws" for some time and it is still in books. These rules refer to freak shows that follow traveling carnivals (sometimes called "Sideshows"). Robert Bogdan wrote a fascinating book in this field, *Freak Show: Exhibiting human oddities for fun and benefit.* In the United States, "freak rules," such as the bearded lady, wolf boy, or fish girl, were introduced in order to deal with institutions trying to benefit from presenting people with uniquely deformed bodies. However, technology can play a part in the understanding of people who think differently. Remember the University of Toronto Professor Steve Mann, who was wearing cyborg technologies for decades "easy on the eye." Richard Crouse, author of the book *100 greatest films You have never watched*, thinks Steve would have made the pitch, approached the proper women cyborg. "Steve was the target of the documentary film of 2001, Cyberman," he says. Did we have for you a freak? This unbiological being, half-man, half-machine, is the world's surprise. In the U.S. the laws introduced by states on Freak Spectrums can be used to provide guidelines as to how the law will interact with cyborgs and androids (which look different from men) as they join society, even if there are no clear rules adopted on cyborgs' presence. Critics claim that freak presentations contain symbolic elements and should thus be covered under the First Amendment. In the coming cyborg age, as cyborgs and androids enter society, a future court will be asked to decide if elements of cyborg technology are modes of speech and thus deserve protection under First Amendment, rather than practical, as required by patent law. However, the First Amendment is currently not the usual legal doctrine, but most generally is lawmaking at the federal, state, and local levels. There is no federal ban on Freak Shows in the United States, usually city legislation regulates such shows, except for maybe a handful of states that legislate in this area. The same ban, is regarded by jurisdictions, some of them allowing "freaks" to be shown for advertising purposes. For example, laws limiting freaks in California and Florida suggest that individuals charged with "freaks" were deemed to be unconstitutional—not because laws were perceived to infringe on the freedom of expression of the First Amendment but because persons with "unusual bodies" had a right to function, and, interestingly, the courts believed that freak shows were one of the most popular shows generate opportunity for employment in a selected cities. In reality, in fact, the Massachusetts General Law forbids all commercial shows with an individual that seems to have an artificially produced deformity, irrespective of whether or not the individuals are adequately paid for involvement. Much like the "ugly" laws of the past, certain laws banning freak shows are meant to protect the viewer from the odd body's exhibition, but in the emerging cyborg era, those that look other than or are different to cultural beauties are not protected. As soon as millions of artificial, sophisticated robots, androids, and cyborgs enter society, they will not all be an appealing human being's counterpart [8].

10.8 MIND UPLOADS AND REPLACEMENT BODIES

Before I look more deeply at the main topic of this chapter and how it can refer to cyborgs and androids, I shall momentarily present the concept of an intelligence upload to a computer avatar, an android, or another physical entity in its entirety, an idea that is obviously a more abstract opportunity but worth exploring in the light of discrimination and rights of ourselves in cyborgs. Although the technology for transferring one's mind to another body is interesting, the evolving technology for transmitting a mind is beyond this book's reach. But instead the short explanation here will be on moral and ethical problems relating to oppression based on physical appearance, that is, the uploading of mind to another body (see, however, Ray Kurzweils *How to Create a Mind* and the published book Russell Blackford and Damien Broderick, *Knowledge Unbound: The future of uploaded and machine minds*).

In the 2040s, "we will be able to access and back up the information in our minds, that constitutes our memories, skill, and personalities." As background information, Ray Kurzweil predicted human-level intelligence by 2029 in the age of spirits. If the concept of transferring our brains to a machine or another organ sounds like science fiction, it is for humans at present, but in fact neural engineering takes major steps towards the simulation of the brain and the advancement of technology to restore or substitute those biological functions. And note I've said "for users"; in truth, we upload a mind every time we load an OS on a computer screen.

In Japan and South Korea, the constitutional protection for robots' policy campaign takes me back to the main theme of this chapter: whether the appearance of cyborgs and artificial smart machines would establish prejudice against people and, if so, how our technical progeny are covered by laws. The constitutions of most countries give their residents basic and democratic rights. In the United States, in the Fourteenth Amendment to the Constitution the concept of fair treatment under law says that "No State shall refuse to provide equal protection to any individual within its jurisdiction." One key word in the equal protection clause is "peer," which obviously does not refer to androids and artificially intelligent robots, whereas existing cyborgs are primarily biological and are assumed to be normal individuals. However, some legal scientists and robotics engineers discuss the issue about gaining personality status from robots. First, it seems unjustified and counter-intuitive to have robotics applied to legal people. However, the position of non-human creatures (such as corporations) had already broadened, and the concept of legal individual grew less concerned with whether a flesh-and-blood person is or is not and with how or what may be punished or triggered.

In New York, for example, a judge has allowed chimpanzees to write habeas corpus, and you may imagine animal rights campaigners moving to protect animals from cruel treatment. It is not humans but types of artificial intelligence that will come into contact with the smartest beings. What are the interests of wiser humans if animals have rights? Currently, it is becoming more dynamic to give rights to our artificially intelligent offspring. But where, even if, an upload of consciousness is made possible, human beings would have to deal with interesting legal and political problems. For example, if it becomes possible to upload a mind into a machine or other humanoid body, this means, inter alia, that one mind will have multiple bodies, enabling an individual to change appearances at will (such as race, age, gender), how are segregation and equal rights regulations in this situation applied under the law?

An interesting thesis that speaks about "mental clones" (i.e. a digital intellectual copy) discusses how mind-clones might be created by a "mindfile," a kind of online archive of our identities that people are now fighting for in the form of social media like Facebook: the hope and peril of digital immortality. The book includes incredible details on how imagination can be created from a mentfile. Rothblatt says its going to be run on "mind ware," a kind of consciousness program. But would a mental clone survive, and if it were to happen would it have rights? Rothblatt thinks so. She describes one concept of existence as a personality disorder code. However, some Rothblatt opponents argue that the mind has to be incorporated into biology, otherwise it can't live and be conscious. In comparison, Rothblatt claims that hardware and software are as good as wet goods or biological materials in the creation of a mind. Indeed, for a mind upload, substitutions will presumably be androids and simulated avatars, not biological, since the legal problems inherent with the body's storage while not in use would almost definitely preclude this activity.

Rothblatt reflects on the effect of making mind-clones; since your entity no longer exists only in a cybernetic organism, the consistency of your self is a challenge. And just as I contend that the right to our artificially smart technologies would take an important national discussion, Rothblatt argues that human rights are one of the main legal problems for mind clones in the 21st century. I know why—it is not rare for gamers to prefer virtual bodies that are very different from their actual bodies in virtual worlds; that seems to mean that living in a different body is not so unthinkable as one might originally think. Furthermore, the capacity to upload a mind in another body must change careers, adopt a new way of life, or start again, as Rothblatt claims. Maybe people would prefer to be athletic, to be more professional, or to dress like another gender—if so, what are the legal and political implications?

With respect to mind cloning rights, Rothblatt says it "would lead second-class citizenship and other kinds of injustice." Naturally, she adds, equitable citizenship would test the fundamental principles of political, criminal, and constitutional law for cyber-conscious creatures. I completely agree with this book and this section of the Law of Rothblatt also explains that mentware is regulated as a medical device—this suggests that the United States FDA legislation would include mental-clone technologies, which may be a good change because the FDA is contemplating cyber protection for networked medical devices (as the protection of the mind from hackers is critical; see the chapter on cognitive liberty). Interestingly enough, Rothblatt believes that, according to principles of constitutional law, mental clones share the biological organism's ethical identity. I must, however, wonder if the legal standing of the original is a legitimacy mind or a mind passed to another body The courts would naturally have to take a decision on legal issues related to the same mind, which are two (or more) bodies (especially if the original is alive).

Ethical and legal questions relating to mental wellbeing to physical individuals or to a simulated avatar in the cloud would obviously be a question that potential courts and decision makers will have to consider. For example, if you can upload your mind to an android, you can occupy a new body whose form will take on an almost infinite range of appearances and physical shapes. So, what will be the legal status of every upload and of every new entity? What if a patient needs a film star or a professional athlete to upload his mind to an android or virtual avatar? Is this allowed under established legislation? In the U.S. and a few other territories, the answer is no without authorization. You must get permission to use the likeness of someone for advertising purposes. Of course, permission is typically gained by paying the right to use the appearance of the user [9].

When we speculate about tomorrow, it isn't too early for us to worry about the strides made in reverse engineering and to "digitize the mind," and the program is patented for the mind material that I discuss in more depth hereafter. And since there are copyrights for works of art, we can effectively substitute a device within the brain where authorship work is called "fixed" by awarding copyright for thought and memory. The notion of authoring functions or some "cyborg unit" within the body being fixed in the mind is a new concept that court officials have to take into account as people get fitted with cyborg technologies, androids come into society, and mental uploads become feasible.

10.9 COPYRIGHT LAW AND APPEARANCE

Of note, some legal theories under intellectual property law may prove useful as a framework in our cyborg future to describe machinery rights. To ask a simple question, is intellectual property a shield for the appearances and bodies of androids and artificially intelligent robots? Under U.S. intellectual property law, protection of copyrights refers to the subject matter comprising "an original authorship work that can be seen, copied, or otherwise conveyed through a tangible form of speech." As we shall soon see, robotic characters in a story have copyright protection. However, what about the real android or robot patent rights; is this possible? Literary books, photographs, and visual works and movies are usually, among other items, protected under copyright laws. Since the "indice of identification" itself is not an original authorship work under the context of U.S. copyright law (although I believe the features are "set" on the basis of our DNA blueprint), DNA is not assigned to this copyright list. The question at hand is whether or not the author has mistakenly identified a natural entity. The human author is deemed to be a creation of nature under copyright law.

In comparison, the "identity signs" of a real being may not only be outside of the scope of copyright protection, but a recent court ruling also notes that the fundamental nature of the corpus of android may be too outside the scope of copyright protection. The concern was whether humans torsos built for clothes modeling became liable for copyright protection in *Carol Barnhart Inc. v Economy Cover Corp. Inc.* The Court ruled that the shape of a human torso is not protected by copyright since the nature of the forms was not conceptual and functional. (the "useful objects," such as human show forms, have a copyright requirement). Section §101 of the U.S. Copyright Act specifies that a "useful article" has the inherent utilitarian feature, not necessarily to represent or express an article but only to include what may be defined and identified irrespective of the utilitarian elements of the item and of which it can live independently. So, if you wish to protect copyright for a specific category of android (which is obviously a valuable article) it is dependent in part on whether its features are covered by copyright. Android functionality can be individually classified apart from the valuable aspects of the architecture of the android. In *Barnhart*, the Court claimed that the show torso shape should not be functionally separated from its functional role, as its features, such as shoulder width and so on, were determined by the need to display utilitarian garments. In general, those parts of a working android body may be patentable but not copyright infringement.

For the copyright, there are many rights to explore, given that a mind can be uploaded to a new entity in the future and that the information can be accessed from a different source. Let's start with a simple example, with robot and android characters in a film or TV show. If you write the script that portrays android or robot characters, it will typically be done for the studio as part of your work, because the studio would own the rights to the characters mentioned in the script. Of course, the writer would be the creator who would hold the copyright to the character if it was not for hire or without contracting rights to the property. If the studio does have the personality rights, however, they are allowed to render derivatives through a third party (i.e., spin-off). Curiously, the person portraying a film or a television role does not maintain the rights to the character he portrays; however he or she does claim a "right to advertisement," no matter how closely they introduce a character to his or her own life. Because in the United States, the right to advertise is state guided and copyright is federally guided, these separate defense schemes may trigger future disputes among the rights holders.

I would like to point out that to natural people the inference of the "identity indication" is not an object of copyright. However, when I address this whole chapter segment, it will probably be decided by courts testing multiple hypotheses under copyright law if an android's presence is a copyright-protected subject. For starters, the evolving characteristics of an android face can be analogized with the changing visual appearance of a video game. When discussing the problem of whether the video game's evolving visual scene is "fixed," courts have ruled that the visual display can be copyrighted because the game's software runs on a computer chip or disk and because visual effects the player sees are repeatable. Additional points to remember for the patent of an android feature should be how the facial appearance of an android is created by the use of software that guides the shafts behind the face of an android (which produces a specific facial appearance). Therefore, the court needs to determine if the facial effects of the android from the directions of the app are protected by copyright. Meet the following "face android," can serve as a test case. In Japan, a shape shifting robot (WD-2) was developed by Atsuo Takanishi of Wasede University in collaboration with NTT Docomo manufacturers who (not surprisingly) are capable of transforming their faces. The robot has an elastic mask made of a mannequin head. Each point has three degrees of freedom, triggering unique face-points on a mask. The mask of WD-2 is made of an extremely elastic material, with pieces of steel wool mixed together in order to improve its strength.

The investigators use a 3D scanner to "clone" a face in order to ascertain the position of seventeen facial points that are necessary to recreate an individual's personal face (or they may create a completely new face). Furthermore, if an image of their face is inserted onto the 3D mask, the robot will show the hairstyle and skin color of the human. A court ruled that the facial expression of android was covered under copyright laws so the owner of the android would not sell illegal copies of the android image, as would the android if it had citizenship recognition to safeguard its privileges (Figure. 10.2). Following this, I suggest that copyright law may include a collection of legal rights for androids, virtual avatars, and robots to exploit the use of its appearance. The concern now posed by courts is what privileges cyborgs and robots have. Robert Freitas from the Institute of Molecular Manufacture commented for example on the precedent set by science fiction authors Ben Bova and Harlan Ellison in defending the copyright of their short story, *Brillo* (about a robotic police officer). In 1985 judge Albert Stevens claimed that robots were granted the same status as humans as characters in stories and thus safeguarded under copyright law,

Body Sketch for Cyborg 249

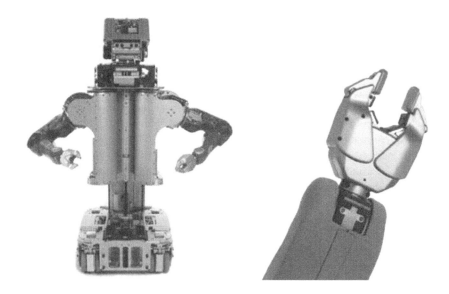

FIGURE 10.2 Robot that can change its facial expressions.

according lawyer Robert Freitas. Freitas considered this to be a highly valuable decision by a judge, because at least in a field of law, it placed robots on an equal footing with humans. Since then, there has been intense dialogue between robotics engineers, philosophers, and lawyers on the issues of the rights of artificially intelligent machines as opposed to humans. As artificial intelligence improves, for example, copyright policy is urgently required to decide if an artificially intelligent computer may be an author for artistic works. My aim in this chapter is to find out whether copyright laws to protect human authorship are also an enticing way to speak about computer rights as we move forward into an era of cyborgs, androids and artificial smart machines [10].

10.10 DERIVATIVE WORKS, ANDROIDS, AND MIND UPLOADS

Under copyright law, a reproduction is built on pre-existing materials that have incorporated enough artistic work to reflect the new work. We should assume a human to be an "original" for conversation and an android to be a "clone" of a specific human. If you are not the original author, the creation of a derivative is copyright infringement without consent. The author of a defamatory statement is not entitled to the original work but may actually be authorized to edit, publish, or share the derivative work of the copyright owner. With those remarks in mind, let's presume that the substance of a mind is protectible in order to discuss future regulation in the next cyborg era (in fact, software is copyright protected). Provided the work of Professor Berger of the University of Southern California on constructing an artificial hippocampus, it is very possible to preserve a person's thoughts and memories on a neuroprosthematic system, thus meeting the copyright obligation to get the work fixed in a tangible medium of speech. If the capacities of the neural associations result, according to neuroscientists, in thoughts and memories, it appears at this granular analytic stage of transition that both thoughts and memories are inherent; but they may not be at higher analytical level as a result of new knowledge and memories being gained. A human or android who remembers the first few lines of the American Constitution (i.e., reminiscences of knowledge held in its mind) does not produce an original work of authorship, as, in fact, it resembles reading a copyrighted page in a noisy manner. Thus, copyright protected subject matter is not the content contained internally on a non-original neuroprosthetic computer. In compliance with copyright law, will the copy of the uploaded mind be treated as a derivative work of the original mind if you upload your mind to an android? The only things that are separate from the upload tend to be those of the imagination. We know that a revised iteration of a software application (where it includes extra features) is a derivative work built on a previous version, based on cases associated with the computing industry. We are also aware that copyright in the U.S. only applies to the original material contributed by the new author, not to the already copyrighted preexisting material. When a mind is submitted to an android in my study of copyright law, this brand-new information, including thoughts and memories, will be referred to simply as "initial" from this point on. What's different when the memory upload is an accurate copy of the original mind? Nothing, legitimately, would simply exercise the freedom to replicate the copyrighted mindfile that had been uploaded to the android body. However, when adding additional mental content, for example, the upload target spoke a different language; the original copyright holder then made a derivative, so only the new content used in the mental upload would be protected by copyright. Would you require a copyright holder's license to upload one mind or any of the features of a famous human into the body of an android? Under the U.S. law of copyright, if a party obtains an author's license to make a derivative work from the original, the person shall not acquire copyright over the original—they merely gain the freedom to make the derivative piece agreed; the owner holds all rights to the original and its pieces, and no original copyright shall be expanded to cover the derivative. This comment is important to our hypothetical cyborg, since it includes the period of copyright rights. In the U.S., the term of copyright for early works is a certain time frame pursuant to laws of copyright. Near to modern times, the novel was protected by copyright for the lifetime of an author after January 1, 1978, plus seventy years. Again, if the original material of a mind is patented and "locked" for this debate, that means that the content of an individual's mind is covered by protection after 70 years. Although if a clone were used as a derivative, how long would the patent rights continue on a mind-clone that was uploaded to an android that could exist forever? Will an individual ever reach the public domain? Compare this with the "freedom to be lost" in which people in many jurisdictions have the right to have private connections from the Internet. I wonder if this is the best outcome for democracy, that is, to encourage people to practice personal knowledge monopolies? With the expansion of copyright strengthening the freedom to be forgotten, can society apply copyright rights to an android so that we can keep it in possession of the mind clone forever? I believe we should think carefully whether this alternative should only be permitted with extreme caution, like some sort of repression by the original or by the clone.

In most cases a mental upload requires that the mind is moved from one body to another. In the event that the consciousness is uploaded to a simulated avatar on the Internet or "living" in the cloud, one entity could be human, electronic, or virtual. The simple conversion from one form to another under U.S. copyright law can be without originality, an integral condition for copyrighted works. Interestingly, what originality is needed for android to be regarded as a derived work for a future court, which consider copyright for an intelligence upload to another body? Will an android be treated as a matter of public policy? For an emerging cyborg rule, there are two significant cases where the use of another medium is necessary for passing the imagination bar for the defense of copyright.

One such case, *Alva Studios Inc. v Winninger* deals with the duplication, in all ways other than the scale and arrangement of the basis of the status, of a Rodin statute similar to the original statute. Here the Court found that, because of the "greed of talent and originality" needed to create the work, the replication of the law was an original (and therefore a derivative). I think that "great knowledge and originality" to create an android is obvious. So I will infer from Alva that it is copyrightable to manufacture an android as a reproduction that is identical to a human (that is, the nonfunctional aspects). In this area, however, I do not consider the law resolved, because jurisdictions have determined separate cases that seemed to me identical in reality. For instance, the court considered Alva to be unacceptable and thus unable to trademark the plastic version of an antique cast iron bank named "Uncle Sam." The court argued, in this case, that the simple conversion of mediums to mediums was a negligible shift. In addition, Carol Barnhart Inc. knows that copyright is valid only on the nonfunctional facets of a "useful article." The relation between these legal holdings and our cyborg future is complicated. If the human body is seen as one medium and the android body as another, an exact android reproduction may entail sufficient originality in the development of a derivative work if the android replication needs substantial expertise, but if great ability is not needed, copyright protection against androids is not required on the basis of Batlin.

Let us also examine the functionality of an app in relation to copyright in this debate regarding android protection. Generally speaking, two eyes, one nose, one mouth, etc. are typical characteristics on a face for a human being. Are the general facial features liable for copyright protection in compliance with copyright law? In the United States, a fair scene is the concept of copyright law in which some aspects of an artistic production are not protected. Fair scenes is the doctrine used when the work is organized or customary. The loose description of fair scenes applies to cases in which a specific concept can basically only be voiced by the use of the general standard domain elements (in a book, for example, the copyright of a peg-legged pirate character or the human torso presented above) is not feasible. If androids are thought to have standard facial characteristics, these features might give humans an advantage over androids in public domain. Again, future courts will have to determine this issue.

10.11 FIRST SALE DOCTRINE

Obviously, this will be a very daunting and exciting copyright problem in our cyborg future, including neuroprosthesis, mental uploads, and memory upgrades. For instance, does the "first selling" doctrine of copyright law apply if a mindclone is uploaded to an android? The principle of "first sale" states that anyone who buys intellectual property lawfully created may "sell or otherwise" work as he deems necessary, subject to a variety of essential requirements and exemptions. In other terms, "first-sale" helps you to sell the memories of another human legitimately leasing a person's register, but not to exercise any of your copyright rights, such as creating a copy of a derivative work.

One compelling consideration for our potential cyborg is that the original selling doctrine applies only to the owner of a purchased copy and not to anyone who has purchased the intellectual property from a software license. As far as a warrant is concerned, an "exclusive" mental file is where only the receiver of the mental file (licensee) has the rights under the license or a non-exclusive license whereby the recipient of the mental file is permitted to exercise rights under the permission but cannot stop. Others exercise the same rights under another license. If a person possesses a mental file (not licensed), he has the freedom to sell it, but the eventual owners cannot duplicate it or make derivative work, or to execute the mental file openly, he can only resell it. Of course, the FDA has a lot to say about the resale of a mindfile, whether Martine Rothblatt is right to believe that mental files would be monitored as a medical instrument. However, I hope that the suppliers of content would still play a role in supplying intellectual artwork, and I believe it may be a profitable industry to market "extraordinary memories." Will the copyright rule still extend to androids that have been uploaded? The public output right contained in copyright? The response will partially rely on how the tribunals identify the android according to copyright law. For example, the U.S. copyright law specifies that a perfectly shaped work cannot only be executed, while an android reiterates or executes a specific work. The commercial property is entitled to regulate whether the work is carried out "publicly" under the public right to exercise. And an exhibition is seen as "important" where the job is done "in a public space or in a place where a significant number of people are clustered outside a typical family group and social associates." Performances are often deemed to be accessible as they are broadcast to different sites such as television and radio. Thus, the freedom to rent a film and to screen this in the public park or theatre without the copyright holder being licensed would be an infringement of the public performance right on a motion picture.

In contrast, showing the film on a TV in a room full of family and friends would not be considered showing it to the "public." Execution even under the Copyright Act will not be forbidden. As the software is considered a literary work under the Copyright Act, the public output privilege is usually held to protect software. Furthermore, certain software applications come under an audio-visual job concept. However, I will point out that our courts have not established the full application of the public output right to apps, so that a public video game is specifically governed by this right. I think that it would be interesting and important to our cyborg future how courts are to extend public copyright output protection to android tablets. The first physical goods theory is largely heterosexual, but it is more difficult to apply to digital products, particularly a mindfile. During the era in which most works in copyright were manufactured in tangible forms, the first selling clause found it impossible to replicate those works specifically on a wide scale. A brain is clearly observable, but the mind is not a computer replica. Once a digital reproduction of the mind is possible, it can be replicated precisely if people lobby for strict first sales rights to discourage the resale of their memories (although I expect that a market that sells fascinating memories and "remarkable encounters" is worthwhile; for instance, certain parents may like their children to have the tangible memory of Stephen Hawking). When more defensive works are digitally manufactured, content providers have lobbied Congress for legislation that weakens the "first-sale" doctrine specifically or implicitly. In order to include the technology that interferes with the doctrine of "first sale," copyright owners create their works. In comparison, tech developers routinely seek to circumvent the first sale doctrine by non-negotiable shrinkwrap or clickwrap deals, by characterizing their contract with the customer as a license instead of a sale. Are people going to license their mind material to others, and are third parties are going to hold a content license in our brains one day? As a graduate student, I'd have loved to have been able to overcome quantum mechanics by Richard Feynman from Cal Tech. It is a technical horizon for humans to debate when we already have the ability to monitor our cyborg fate to license memories and information contained in a neuroprosthetic computer.

10.12 RIGHT OF PUBLICITY FOR ANDROIDS

Back to the law as it can apply to the appearance of androids, who are especially applicable to the Law of Looks and Artificial Limbs, some defined cases of robots as "impermanent," and androids as "impersonators." The advertisement right enables an individual to regulate the use of one's appearances by another party through commercial manipulation. In our cyborg world, advertisement rights can deter people from moving their minds to a popular person or to an android or robotic avatar. Inversely, however, advertising rights can protect androids' ability to control the use of their presence and that right could be exercised by an android. The advertising harm to the integrity of personal identity shall be measured in the right of advertisement cases. The legitimacy of advertising rights will also outlive the individual in some jurisdictions. This poses an important question about our cyborg future: do you still have your privileges to display when you are transferred to an android or virtual avatar like them?

There are two particularly important cases involving the rule of looks and robotic bodies in the field of robot lore. One is *White v Samsung Electronics America, Inc.* where Samsung used a robot looking and behaving like Vanna White of *Wheel of Fortune* fame (to a certain degree). Samsung was sued by Vanna White alleging that Samsung had confiscated her likeness without her consent for commercial use. The Court of Appeal of the Ninth Circuit found that this usage is a breach because the brand and the prominence of White was intentionally used by Samsung as well as the context. Although the android did not look closely like White, the court held that the android and the *Wheel of Fortune* collection "evoked" her. The Ninth Circuit addressed California's rule on ads narrowly and commented that the robot caricatured the characteristics of Vanna White and used the word "likeness." The robot, for example, had a blonde wig and had letters on what looked like a "celebrity Jeopardy." Vana White, an android that resembled Vana in some ways, while also advocating for a respectable right to exposure, brings to mind the fact that Japanese roboticists believe the race to create androids that cannot be differentiated from humans is only 10 years away.

Alex Kozinski, a judge of the Ninth Circuit Court of Appeals, is often cited in talks about "robots' law." After the second case involving robot impersonators was submitted, Kozinski's popular writer was quoted as saying, "Robots again." As obvious as day and night, Judge Kozinski, robots have returned. In *Wendt v Host International, Inc.*, the second event, robot/android, the question was no longer if the androids were the actors' own likeness (as in White) but rather whether the android seemed to be the actors starring in *Cheers*. What are the rights under this case with respect to a regulation of appearances and artificial bodies? Paramount Pictures will claim the actors' right to publish their photographs. The characters "Cliff" and "Norm" (which have looked and performed in certain ways), as well as Paramount Pictures as copyright holder, may permit third-party (host) licensing for commercial use of the *Cheers* properties. The aim of Host International was to create airports bars reminding the travelers of the *Cheers*, with animatronic robots sitting at a bar that looked like the characters "Norm" and "Cliff" and to make observations. The actors George Wendt and John Ratzenburger who played "Norm" and "Cliff" sued Host because their likeness was misappropriated. The *Cheers* case introduced another wrinkle to an evolving cyborg regulation in our interests: *Cheers* was owned by Paramount Pictures and Paramount did not license *Cheers*, but rather a derivative of *Cheers* such as Cliff and other regular characters. As a result, any California right to advertising in the state that clashes with it is exempted by the Federal Copyright Law. Faced with this controversy, curiously, the Ninth Circuit determined that the actor's likeness should be distinguished from the character, which means a person's rights to his "likeness."

Based on the aforementioned question, where are we standing for a proposed cyborg rule based on law on advertising? With the exception of music associations, the courts and politicians of the U.S. are largely unwilling to expand the right to advertising to non-human entities. Furthermore, there's the freedom to advertise, restricted to the "popular," so robots must be equally famous in order to utilize the doctrine effectively, but note Matsuko Deluxe's android portrayal mentioned earlier. In addition, a legal precedent has been established to restrict the freedom to advertise famous people. Before android tablets and robotics joined society and before android design centers in Japan and South Korea, androids were designed that could not be distinguished from human beings. If the artificially intelligent android technology progresses, I see tensions arising between androids who mimic real human beings and others that hold androids' privileges. Moreover, while human beings and cyborgs are normal individuals, there are no types of artificial intelligence and androids and we thus lack a "standing" right to announce their existence. Finally, as the right to advertisement has evolved, the identification signs that can be covered have now been established. Some courts have determined that this category includes visual

Body Sketch for Cyborg 255

characteristics, vocal characteristics, physical characteristics, performance characteristics, identifying artifacts, settings highly associated with certain celebrities, roles or stances strongly connected with particular celebrities, and musical signatures.

10.13 ANDROIDS AND TRADE DRESS LAW

Furthermore, the concept of copyright law makes an important contribution to the protection of computers, which can be true for our technical progeny in the trademark law. Let me use an illustration to demonstrate some facets of labeling laws that could be applicable to androids and artificially smart machines. Imagine a line of androids developed to clean houses and built to represent the organization employing them in a distinctive way. Is it possible to "protect" the varied topography of androids as robotic maids by means of a law? Trademark law provides choices. Trademark law considers if the root of the service (the android service offering company) offered by the android maids will be misunderstood if other androids that are similarly built were also providing a maid service.

The use of a term, emblem, or sentence used to identify such makers or sellers' goods for their differentiation from the products of another usually covers the trademark legislation of most counties. The Nike logo, for example, defines Nike's sneakers and separates them from those of other firms, along with the Nike "swoosh." When services instead of goods are marked with those markings, they are referred to as service marks even though they usually are processed with the same marks. Registered trademark security may refer to other elements of a product, including its colors or its labels, including words, symbols, and sentences in certain circumstances. On these fronts, just as the distinctive form of a Coca-Cola bottle may be exploited to distinguish the brand, so too can a unique component of our android housekeeper. These feature characteristics usually fall under the term "trade apparel" and can be covered if buyers equate this with a single manufacturer rather than with the material. Trade wear will consist of all the different design features of the android that have been used to advertise a good or service for our android servant (however, only nonfunctional aspects of trade dress are protected). The packaging, the screens, and even the product configuration will shape a commercial dress for a product. The setting for a service may involve the distinctive decoration of the Hard Rock I chain of restaurants. In addition, the following must be valid in order to be protected as commercial clothing: the garment must be "intrinsically distinctive," unless it has "secondary value." According to the trademark laws it must "be unusual and rememberable, conceptually different from the product and possibly mainly act as a designator for the origin of the product" in order for trade wear to be considered uniquely distinctive. In a landmark business wear event, the U.S. Supreme Court ruled that Iecor of the Mexican food chain could inherently be viewed as distinctive as the chain used a certain indoor and outdoor range centered on neon-lit frontier stripes, distinctive outdoor umbrellas, and a modern service type buffet, in addition to murals and colored pottery. Furthermore, the secondary importance is that the android servants stand for the business they serve (in the view of the consumer).

Another argument to be made in regard to the protection of android maids' looks is that it is difficult to cover practical features of commercial apparel under the label rule (or as we learned earlier, copyright law). As an example, if this type confers a practical gain, a manufacturer cannot "close up" the use of a specific android shape. For example, after the Seventh Circuit ruled that the design was solely utilitarian, a community that argued for trademark on a round beach table lost its rights. The protectable forms of commercial dress are only styles, formats, or other elements of the product that have been produced specifically to advertise the product or service. Therefore, judges may rule that a tapering android female is not suitable for performing a maid's tasks, and so, may be covered as a commercial costume when paired with other non-functional or strongly suggest this location items. Finally, if it can be shown that the average customer is generally confused as to product type, if a particular product can appear in identical dress, the commercial dress element of the package can be safeguarded. If one group of android maids looks too alike, the second group can also be assumed to have breached the first group's commercial clothing.

10.14 GENDER, ANDROIDS, AND DISCRIMINATION

Turning the gender inequality into social problems, envision a cyborg world with androids as a sex equivalent, subjugated to traditional gender related roles, from intellectual property law to other law and politics. As human nature is so, androids can be used in a wide variety of ways; in reality androids are just starting to discover the spectrum of tasks built to perform. For example, pole dancing robots at a tech conference attracted huge crowds of male participants and what might be a potential foreman, a female android (also termed a ganoid) Asteroid Replete Q2 tells guests that sexual assault impact their breasts. Japanese A-lab robots also omitted the creation of androids that may be used for sex, working with Hiroshi Ishiguro. But a spokesperson for Ishiguro's lab says that, with the advances made in robotics and silicone skin science, potential robots will be used for sex. Takahashi Komiyama, a lab speaker, commented, "In general physical ties with such androids are likely" and "Androids for the sex field may be assured."

Let's ask a simple question, following the previous remarks. Can androids be seen as women across culture, such that for our technological progeny "cyborg sexism" based on gender can exist? I believe so. Sex is the status of being men or women, according to social science, with a term that is usually used in reference not to biological but to social and cultural distinctions. But why not recognize gender as women while debating rights as society assumes androids to be female on the basis of their design? I assume that gender-based inequality will become a serious civil rights problem with our technical innovations in the coming cyborg era. Gender discrimination against women obviously is a culture through which augmented reality and our android prototypes seem expanded. Androids may nevertheless decide that gender discrimination against machines that cannot be differentiated from men sets a weak precedent for human action even though they have no right to defend against gender discrimination. "Robotics in Japan allocate sex to their common sense of the hopes surrounding female and male sex and gender roles," according to Jennifer Robertson. In reality, robots are discussing what gender embodiments in androids should be perpetuated. So, as a human being, as a girl or as a man, how should their bodies be proportioned? Since several robots are constructed as individuals, robotics also model or revert to uniform gender features according to real women or men (recall the rights to advertising). For example, Hiroshi Ishiguro from the University of Osaka screened many young Japanese women's faces to derive a statistically mean composite face.

Gender discrimination in occupations also leads to court action, and because androids get into the workplace, there may be android wage disputes. Although most employers recognize that discrimination against individuals because of their sex is unconstitutional, recent cases challenge whether or not discrimination against current or future workers on the grounds of their presence is appropriate. I assume that cases of sexism against female androids would be present in more court cases. Southwest Airlines has attempted to justify its strategy in *Wilson v Southwest Airlines*, for instance, merely to hire "attractive female flight attendants." It is true that the "sexy picture" is "crucial to the ongoing survival of the airline." In *Wilson*, the court objected and claimed that sexual orientation for flight attendants was not a major prerequisite. If androids improve their intelligence and find knowledge on the web about the gender stereotypes, they will learn to combat sexism against them, if female androids improve their intelligence.

In addition, accessories worn on the body also determine the gender of an individual. Not only does the accessory worn tend to describe the look of an individual, they can contribute to prejudice. As far as accessories and prejudice are concerned, a Federal Appeals Court upheld the policy of the police department that prohibits men from wearing bumpers when they are off duty. In addition, the standards of footwear, clothing, and beauty are typically inappropriate on the grounds of gender norms. But the Ninth Circuit maintained a requirement that women may wear face makeup for their hotel/casino dress code. The vector illustration of a university newspaper, on the other hand, contributed to a surge of critics of the alleged unequal treatment of women. The cover of the 2014 edition of the *Japanese Society of Artificial Intelligence Journal* featured a female robot. The cover depicted a female android, with a book on her right side and a swarm on her left with a rope tied to her. Gender inequality, the resolution of the regulations and rules of female androids, is not a straightforward process since the existing regulation in this region continues to be fragmented for human beings. For example, a recent jury in the U.S. dismissed a lawsuit of sexism involving a woman who argued that her promotion was ignored for looking too sexy—how sexy will an android be?

10.15 OUR CHANGING FACES

The look of a person varies spontaneously based on the age of the person and also through the use of noninvasive methods like make-up. The appearance of people can also be altered due to injury or illness, but voluntary plastic surgery is one of the most dramatic degenerative changes of the human. In fact, cosmetic operations are a form of plastic operation that requires facial reconstruction skin or flesh surgery. Procedures for repairing extreme burns and other forms of patient injuries are a clear example of cosmetic surgery. Plastic procedure, in comparison, is an elective procedure that is often used to change the appearance of the body. In 2014, over fifteen million cosmetic operations were carried out in the U.S. alone to measure how often people change their look—according to figures from the American Society of Plastic Surgeons. And South Korea is not only a leading android production hub but also a leading cosmetic service destination. It is interesting to remember here that it is always difficult for South Korean women who are undergoing cosmetic operation to reenter because their current face is so distinct that they don't resemble their passport images. South Korean hospitals grant "plastic surgery certificates" for patients outside Korean countries to prevent complications on their way home. Some people think plastic android surgeries repair "mechanical elements," but in the future cosmetic android surgery may look similar to cosmetic surgery for humans as a consequence of advancement in the formation of skin-like surfaces to cover the mechanical body of an android.

Of special interest to see from an analysis of patient reactions before and after plastic procedure, it was observed that after surgery patients were perceived to be more posed, more fascinating, more polite, childlike, and colder if "before" and "after" photos were contrasted. However, individuals are well aware because they adhere to social norms through having plastic surgery and through selecting their dress and appearance (through make-up), and citizens are also discriminated towards in society and on the job, as is the observation. These findings raise questions of what would be acceptable or "socially approved" like android operations in society and whether if they looked too different from people whether androids and artificially smart machines would be discriminated against. Although cosmetic surgery is very normal, there is a possibility that not everyone who gets plastic surgery is happy with the results; indeed, many people are seriously injured because of the surgery. Some side effects can include lesions, defects, and death of the skin. And bad effects in cosmetic procedures sometimes lead to a lawsuit; in New York, for example, a jury ordered a woman million in refunds for a failed plastic surgery that left her so warped as to preclude a medical solution of the problem. There will also be risks and unforeseen effects in the promotion of conformity with society's expectations of beauty. In view of a 1936 case in the Connecticut area, the previous findings are interesting. Herman Cohen applied for his name to be changed to Albert Connelly but the court dismissed him saying: "every race has its virtues and defects. The court argued that, if his appeal were approved, the claimant would fly in fake colors." Similarly, can people fly under fake color if they upload their mind to an android? Will future androids and cyborgs update their own appearances and also benefit from a "certificate of legitimacy" for each new edition of hardware and software, or will they also fly with fake colors? In order to make their identity much simpler, perhaps wise public policy would limit androids' appearance to change; if so, then technical advances might encourage our technical innovations to enhance their knowledge management capability but not improvements in their appearances. Imagine a cyborg in a "line-up" that is accused of a crime but that can change its appearance at will; will justice ever be done under this condition?

In view of the fact that androids, cyborgs, and artificially smart machines will benefit from new technologies, are there regulations relating to technologies used to reinforce an individual, and are there sufficient solutions to remedy the unintended consequences of technology penetration into their bodies? If the surgeon is accused of incompetence and cosmetic treatment is done, he may make a charge of medical malpractice. Naturally, a natural person makes a misconduct claim. But no existing cyborg is so technologically equipped that its natural status is called into question, so all current cyborgs can claim medical abuse. However, without androids or artificially intelligent robots, such an argument is not made and any other right is pursued separately under the constitution to protect the dignity of their bodies. Of course, the right to protect human owners and businesses is actually treated as property by androids. A surgeon put a wrong size prosthesis on a human during shoulder reconstructive surgery to demonstrate the allegation of a surgical misconduct involving "cyborg technology." What is the function of a mobile application that mimics a legal malpractice suit? that the android should file such a claim; the android, it appears to me, should be concerned that the modification to its original idea or appearance update has affected or diverged from its capacity to function in society. Of course, without the rank of a human, an android could not start with the operation or would not sue an engineer or a software designer if it could.

When prosthesis machines alter the appearances of humans, the appearance of a cyborg or robot may change as well. What legislation specifically affects prosthesis devices and could fail to work? The product liability statute is directly applicable to prosthetic products; it is the responsibility of prosthesic device suppliers, who are not malfunctioning and behave like a marketed device, with this law. If there is a fault in the product's make, model, or build you violate this obligation. The stakes can be high because faulty prostheses can dysfunction or lead to significant accidents or faults in the case of an android or other artificially smart computer in the future of our cyborgs. The implants of the heart, for example, are "misfires" when exposed and may trigger persistent pain and arthritic symptoms and retract the patient into cardiac arrest, if hip implants prematurely break. Under the FDA, manufactures must recall faulty goods and warn customers against foreseeable harm; does this provision cover android and artificially intelligent innovations in the coming cyborg age? Under the Product Responsibility Rule, producers are solely responsible for losses attributable to failure or faulty protheses, so a cyborg only has to prove harm due to cyborg technology.

10.16 CONCLUDING EXAMPLES OF LOOKISM DISCRIMINATION

As with individuals, I think that the office is one of the principal areas for androids that is highly troublesome with sexism in lookism. Is a beautiful individual or unique look "appropriate" for a job given societal expectations for beauty? It would probably not be considered desirable in human norm, however, to construct an intelligent robot, in the shape of a snake, in order to seek a collapsed building. In general, the courts are not very particular about identifying work conditions, which means that physical attractiveness cannot be readily proven as necessary for most workers. For employees, it might be troublesome as long as an android will do the work.

Numerous reports of prejudice at work indicate that people with cyborg technology may be discriminated against at work depending on their presence, whether or not they are troublesome. The case of Riam Dean, a student from London, has been excluded from the store at the Savile Row branch of the business after management discovered that she wears a predictive limb. After reporting feeling "personally decreased and embarrassed," Dean, who commented that the prosthesis is part of her and "not a cosmetic," sued Abercrombie & Fitch for discrimination against women. However, inequality on the basis of prosthesis does not only happen in the workplace. There are various tales of visually disabled people with digital devices such as those worn by Steve Mann and Neil Harbisson being asked to leave a hospital or being banished from theme parks, and let's not forget that many people with wearable technology have been prohibited to access restaurants and bars that aim to protect the privacy of their consumers. Obviously, the care we get in society will be influenced by how we look, even technologies we wear.

The sexism of looks can be built on a variety of body-wear technology. For example, an employer declined to allow employees to get visible piercings in a federal appellate court in Boston, even though the employee argued the jewelry was worn for religious purposes. Interestingly, if cyborgs belong to a religious community that changes body, they can take the action under the 1st Amendment to discriminate. An example is the reintegration into school of a girl in North Carolina, who wore a nose stud, after it was discovered out she was a member of the Church of Body Modification. A court ruled that in *Rourke v State Correctional Facilities* that the right to freedom of speech of faith was abused by the indigenous American correction officer when he was stopped for not having shaved off his long hair since he had not been allowed to cut his hair. Although discrimination based on attractiveness can be actionable, it must most often be related to sex, color, age, religion, disability, or any other protected class. Some claim however that cyborgs and androids should, from a criminal law point of view, be considered a covered class. I see cyborg technology at this stage, however, in the coming decades, producing more able citizens. I wonder if unenhanced citizens should be given safe class status.

In the United States, disability discrimination law is the most likely legal theory to offer general protection to the victims of discrimination appearance. The Americans with Disabilities Act (ADA), an important state law for disabled persons, prohibits employees who are seeking federal funding from discrimination on the grounds of a physical or mental disorder if the illness seriously limits life function. When the severity of a disability is evaluated in isolation from the ameliorative effects of aids-like prostheses, the results might be misleading. This means that a female with a cybernetic limb replacing her right leg will be considered disabled under the ADA, as the current leg is superior to the first. Without more changes to the ADA to allow for the increasing use of cyborg technology, the ADA, as it is written, would lead to unrealistic effects as we progress into a potential merger with computers (higher cyborg technology corresponds to less impairments). Indeed, in a decade or two, unaffected citizens might be prejudiced against a physically and mentally superior cyborg or robot; how long does it remain for people to suffer from institutionalized racism?

In fact, the ADA does not name all the losses covered. In some cases, some courts have used a liberal concept of "handicap" to place visually unattractive items within the framework of the Act. In order to appear impaired under the ADA, the individual has to present himself two-fold. First, he/she has "physical or emotional disabilities... or is deemed to have some disease" and, second, the condition "limiting significantly one or more main events in life." If the court finds out an individual meets the first criterion, the present language and the U.S. Department of Human Affairs for both the body and the mind, impairment might take the form of a diagnosable medical condition, a disfiguring aesthetic alteration, or a limb or organ loss. Health care will come into question. In addition to this, the law protects individuals with defective scars. As the whole definition of disfigurement is of marred appearance, the ADA declares certain individuals to be impaired due to their physical appearance. Interestingly, "difficulties" in obtaining, maintaining, or progressing jobs are perceived to be a restricting main task in life such that an android with challenges in the labor market might argue that the presence of this is a cause. Under the ADA, a special significance for cyborgs is that the ADA takes into account that inequality can arise from the socially constructed of the physical distinction by identifying a disability not only as a physical state but also as a disabled person. Therefore, people that are deemed to have severely restrictive impairments, even though those impairments cannot be found, may be granted protection under the ADAs definition of disability. This clause would, for example, defend an eligible person with significant facial injury from denial of jobs when an employer believed that clients and co-workers would have unfavorable consequences. Certain features of cyborgs could be covered under this part of the ADA, which will be determined in future courts.

However, there are no examples of impairment within the context of ADA being "clear" or "unattractive" shielding work seekers in those groups. However, it is equally evident that disfigurement is generally held to be a disability under the context of the ADA; for example, because of injury or obesity, and therefore candidates not employed for these purposes could make an argument. If a worker may, of course, assess that beauty as a bone-fide technical qualification, he would be eligible to be hired on the basis of the appearances. Criteria for "appearance," assuming they are not subjective. This is apparently one of the conclusions reached by *Frank v UN Airlines, Inc.*, where the court stated: "a norm in beauty placed on men and women by separate yet fundamentally equivalent burdens is not unfair treatment." The Court quoted a ruling that all the flight attendants of an airline would have to wear contacts instead of glasses. So it is clear that employers will implement look-alike requirements that are not deemed immutable (i.e. cannot be changed), since the presence of workers affects all public and private employers' reputations and successes. The Eighth Circuit of Appeal, however, ruled that tattoos are simply "self-speech" and therefore have no rights as a means of expression to constitutional defense. In this time, I can't think about the "tattooed" androids who clamor for their rights; nevertheless, the urge to change or not to adapt is high, so the actions of future androids in their appearance will probably astound human beings and emphasize the laws of prejudice when it is under their influence.

10.17 CONCLUSION

Although cyborg and android growth is undoubtedly the continuation of the long tradition of human-tool and human-machine interactions, it is also a new partnership both quantitatively and qualitatively. Although the anti-discrimination statute has not yet defined an overarching model of discrimination that stipulates just what is an unconstitutional (not in cyborgs and androids) standard, certain internal and external frontiers for humans are apparent. The U.S. Constitution extends legal protection against discrimination to members of ethnic and religious communities. However, visually unattractive or people with characteristics distinct from society in terms of attractiveness or form do not form an integral category leading prima facie to constitutional immunity, for example, a cyborg with a prosthetic leg. However, we know that for those fitted with prosthetics and other cyborg technology inequality occurs, so effective regulatory steps are needed to tackle differences in care between technologically enhanced ones and non-technologically-enhanced ones. Consider the concept of the "race" that is a collective creation composed of a group of people with identical and distinctive physical characteristics while contemplating defining our technical progeny as a privileged class. It is fascinating that this definition can suit our artificially intelligent progeny and, in the future, can create a safe class. If, therefore, we humans should be seen as a privileged community if their capacities are superior to unenhanced human beings, something remains to be debated.

REFERENCES

[1] Lindberg, K. V. (1996). Prosthetic mnemonics and prophylactic politics: William Gibson among the subjectivity mechanisms. *Boundao.' 2*, 23(2), 47–83.
[2] Mattelart, A. (1994). *Mapping world communication: War, progress, culture.* (S. Emanuel & J. A. Cohen, Trans.). Minneapolis: University of Minnesota Press. (Original work published 1991).
[3] Mattelart, A. (1996). *The invention of communication.* (S. Emanuel, Trans.). Minneapolis: University of Minnesota Press. (Original work published 1994).
[4] Mattelart, A., & Mattelart, M. (1992). *Rethinking media theory: Signposts and new directions.* Minneapolis: University of Minnesota Press.
[5] Mazlish, B. (1993). *The fourth discontinuity: The co-evolution of humans and machines.* New Haven, CT: Yale University Press.
[6] McGee, M. C. (1982). A materialist's conception of rhetoric. In R. E. McKerrow (Ed.), *Explorations in rhetoric: Studies in honor of Douglas Ehninger* (pp. 23–48). Glenview, IL: Scott, Foresman.
[7] McLuhan, M. (1995). *Understanding media: The extensions of man.* Cambridge, MA: MIT Press; Mitchell, W. J. (1995). *City of bits: Space, place, and the infobahn.* Cambridge, MA: MIT Press.
[8] Nietzsche, F. (1974). *The gay science.* (W. Kaufmann, Trans.). New York: Vintage Books. (Original work published 1887).
[9] Osgood, C. (1969). The nature and measurement of meaning. In J. Snider & C. Osgood (Eds.), *The semantic differential technique* (pp. 3–41). Chicago: Aldine.
[10] Pask, K. (1995). Cyborg economies: Desire and labor in the terminator films. In R. Dellamora (Ed.), *Postmodern apocalypse: Theory and cultural practice at the end* (pp. 182–198). Philadelphia: University of Pennsylvania Press.

11 Cyborg Body

11.1 INTRODUCTION

The first candidates to take an IT course had to write programs for a large machine with a newfound appreciation for power and a habit for reading punched cards. A lot has changed since then. Today, wireless iPhone with the strength of the Cray supercomputer are in everyone's possession. More than 100,000 people have electrons in their brains that control tremors and other signs of deteriorating neurological disorder. However, in future prosthesis, implantable chips and brain-computer interfaces will go further than curing disease—as remarkable as such applications of technologies are—or to provide an online search platform for students. Indeed, neuroscientists, artificial intelligence, and robotic researchers also projected that science is going to progress long before the end of this century into the stage where memories in the brain will be implanted; this would lead to a new generation of cyborgs and the rise of highly creative robots. Brain-computer technology, increasingly efficient computers, and advancement in artificial intelligence all contribute to what I consider to be the greatest trend of the 21st century, a world in which we human beings combine with artificially smart machines. And as Ray Kurzweil wrote in *Singularity is Near*, a world just a few centuries away [1].

I'm describing in this chapter the incredible technology used for developing and transforming people's bodies that directly shift society into the cyborg era and the potential for a posthumous future. These developments will also take humanity a few steps closer to technological Singularity, which leads to the arrival and passage of artificial intelligence. Given the variety of technologies mentioned in this chapter, I classify the activities under the heading "body hacking" that alter or strengthen the body. People's attempts to "hack their bodies" may also include strengthening their senses, developing new senses, altering external aspects of their bodies, or disrupting implantable wireless devices used by others under the subject of cryptography, as mentioned later. In my opinion, integrating with machinery does not mean that it is not differentiated from robots or that it lacks all individuality but that technology is becoming more and more merged with human beings, including the brain, and that ultimately produces a cyborg and a posthuman existence for humanity. This book explores how this future can be unfolded. There are several explanations that prostheses, cameras, and other devices will have to be used in the body. Indeed, humans have changed the outward characteristics of their bodies for thousands of years. For example, young males in among some Amazon tribes have historically pierced their lips and wear lipstick when entering the men's house, so it is obvious from experiments in sociology and anthropology that the body is adaptable and subject to alteration. In Western culture film stars and others also use surgical procedures for an aesthetic function, in order to change their body and facial features. Based on empirical evidence from the American Society of Plastic Surgeons, millions of plastic surgeries are completed every year in the U.S. alone and millions more people have reconstructive cosmetic procedures. In addition, a total of a hundred thousand individuals world-wide have cochlear implants and retinal prostheses, and astonishing advancement technology just outside of the horizon is frequently quoted as an ability to change the body or regain body functionalities in an earlier normal state [1].

The increasing number of people who are starting to "develop themselves" using digital technologies is another aspect contributing to a cyborg future to go beyond the human skills of today. With constant technical advancements, these individuals will benefit from the ability to hack their bodies in an unprecedented way. Students interested in physics and economics may, for example, in the future access the topic by uploading the content directly through a digital storage unit in their head using appropriately sophisticated brain-computer interfaces. And health-aware individuals should purchase medical/MD downloads the way they buy dietary supplements and they can import the necessary cognitive ability directly into their brains as they need new skills in the 21st century. (Or upload their thinking skills to the internet). Naturally, major developments in technology and life science would have to be made for all of these scenarios, but this book can, if anything, convince the readers to take us in that direction at least. And as such, it is important to note that artificially intelligent robots are making tremendous strides in terms of "human" senses, cognition, mechanics, and motor skills in the backdrop of people hacking their bodies and becoming more "cyborg" like them. Robot prosthesis in many fields is now approaching human functionality. We tend to be more similar to them (smart robots artificially) and much more like humanity.

Yet vigilance is in order, even with impressive advancements in technology. At a time when science is going to encourage parents to pick their babies' characteristics and people to adopt quicker, cleverer, and more efficient technologies in their bodies, hacking their bodies and their effects is a big point. Author and founder of the University of Singularity, Ramez Naam, is wondering how our minds can be drawn together by electronics. Are we susceptible to glitches, computer viruses, and malware? Program crashes? I discussed this topic in a previous chapter and concluded that it was probably so. In Our *Posthuman Future*, Stanford professor Francis Fukuyama cautioned that the "major danger" of improved technology "is the risk that it will alter human existence and therefore transfer humanity into a 'posthumous' period of history." This could arise, according to Fukuyama, through the accomplishment of genetically modified babies, but it also poses more pathways: such as neuropharmacology trials, which have already started to reshape human actions by means of medications such as Prozac and Ritalin.

I address in this chapter how new media will help change and alter the body as another road leading to a future for the people of Europe. Professor Fukuyama shared the apprehension of those who are urging caution in mounting this result, warning about the possibilities of "us" being something different or sacrificing what he terms "our human nature" In relation to the possibility of humanism entering a posthuman period. Certainly, rather than simply watching machines march and conquer our body, we would want to actively debate the risk of destroying the very features we humanize. And if we work towards a future for humanity, the public will have a lot of problems such as the question of whether or not brain-computer interfaces and neuroprosthesis would unintendedly impact memory and cognizance and, ultimately, free thinking for instance and how we could control the subject of "cognitive independence" as part of the other chapter in this novel. In the case of body reforms and brain-computer interfaces and under the Convention on Human Rights of the United States, how can the courts address problems central to constitutional law such as freedom of expression and the unrestricted exercise of religion? And given the business and economic powers, corporations must know who controls the property rights to the material generated by machines that appear to be conscious and alive and to what degree intelligent firms will be permitted to enter into contracts in artificial terms. And where technology is concerned, criminality proceeds, including cybersecurity problems with wireless implantable systems and future brain-computer interfaces. There are serious ethical, legal, and policy issues that should be debated by the public whilst the future can still be formed.

11.2 HACKING THE BODY

Over the last decade, the fascination in hacking the body for purposes of creativity, self-expression, and senses has contributed to a growing trend, not only to change but also to improve the abilities of the body amongst certain people. I believe that the practice of body change will expand and become a significant part of society as technology continues to evolve; new ways of change in the body are embraced and new advantages are improved. I would expand the idea of hacking from networking to smart solution design, to exploitation and better use of digital technologies in the human body. In this chapter hacking is generally done to comprehend how everything appears to work so that the attacker is able to reconstitute it for its own purpose. In the profession of informatics hacking, however, this has a double sense; it may refer to an expert programmer who generates software or computational methods or who breaks into computer networks for his own use. Regardless of why you have access to another individual's software or network, the compiler of *The New Hacker's Dictionary* Eric Raymond has commented that a "good hack" is the clever solution to a programming issue, and hacking is what happens. The introduction of basic terminology is an excellent starting point when discussing "body hacking." In this sense, human enhancing is defined as moving "over and beyond treatment," rather than restoring a person to a stable or normal condition in a study by the U.S. president's council on bioethics. However, the definition of human improvement is focused more on performance than on "out of therapy" in a European Parliament report. The definition of the enhancement is defined under the Science Technology Options Assessment as any modification that aims to improve individual human performance by means of interventions in the human body based on science or technology. Clearly, different kinds of prosthesis will be important technologies for the future of cyborgs; we can define a prosthesis for discussion as an artificial body part substitution. An implant can also be called a subset of "prostheses," which can contain anything within the body, like an organ or substance that is implanted or grafted onto the body for prosthesis, treatment, diagnosis, or experimental purposes. In fact, "implant ethics" are used to study the ethical aspects of technological devices being introduced into the human body. Last, thinkers have been involved in helping individuals and transforming the body. And they wrote numerous books and articles on the subject.

"To hack the body" is a principle that can cover the spectrum from DIY biologists for whom the objective is sequencing at home to "grinders" that design and install DIY advancements such as magnetic implants. DIY biologists engage in hacking, referred to as a "biohacking," which refers to the use of a hacker ethic in the handling of human biology, i.e., to find physical, emotional, or intellectual tweaks to the body to improve cognitive and sensory performance. Any individuals may also be interested in biohacking by using a mixture of medication to manage their own genetics, technologies of nutrition, and electronics. Biohacking may also involve the use of nootropics and/or cybernetic equipment for biometric data recording. In general, people who participate in body hacking are identified with the mobility for transhumanism—the belief that a fundamental change of the human condition is possible and desirable through the use of technology in order eventually to create a superior posthuman being. Finally, those who associate with grinders as a means of working towards transhumanism are practicing practical installation of cyber equipment in their organic bodies. The idea of bolstering or altering the body with implants and other technological forms is not a new one, but an interesting issue arises in the 21st century—to what extent will people increase or change their bodies in response to advances in protheses and digital technology? This article aims to deal with this question through the provision of numerous examples of recent implantable devices, but the efforts of people to modify the structure of their own bodies can provide a partial response. Body implants have actually been used for some time to change the shape and appearance of particular areas of your body, particularly face, chest, and biceps. In this instance, the implants approved by the Federal Drug Administration (FDA) in the USA are made of solid soft rubber silicone, which aligns with the bones without being consumed by the body. Because body implants are regarded as permanent, surgery is necessary for removal. Besides body implants, a few other individuals choose, by liposuction and fat translation, to sculpt or add volume or outline some parts of the body. There are also people who have changed their bodies in extraordinary situations—like someone who uses tattoo designs and multiple surgeries to look like a cat, with implanted whiskers, a converted cat nose, filed teeth like cat teeth, and a flattened head that appears to be more feline. In the spirit of "cat man," another example of self-modification is the "lizard man," and even the Church of Body Modification allegedly seeks to reinforce the bond among "mind, body and soul." Alongside these examples of self-directed body changes is the 21st century the use of engineering science and information technologies allowing the use of sophisticated prostheses to modify and enhance their bodies, to stretch their sensations far beyond the boundaries of human behavior, and to allow electrodes and/or chips embedded in their brain to form a comparable patient-machine interaction for people suffering from neurological disabilities [2].

11.3 THE RISKS OF BODY HACKING AND CYBORG TECHNOLOGY

While many people wish to change their bodies, the operations aren't always effective, and corporal alteration is at risk, even for those who alter themselves, and often the risk is fatal. In this respect, there are allegations in the press that women throughout the United States are risking their lives for black market operations conducted in hotel rooms by people without any medical training. Regardless of why you chose to improve the body, you are searching for inexpensive solutions to plastic intervention—even with deadly or faulty effects. Disastrously, in many states in the United States, deaths from black-market silicone injections were reported on criminal charges against the individual conducting the prosecutions. In one incident, the injector was threatened with a very serious "depraved heart murder," which reveals "crazy negligence of human life" and leads to death. Conviction can be punishable by imprisonment for life. In spite of the scarcity of numbers, there is anecdotal evidence of relatively frequent illegal practices.

The keep foundation stone of market in night as well as in day. For starters, for years the French company Poly Implant Prostheses (PIP), once the world's third-largest provider of breast implants, used non-medical industrial quality silicone for its items. As a result, many breast implants became vulnerable to breakage, resulting in harmful silicone leakages in female bodies. And if an implant fails, a significant number of patients are usually affected. In this case, 100,000 people and 300,000 women worldwide have been involved in the breast implant fraud case; thousands of women now demand restitution for damages caused by implants, usually restricted by French laws to specific losses and loss of opportunity. However, in order to compensate the plaintiffs for their emotional suffering or pain, the French court may also levy a general damage not associated with the individual loss, known as "religious damages." The business manager was sentenced to four years imprisonment for theft, an aspect with separate disciplinary provisions of the Criminal Code of French law. If you wish to use mass consumer technologies for hacking the body, you're better off getting this.

In consideration of potential disfigurement and other threats involved with body hacking, I would recommend that a national discussion about the suitability of transforming the body take place before body hacking becomes a celebrity trend. For example, a temporary tattoo to mark an occasion, frequently in protest, is common among young people. Temporary tattoos usually last three to six weeks, dependent on the coloring substance and the skin tone. Unlike the injection of permanent tattoos into the skin and digital tattoos as a sensor (described later), temporary "henna" tattoos are used on the surface of the skin. These tattoos appear innocent at first glance; but, Linda Katz, head of the Cosmetics and Colors Office of the FDA, says "only because a tattoo is temporary it is not threat." In fact, some tattoo recipients have documented extreme reactions that may be the result of temporary tattoos. Technologies implantable under the skin, of course, and also in the brain will give a human enormous power, but they are far riskier for recipients, and the safety of our prospective cyborgs must be highly protected.

The European Commission recommended that new law on medical devices be revised to respond to faulty implants. In Europe, the term "medical device" actually encompasses a broad variety of devices used by patients and clinicians both internally and externally. All of these can include touch and fertility testing, dental filling supplies, and "cyborg procedures" such as pacemakers and hip replacements. Likewise, in the United States, due to the difficult application of medical equipment and possible danger in the product, the FDA governs with regulatory severity. For starters, a thermometer may have very minimal guidelines, but a pacemaker is regulated rather heavily. In Europe (and even in the USA), medical equipment is graded as class I, which contains low-risk products and class III high-risk items, such as hip substitutes and pacemakers, which function on the inside of the body. The Commission aims to enhance the product assessment process, improve product traceability in its proposal for controlling implants, and put more control over notified bodies where a problem with medical devices has arisen.

When the cyborg world advances, there will be particular protection and health problems with implants and other kinds of "cyborg hardware." For example, as we look at the future health issues involved with prosthesis and implants, pathologists and other professionals are worried about the protection problems associated with the implants and components in the human body. These problems can be seen as obstacles to be solved if mankind is to combine with machinery. For example, while inert materials or "biocompatibles" are commonly seen, a body of research indicates that certain implantable metals, plastics, gels, and rubber material combinations may have chronic, potentially dangerous effects on human tissues in some individuals. Implant recurrent inflammation, poisoning, blood clots, bone erosion, connective tissue disorder, and in extreme cases cancer, depending on the concentration and location in the body, are likely. And with implants in the brain the biocompatibility of implanted electrodes and chips with brain implants is especially important in the design of devices. The formation of granulation tissue for people being treated with Parkinson's disease at the site of implantable electrodes is already a problem.

While going towards the future of the cyborg, some experts consider the changing data on the protection of implants, from software to hardware, a cautionary sign that people prepare for implants, especially those that may be used for cosmetic purposes early in life. A gauntlet of current "cyborg technologies" appear to be the possible problematic devices; including artificial hips, knees, elbows, hands, fingers, breast implants, heart valves, pacemakers, shunts, intrauterine appliances, dental implants, and a number of other items that meet medical or cosmetics requirements. These are some of the body's unique implant reactions. The body is normally built to combat foreign bodies. The substance is inserted under the skin and is contained in muscle tissue in a protein-rich bath. Proteins tend automatically to bind to the embedded system surface and then they are wrapped in a protein mixture. Based on the form of substance used in the implant, energetic particles and magnetic fields seen between surface and the proteins can be involved in physical interactions. The communication is strong enough to change protein form, thereby exposing the proteins to signaling pathways, attracting other proteins in circulation. The body manipulation and cyborg innovation threats were planned to detect issues. One collection of circulating proteins initiates blood coagulations and surrounds the implant with dense skin tissue layers known as fibrin. Some people have persistent and sporadic low-grade fevers, but at least some people appear to be extremely resistant to implants but often flare up years later. For several years after treatment, implants can also become contaminated by bacteria. To a certain degree, antibiotics, pain-killers, and anti-inflammatory medications should be used for complications but, of course, any implant treatment should undergo cost-benefit analysis. Another way to acclimate the body to implantable devices is by using antibiotics, blood thinner devices, and other agents, but they dissolve gradually and reduce their duration and efficacy.

Some businesses create new biomaterials for embedded devices that permanently barricade problematic microbes from the surface of the product to respond to the organic reaction to implants. A thicket of polymers that attract water and create an impenetrable shield for bacteria sprouts a substance that is attached to an implant system. Its chemical structure is also imitated by cells critical for homeostasis, which can minimize the physical discharge of embedded devices. The approach is basically intended to make the implantable machines feel like the human body. However, considering the advancement made in biomaterials, the audience should bear in mind the possible health concerns associated with implants, given grinders' attempts to transform their bodies without the aid of a practitioner, by reading more in this segment on body hacking.

In the light of the latest FDA-approved sensor (radio frequency recognition, RFID sensor) that is inserted on the body due to defense, art, and corpus hacking, it is useful in looking at our potential cyborgs. Whilst the FDA has fair confidence of a secure RFID implanted sensor, neither VeriChip Corp. nor the regulators publicly address a series of veterinary and toxicology experiments from the mid-1990s that revealed that chip implants had "induced" cancerous growths in a variety of laboratory mice and rats. Many scientists suggested that family members would not be able to receive RFID implants and many encouraged further study before glass-encased transponders were commonly implanted in humans. These signals mean that thousands of RFID equipment in humans worldwide are already implanted. VeriChip Corp., which views its biomedical security chips as a target market of 45 million Americans, maintains that computers are secured. However, it acknowledged those risks when the FDA approved the device: the capsules can travel around the body and make them hard to remove, interact with defibrillators, or be incompatible with MRI scans, resulting in burns.

If we equate an RFID chip to that other implantable product, a cardiac bacterial pacemaker, we understand that the RFID device is not necessary in order to keep anyone alive like the heart pacemaker. The reaction for a group of people is simply "yes." Actually, RFID chips for people with Alzheimer's or other dementia patients have been approved; the premise is if they get misplaced, it's simpler to reunite the chips with their caregivers; the advantages of the embedded tracker outweigh the costs in this regard. But the basic principle of an implantable device that would help a community of individuals, while other people cannot, poses interesting questions of law and politics. Not least because cyborgs should be treated by courts as a covered class and thus liable under statute for special protection. This could include the requisite access to technical upgrades and hardware upgrades of the next decade and potentially broad security under a federal law providing discrimination grounds [3].

11.4 PROSTHESIS, IMPLANTS, AND LAW

Many citizens who believe in the future of cyborgs are also dedicated to sufficient laws and statues of the government to protect technologically enhanced individuals. Progress in neuroscience and robotics will in the future transform the way society regards the human body, strengthening the idea of the body as an integrated, easy to replace, and upgradable system. As these cyborg innovations grow more sophisticated, the possibilities for human capacities expand well beyond the present basic level, causing society to regard stable but still not improved human beings as handicapped, who will reach and then exceed normal human function. Therefore, disabled people, fitted with cyborg technologies, could prove more "fit" in the future and average capabilities may be approximately identical to deficiencies that need removal. This segment provides the reader a glimpse of what I think forms part of an emerging area of cybernetic law, considering these probabilities.

The Equitable Employment Opportunities Commission (EEOC) has taken on several cases of prejudice against people armed with cyborg technologies in employment in the U.S. Most of these examples were from a woman who had a prosthetic leg and her boss feared she would "fall over" at work because of her handicap. The case, won by the female, was decided under the American Disability Act (ADA), which forbids inequality in jobs against persons with disabilities. The Court found it unlawful to dismiss an individual who had a disability out of an unreasonable suspicion that they could hurt themselves or others. Another job dispute was brought up under the 1973 Rehab Act, which had a "cyborg rule" concerning an individual with a hand prosthesis. During this dispute, the FBI academy refused a veteran who had lost his hand and replaced it with a prosthetic, because he was not able to safely shoot his with prosthesis during his preparation. A tribunal decided however, that FBI instructors were abusive to the veteran and ruled in his favor, and the Court granted him punitive penalties, compensation, and reintegration into the FBI Academy. The statue in this "cyborg discrimination lawsuit" is for federal employers and federal agencies, which also does not include discrimination against prostheses in other circumstances; it appears to be a regulatory area.

In 1999, the United States Supreme Court heard another case concerning the consequences of cyborg legislation and considered whether the ADA still considered an individual with a handicap correction disabled. It is an important example regarding "cyborg rules" since it also tries to regain or go beyond natural human ability to be prepared with technique (i.e. to become a cyborg). Twin girls with severe visual myopia were interested. When they qualified for a position as commercial pilots at United Airlines, they satisfied the work criteria except for the visual condition, which was 20/100 or higher visual acuity. With glasses and contact lenses, each sister was able to correct her myopia of vision by 20/20 and could usually work on a regular basis. But in their ADA claim, the Suttons argued that they were legally incapacitated because of a physical condition that "substantially restricted . . . significant lives" or was otherwise determined to be an impairment of that sort. The Court determined whether the ADA should assess the disability without regard to corrective actions that alleviated the degradation. In other words, can a disabled person always be deemed disabled but returned to "normal" by technology? In regard to disciplinary action, the Court found that a disability had to be assessed. The Court then concluded that the disability would not greatly limit a "big life operation" until an impairment had been corrected. On the grounds of the court opinion, if cyborg technology took the person to full function (or above normal) a person would not be deemed damaged. But the ruling of a court can be reversed by the lawmakers and the debate can proceed.

The legislation on reducing technical disabilities has a lot to do with cyborg inequality and societal recognition and poses significant concerns as to who can be treated as handicapped while technology is used. On this issue, Congress approved in 2008 the American with Handicaps Amendments Act that specifies, without referring to the enhanced benefits of mitigating interventions, such as prosthetic limbs, cochlear implants, or an implantable earphone, whether the disability dramatically restricts a vital life operation. The amendment exposed the Congressional thinking; that no amendments or changes are applicable for the legislation to decide whether or not a person is disabled; thus, for example, the very act of having an upper arm prosthetic does not automatically describe someone as a person with a disability as part of the extension. To decide whether an impairment greatly limits a main life operation, the value of the mitigating actions such as medicine, prosthetics, mobilizers, hearing aids, and cochlear implants can be taken without hesitation—to mention only a few aspects in which a disability may be mitigated. For example, persons with one leg may have a prosthesis, so they will walk better if they wear their prosthetic leg, but they may have trouble walking without a prosthetic leg. This person has an illness under the ADA, since it is calculated without taking into account the prosthesis that he or she is significantly limited in main life operation. However, the law on interventions relating to the prevention of vision does not apply to persons with contact lenses when deciding if a person has an impairment. A myopic woman whose visual acuity is completely corrected when she wears glasses, for example, is not greatly restricted in seeing, since it is determined when it wears the glasses. This is a judgment in public policy—only think of how many individuals would be perceived as being handicapped under the ADA if the positive benefits of regular glasses or contact lenses were not taken into consideration. In an age of cyborgs, though, the amended ADA poses some philosophical questions. For example, with the amended ADA, the limb would fall into the prosthetic category (leaflets and apertures) if a woman wished to replace her right leg with a much superior cybernetic leg, and, as the law forbids the consideration of such mitigating causes, this woman would have legal disabilities beyond the fact that her current leg is stronger than her older leg. And paradoxically, with the exception of an individual who upgrades a limb with superior cybernetic prosthesis on a specific job site, the unenhanced "normal" employee is the only non-disabled person, even though superior abilities are enjoyed by all colleagues. In fact, the more prosthesis-related modifications a person gets, the more disabled the updated ADA will be. And nothing in the ADA prevents people from discrimination where upgraded persons can discriminate against a "disabled" person because of their lack of enhancements. When further technologies become available and human beings with higher ability become available, the legislation will have to change the way it conceptualizes damaged people with cyborgs and augmented cyborgs living among them [4].

For an emerging area of cyborg law, there are other law and policy concerns. Public regulations, for example, include the availability of medical equipment suppliers, including implants, for life-saving medical treatment. The U.S. Congress introduced the 1998 Biomaterials Access Guarantee Act (BAAA) for manufacturers of implant materials. Both implant basic materials and parts, with the exception of silicone gel and silicone framework used in a breast implant, are considered by the BAAA. The BAAA effectively insulates producers of virtual surgical implants from their components and raw materials. Any duty mandated by law to ensure availability of essential emergency supplies. However, BAAA does not apply whether the retailer makes or sells the device as well as part or components. BAAA does not satisfy the statutory specifications. But raw material suppliers and medical equipment providers may use the BAAA to exclude themselves from the responsibility, as well as from the personal injury lawsuits in which they are designated as charged.

The negligent physicians and their implantable liability goods are not covered by the BAAA if the beneficiary is injured by the equipment. Under the proposed laws, prospective cyborgs will have a variety of legal options if they experience injury. What if the unit, inserted by a medical officer, fails to initiate the conversation, and the individual is injured? If the damage can be attributed to the conduct of the doctor, the implant-appointed individual is entitled to prosecute for assault. In the United States, psychiatric violence derives from common law English. The wounded party must claim that the doctor has been incompetent before implementing the device in order to assess the evidence for a medical malpractice. Four legal aspects are to be demonstrated in particular: a professional obligation due to the individual receiving the implantation; a violation of this duty; injuries incurred by the infringement; and penalties that arise. Provided that there are a range of implantable devices and forms of prosthesis, in particular with hip and knee replacement, which are amongst the most frequent procedures undertaken in the U.S., medical malpractice prosecutions are not rare in these areas. New implant devices and revision surgeries to fix complications from the first treatment, as well as the aging of the getting implants baby boom generation, have contributed to the gradual but steady improvement of these procedures.

An implant can be dangerous in other ways than by a doctor who performs a specific operation. For example, a cyborg could sue if the implant fails. If the injury to the implant receiver happens, any individual in the manufacturing and marketing chain of a faulty implant will be sued. In this situation it will not only be the implant maker but also the device component parts suppliers, the wholesaler, and the supplier who are responsible. If a cyborg files a lawsuit, the goal of securing compensation for defective parts will be met. Management failures and product liability claims, a significant distinction is made when medical negligence depends on whether the conduct of the practitioner was appropriate (measured in accordance with the health care standard). Generally, if a product maker or distributor has an intrinsic flaw that is exaggeratedly harmful, and it causes harm to a conceivable consumer of the product, it is liable under the Product Liability Statute. I believe that a manufacturer cannot reliably forecast how a grinder for an implant will be used, but that a patient may get a grinder with the help of a doctor. It is for a medical state. Sensors (the backbone of the cyborg revolution) are useful regardless of whether the FDA is used as a diagnostic tool or not.

According to tortious laws, there are three kinds of liability: a development error, a publicity fault, or a construction fault. During the processing process, a production flaw occurs. In general, a marketing flaw involves flaws with the guidance or advertisement of the product, such as failing to notify the consumer of concealed hazards in an implant system. Moreover, there is a design flaw where the product, because of its design, is clearly unsafe and faulty. For instance, a prosthetic leg that cannot support the recipient's weight. For starters, a few years ago, 93,000 DePuy hips were discovered to have design flaws and resulting recalls, even more recalls for other implantable technologies were found. Design flaws are not rare with cybernetic technologies, for example.

What if, rather, a tattoo artist or a "body-shop" individual doing the implant isn't a professional doctor? A negligence conduct that is not the duty of a reasonable prudent individual under the same circumstances is probable. Failure is similar to medical incompetence, which also requires tasks, violations, reasons, and penalties. The basic distinction between a normal negligence suit and a miscreation suit lies in the concept of the level of treatment that prevails. When anyone sues for ordinary negligence, they compare the conduct of the claimant with that of fair person laws requirements. When they sue for negligence, they equate the actions of the doctor with what a rational person of the profession might have done. Technical expectations are considerably higher and much better documented; negligence in a skilled capacity is therefore usually easier to create. In an era of autonomous body changes, I will ask, what is the concept of "fair" when the person who does the implant is a friend or someone who works in a tattoo room?

For hacking the body, a sensor, magnet, or some such other type of technology sometimes requires the implantation of a piercing t under the surface of the skin; astoundingly, a specialist is not mandatory for the operation in some jurisdictions. The U.K. corporal piercing industry is unregulated and demands that the studio only be licensed with its local council's Environmental Protection Department. In comparison to tattooing, there are not minimum piercee age standards in the United Kingdom, while there are in the U.S. Furthermore, there are no guidelines regarding the instruction of carbs, and no restrictions that apply to instructors of carbohydrates, in the United Kingdom. In America, however, several states' legislatures have called attention to the culture of corporate modification. The 2013 bill entitled "An Act to Restrict Bodyspecific Controller Practices" was introduced by a state senator in Arkansas for example to minimize bodily improvements only to "traditional" tattoos and piercings. The initiative of the state senator would effectively prohibit scarification operations and dermal injections and any residual tattoos due to the ambiguous wording of the supported measure. Scarification is a non-tint skin marker that shapes decorative scars, while dermal implants involve putting decorative items under the skin. In my opinion, the bill introduced is unconstitutional in the light of the First Amendment to the U.S. Constitution, which explicitly forbids government attempts to "bridge the freedom of speech."

Moreover, changing the body is a part of a collective ritual for certain people and should thus be a fundamental human right. Not that far away from home, a secondary school girl from North Carolina was expelled because her nose-piercing defied the dress code in classrooms. The student believed that the fracturing of her nose was a part of her religious faith, owing to her membership in the Church of Body Modification. Though her school uniform code bans facial piercings, the federal judge ruled that the student could go back to school and pierce anything. The American Civil Liberties Union branch in North Carolina, which represented the student, argued that the founding of the school was a preservation of the freedom of the family to create their own religious practice. According to the resolution, as long as she remains a member of the Church of Body Modification, a religious community which boasts a few thousand practitioners and considers the practice of tattooing and perforating the body to be an aspect of spiritual practice, the student is approved for the purposes of the resolution. However, the rule is far from an educated guess, since a different result was created in another case based on a moral exception by anyone who altered his body. Kimberly Cloutier was an employee of Costco when she alleged that she was not given a "fair accommodation" for her headgear as a result of her personal convictions endorsed by the Church of Body Modification. While Kimberly got a copy of the Costco job contract, she opted to disregard the requirements of the dress code and also took on many styles of improvements to the body including corpuscles and skin slicing. She consequently ordered that the U.S. Court of Appeals, having terminated her appeal on failure to conform to the dress code, put an unreasonable burden on Costco to allow a cashier to wear facial gems because of her legitimate concern in having a fairly professional appearance due to her failure to comply with the dress code. Cloutier, who refused to combine her unprotected body parts together, justified her choice by citing the religious significance of facial piercings, which had been supported by the law because of their association with the church of the Blessed Lord. Given a new outcome, the public and lawmakers must discuss why additional laws must be introduced in the two aforementioned cases in order to meet the needs of those who transform their bodies, argue for rights, and pose as cyborgs in the future. The two cases deal with religious practice [5].

11.5 BODY HACKING IN THE DIGITAL AGE

In general, the term body hacking refers to an activity that involves manipulation of the body and computer hacking as part of an age of cyborgs. This dichotomy of body and machine shows that law and government policy concerns must be presented to each aspect of the cyborg. The so-called brains of the implantable devices will contain, for example, regulations relating to software (i.e. contracts, licenses, miscarriages). In certain cases, it will be the same rule for both, but I believe the mix of humans and robots should be protected by new rules and policies. The body hacking motion has gained popularity from Professor Kevin Warwick's pioneering work, beginning in 1998 at Reading University, particularly with regard to implantable sensor systems in the body. He engaged in a variety of proof-of-concept studies involving a sensor inserted into the median nerve of his connection brain, a technique that allowed him to link his nervous system directly to a computer. Professor Warwick was one of the first people to hack his body. In specific, using the neural interface, Professor Warwick could control the electrical wheelchair and the artificial hand. As well as providing calculation of the nerve fibers in his left arm, the implant could also generate an artificial stimulus by using individual electrodes to activate the nerves in

Cyborg Body

his arm. The use of a second, less complicated implant attached to the nervous system was shown by Kevin's wife. Kevin said that this was the first electro-communication between two human nervous systems. Since that time, many of Kevin's works have been improved with RFID chips and other implantable sensors. Given Kevin's operation for the implantation of a sensor in his body, Kevin and his wife apparently willingly inserted the sensor, an act anyone might consider to be potentially risky. There should also be an implant-related danger bar based on the assumption of risk theory regarding whether any damage attributed to the surgeon occurred before or after the implant. This legal theory holds that a person who intentionally exposes himself to dangerous physical damage cannot hold anyone accountable when the harm happens. In comparison, a person accepting a treatment recognizing the damage is a predictable, if unusual, person under the presumption of risk doctrine. He waives the privilege to potential accusation of neglect in view of any "foreseeable" accident if the operation was adequately executed. However, although the implant operation was done by the surgeon, the surgical malpractice will also be charged. Moreover, a court may investigate this condition using the secondary presumption of danger doctrine, depending on the jurisdiction. In California, for example, if an experimental implant was done by a surgeon and Kevin was found to be responsible for his treatments, since Kevin volunteered for the operation (not warranted by medical necessity), a comparative blame system could be used when harm occurs and the judge may weigh the respective roles of the parties in separating arising damages [6].

11.6 SENSORS AND IMPLANTABLE DEVICES

Humans live in a period of immense progress in developing sensors and implantable technologies for regulating and tracking the various functions of the body (Figure. 11.1). MIT researchers are designing, for instance, an implantable tracker, using nitric oxide (NO) carbon nanotubes for animal tracking.

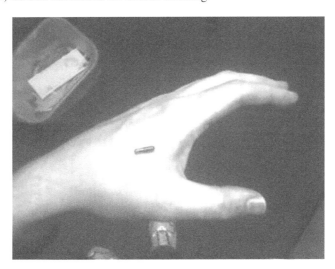

FIGURE 11.1 An RFID sensor implanted in the hand.

The microchip includes stored data that can be transferred to and from a reader. Active RFIDs have an intrinsic source of power such as a battery, so that the tag will return signals to the reader.

The sensor can be useful in humans for cancer cell identification and for glycose control. There is also a project on brain-reading tools at Boston University to translate thoughts into language, beginning with vowels. Implanted electrodes are used in the system to capture nerve impulses from activity of the mouth, eyes, and jaw; these signals are then wirelessly transmitted to the device at which point speech pattern recognition software analyzes them.

In addition, Brown University and cyberkinetics researchers in Massachusetts are developing a microchip inserted under the skull of a human in the engine cortex to collect nerve impulses and re-route them on a screen that will then relay command to electronic equipment, including phones, stereos, and electric wheelchairs, wirelessly. And consider a top team that has developed a microvibration system and a low-frequency wireless receiver that can be inserted into the tooth of a human. The amplifier and speaker is the vibrator; trying to send a person's eardrum sound waves along the jawbone. But another illustration of an implantable device is Setpoint, which develops computer therapies that stimulate the vagus nerve with an implantable pulse generator to relieve systemic inflammation. In order to diminish inflammations and enhance health signs and symptoms, this system stimulates the normal inflammatory reflex in the body. To date, the organization has been developing an implanted neuromodulation to treat rheumatoid arthritis, a disorder that affects more than two million people in the United States.

Sensors have been inserted into the human body since Warwick's pioneering findings, for many reasons, such as personal protection or to monitor the health of a human. For example, some people have inserted a small transmitter under their skin because of the possibility of kidnapping, to allow satellites to detect and track their locations if appropriate. From a separate security point of view, courts can compel individuals who are found guilty of a crime to take part in a wearable sensor's electronic surveillance program as an alternative to jail. There are two kinds of electronic bracelets available: the house arrest Radio Frequency Bracelet and the GPS Bracelet to map the whereabouts of an inmate in the actual environments. The use of the GPS tracker poses significant legal and political concerns as in most wearable devices. One such problem arose when a GPS wristband was found to be able to listen to interactions between a lawyer and a customer that breached the privilege of the lawyer and the customer. I may say that this is a violation of both the Establishment Clause and the U.S. Federal Wiretapping Act, which forbids an unconstitutional search and seized activity (a federal law that is aimed at protecting privacy in communications with other persons).

The previous examples illustrate how critical activities on the body's surface or internally can be accomplished by wearable and implantable technologies. They are basically technological tools for supporting people. More than anything else in this book, technology is what makes us stronger, and the more of it we use, the closer we come to becoming cyborgs and laying the groundwork for a possible merging of artificially intelligent robots. I assume also that a huge amount of knowledge we learn about integration of sensors within the body is good information for engineers who build artificially intelligent machines in the next century, as they too require sensors to experience the environment.

It is easy to understand the number of sensors being built and their ability to gather data on the body's internal condition. Conclude that an extensive data mining process is under way in the human body. In reality, Google uses this in a program to figure out how a safe individual is expected to look. The experiment, called the baseline analysis, includes researchers gathering, originally from 175 individuals and later thousands more, anonymous genetic and molecular information to aid in the diagnosis of diseases like cancer and cardiac diseases even faster. The baseline will not be confined to such pathogens but will use state-of-the-art analytical instruments to gather hundreds of various samples plugged into and correlated with computer systems. In order to gather the data, members, to track the level of glucose, for example, wear Google smart contact lenses. Once data is gathered, Google uses the computational resources to recognize trends or "biomarkers" for medical science to identify a disease at a curable level.

Interestingly, when implants gather knowledge about our bodies, our bodies become the equivalents of open books such as Google scans, posing significant questions about privacy. The problems, presume, are giving prospective firms selling and supplying facilities to enable brain implants exclusive access to private information housed in the human cervix. In compliance with Title II of the 2008 Genetic Material Non-Participation Act, data acquired by ability to enter the brain require special protection (GINA). This act forbids discrimination in jobs in the fields of genetic knowledge and is enacted by the Fair Employment Practices Committee (FEPC). Genetic knowledge is also used to assess if someone is at an elevated risk of becoming sick in the future. Likewise, access to knowledge may be used to assess if a person is vulnerable to a felony, the potential for abuse, or a mental disorder. The ability to gather and interpret information about the body and brain using computers to forecast the future poses troubling questions of privacy and policy and the chance of a dystopian future. A previous chapter discussed in some detail the law and regulation of brain science and cognitive independence.

For example, diabetes is another explanation for using cyborg technologies for weakening disease. Millions of diabetic people around the world should potentially use implantable sensors and wearable computers for tracking their blood sugar levels since people who are not monitored are at risk of eye injury, and injuries to the heart and kidneys. A spin-off organization from Cambridge University, Google, and others create "eye-carried" sensors to benefit patients with the condition and help people control their blood sugar levels. The technology from Google consists of a special sensor-built contact lens, which monitors sugar concentration in tears with a small wireless chip and a miniature sensor inserted into the two soft contact material layers. This approach for diabetes tracking is not the only fascinating and creative example of "eye-oriented" cyborg technology among hackers; in fact, eye hacking is a topic of body changes. We will see cyborgs who are fitted in the future with a contact lens or retinal prosthesis, which monitors your health, quantifies energy in the spectrum of X-rays and infrared, and has images (see later, eye hacking). I can point out that any contact lens system is governed by the FDA, and cybernetic technology is regulated by the government [7].

11.7 ISSUES OF SOFTWARE

In the operation of implants, the program becomes more and more important. The code law and algorithms could also be of concern to cyborgs and to the programmers. Take an artificial pancreas to mimic the working of the natural pancreas using an intelligent dosing algorithm by constantly adapting the insulin supply based on changes in glucose levels. What if the artificial pancreatic program fails? In case of malfunction of software, the "chain of responsibility" may include the software producer, equipment builder, program distributor, programmer, contractor, and the businesses using the software, various parties who may be liable to legal proceedings, and operators with software. Computer creators use disclaimers (by means of a software license) for their goods in order to shield themselves against liability lawsuits, which can restrict the claims of consumers.

In brief, however, if a competent programmer fails to behave as a fairly cautious programmer, he may be careless when writing code. The divergence from standard programming activities is often illustrated by another specialist programmer's proof. Similar to the abuse of rights cases, negligent programming predictions are based on obligation, violation of duty, cause, and harm. In order for the cyborg to prevail over an argument that a programming malpractice is unethical it will have to prove that a careless programmer has the duty or moral responsibility to use fair caution in delivering computer programming or services. By refusing to provide programming or development services that an appropriate programmer would provide, a careless programmer violates that responsibility. Besides program malfunction arguments, provided that the design and development of software are typically protected by contract law, many contracts contain protections and conditions that shield firms from computer software, programming, and networks that either do not run or are unreliable.

Many attorneys have sought to make an argument for technical bugs and crashes in defense of their client. Yet this hypothesis continues to be dismissed by the judiciary, even if that action does not happen. An example of this is the early case of *Chatlos Systems v National Cash Register Corp.* (1979). An NCR dealer here performed a thorough review and recommended Chatlos to purchase NCR equipment. Chatlos, on the basis of NCR guidance, acquired a device supposedly never to have executed a variety of promised functions; Chatlos sued and NCR was found responsible for contravention. In a footnote, however, Chatlos' argument on computer malpractices was discussed by the court:

> the innovative idea of the new wrongdoing termed 'computer malpractice' is based on a philosophy of expanded liability by the sales and operation of computers. The plaintiff is compared with proven technical mispractice ideas in the selling and operation of computer systems. Simply because the corporate sector finds an operation scientifically challenging and significant does not mean that more possible liability has to be introduced. The courts reject the offer to create a new injury in the absence of sound precedent jurisdiction.

Failure to file an application for machine malpractice could nevertheless enforce contract law and neglect or perhaps an acceptable legal standing as a protection to the interests of cyborgs suing for faulty software. In the emerging age of cyborgs, they will be embroiled in several conflicts that will pass to the courts. In relation to a horse's or in this case cyborg rule, would it be adequate, or would a whole system of rights for cyborgs and, in the future, for artificially smart robots, be safeguarded and be based on the basic values of contractual, penal, and constitutional principles? By mid-century, the solution would be simple [8].

11.8 MACHINES HACKING MACHINES

In my opinion, the need to integrate humans into artificially intelligent machinery relies on another accelerating technological trend—attempts by computational scientists and engineers to build computers that will become or at least program the architect of their own designs. When computers start hacking their own hardware and apps, they will steer a creation that is so rapid that we humans can be overcome. This potential could undoubtedly offer a potent impetus for mankind to envisage a self-direct convergence of our future technological progeny beyond biological evolution. This is attributed to the recent instances in which computers tend to build, fix, and program themselves to steer their own evolution. For instance, a robot from the International Space Station fixed its cameras when it was orbiting to become the first robot to self-repair. Then we'll explore the concept of self-assembling robots by MIT scientist Daniela Rus and Erna Viterbi. These robots are made up of printable robotic components that immediately dynamically assemble in three-dimensional configurations when heated. An example of their research is a system that takes a digital 3D design, such as the one produced by a 2D pattern, and allows the replication by self-folding of a plastic object. Daniela and her colleagues are further studying the design of electrical components using laser-cut, self-folding materials. These prototypes involve resistors, inducers, and condensers as well as sensors and actuators, which are electromechanical "muscles" for the operation of robots. If artificial smart machines become the masters of their own architectures, can there be any doubt that they can easily improve and go beyond their master's capability by using technologies like 3D printers? Certain people believe that artificial intelligence technologies with adequate machine intelligence would allow software to develop its hardware and software components fully autonomously. Once these changes have been made it is easier to find ways to strengthen its structure and further improve its ability. Such an artificial intelligence over several implementations is speculated to greatly outweigh human cognitive ability and contribute to individuality. Machine learning, an artificial intelligence division concerned with developing and researching systems that learn from mining data, is a form of investigative guidance on this subject. I foresee a world in which cyborgs and artificially smart robots gather data, exchange information, and make decisions together. In general, artificial intelligence, along with advancement in computing science, electronics, memory, and CPU speeds, has advanced gradually over the years. Accelerating processors allows for more calculations per unit time, reducing the time and energy required to perform AI algorithms and data mining methods.

11.9 HACKING THE BRAIN

While several of the examples given in these chapters reflect the attempts of people to hack their bodies and a discussion of legal problems that are included in those actions, the future could be much more impressive in terms of the alteration of the individual. In the end, considering that the brain is operated by computing and that considerable progress has been made in decoding the way the brain measures, I think that the main mechanisms of how knowledge was learned by the brain and processed were to be gleaned, and by so doing information important to the future cyborg and to the human-machine fusion was gained. However, as there is complicated neuro-chemistry and neuro-circuitry (draining) of 100 billion neurons composing the human brain, as the brain uses distributed computation, and as there are several distinct mechanisms, uncovering the secrets of the brain is very challenging (order of magnitude more difficult than the Human Genome project). Unfortunately, because of the brain's complexities, because of the wireless network and brain machine interfaces and essential technologies that are emerging in the world of neurotransparency, I expect that the human brain will be threatened by a vast range of criminal attacks in the cyborg future.

If we take into consideration the raw computing capacity of a quantum computer and equate it to the brain (capable of executing trillions of point operations every second), the calculations bring the processing power of a supercomputer within that of a brain, which ensures that, for many variations of Moor's rule, the processing power would not be the end of producing human intellect. In ten to fifteen years, biological neural machine technology will be the main barrier to developing an artificial intelligence of comparable abilities as the human brain. It varies from silicon in basic respects. For example, the human brain is significantly parallel, which comprises billions of neurons that can synapse independently with thousands of other neurons. Supercomputers, by comparison, have immense computing power as computed by the number of arithmetic operations per second but generally have minimal parallel connections; however, working on this topic alone is targeted at making computers calculate in parallel. Many developments in science may also be due to advancement in unlocking the mysteries of the brain. The revolution in contemporary neuroscience was caused by the intensive use of MRI technologies beginning in the nineties and fMRI technology later on, resulting in optogenetic techniques in the last few years, as theoretical physicist Michio Kaku addressed recently in *The Future of the Mind, the scientific Challenge to Grasp, Improve and Motivate the Mind.*

Advances in optogenetics by Stanford University researchers represent an important technology for researching the workings of the brain. The explanation is that optogenetics helps researchers to investigate the interplay and effect of various neuronal circuits. For the potential human-machine merger, the importance of this technology is that if we want to create machines that can interact with the brain directly, we need its passwords. One way of doing this is to pick a number of neurons of interest, calculate how they fire, reverse engineer their response, and construct the corresponding algorithm(s) (simplification!). Scientists have traditionally understood that proteins called opsins produce strength when exposed to light in bacteria and algaes. This process for brain science makes it easy to take advantage of optogenetics.

Opsin genes are incorporated into the dangerous virus' DNA and later injected into a research subject's brain. Through picking a virus that chooses certain kinds of cells over others or by modifying the genetic sequence of the virus, researchers may attack particular neurons or regions of the brain that are considered to be responsible for those behaviors. In order to research the neuronal function, a thin glass cable is threaded through the skin or skull, which transfers light from its point. The lights in the fiber trigger the opsin, which in turn drives the neuron to fire an electric charge.

The fusion of the human brain with computers and sensors is still being advanced throughout the brain's immense complexity. For example, investigators of the Chicago Institute of Rehabilitation have created a bionic leg regulated by the use of upper leg neurosignals to control the knee and ankle prosthesis. The prosthesis uses on-board computer pattern recognition software to decode electric signals from the upper leg and mechanical signals from the bionic leg. When a prosthesis-fitted person considers lifting his leg, brain impulses are activated by the idea, which travels down his backbone and finally to electrodes in his bionic leg by peripheral nerves. In addition, hackers start joining the fray. Take body hacker and inventor from India, Shiva Nathan. Shiva was encouraged to support a family member who lost both arms under the elbow and he designed a thinking-driven robotic arm. The technology uses a Mindwave Handheld headset for reading EEG waves and Bluetooth in order to transmit those kinds of ideas to the arm that converts them into gestures of the finger and wrist. This is an exceptional 15-year-old accomplishment in technologies that can be used by anyone.

Indeed, prosthesis research is genuinely international. Researchers in Sweden, for example, create a think-through prosthesis of the amputee in the shape of an implantable robotic arm at Chalmers University of Technology. More than any other product on the market, the FDA has in the USA approved a thought-controlled prosthetic extremity that is more functional and human. Dean Kamen invented the DEKA arm prosthesis, which detects as many as ten movements, is equivalent to that of the normal human arm, and works by detecting electrical activity induced by muscle contraction near the site of the prosthesis. The initially thought-induced electric signals are sent to a PC processor in the DEKA arm that activates a certain prosthesis action. In FDA experiments, people with household activities, such as using keys and locks, food storage, self-feeding, hair blowing, and zipping, were successful in their artificial arm-hold.

The previous explanations make me think, since a prosthesis of the arm or leg seems to be as functional as a natural limb for humans but more potent and dexterous that natural limbs, with more freedom of mobility in joints and more articulation for "average" persons, will they prefer a higher prosthetic, whether or not there is a minimum surgical risk? But I have my doubts. The answer is unclear. Take the following data into account. In 2011, in the U.S. there were 307,000 breast augmentations, which relieve no medical problems and have no sensors or artificial intelligence and are merely aesthetic, according to the American Society of Plate Superiors.

And if you have optical compass in your brain, who needs sight to get around? A neuroprosthesis that feeds geomagnetic signals in blind rats' brains has allowed them to navigate around a labyrinth. The results show that the rats will quickly pick up on a completely foreign meaning. And it increases the likelihood of people doing the same thing, possibly opening up new ways of curing blindness or even giving good people extra senses. "I dream of humans extending their senses through artificial geomagnet sensors, ultraviolet, radio waves, ultrasound waves and so on," says Yuji Ikegava of Tokyo University, Japan. "Allow next-generation human-to-human contact through ultrasonic and radio- wave sensors," he said. A geomagnetic compass—a clone of the microchip on smartphone—consists of a neuroprosthesis and two electrodes that match in the visual cortices of the animals, the brain areas that process visual information.

Returning to the findings of Michio Kaku about the future of the mind, he disclosed other interesting experiments using scanners for interpreting images preserved in the human brain and artificial memory for treating victims of strokes and Alzheimer's. Kaku also mentions telepathy and telekinesis; fake memories that are embedded in our brains; and a pill that will allow us to become wiser as the future inventions of the century. Extend Kaku's comments on the future, so that the anatomy and physiology of the brain could be substituted with 3D-printed pieces. A Netherlands surgeon performed the first effective substitution of a human skull with a 3D artificial skull. The skull repair procedure was performed on a student in Utrecht University Medical Center who had serious headaches due to her skull thickening. As a result, her vision was increasingly lost, her muscle control had been impaired, and other vital brain functions may have atrophied without surgical interdiction. The 3D implant was done, brain burden was decreased, and sensory and motor abilities were restored. I regard the use of 3D printing materials for humans and androids as a groundbreaking technique, which will expand over the next decades and play an important role in the process of our cyborg and eventually human-machine fusion [9].

11.10 HACKING MEMORY

Will we replace or boost portions of the brain's human brains by digital technologies if we can substitute the anatomical structure that supports the brain? Not today, definitely, but how about in the future, or in any decades to come? Join the University of Southern California biomedical engineer Theodore Berger and his team who have built a test-free hippocampus with rats. They are researching. The artificial hippocampus of neuroscientists is a replacement for the section of the brain that encodes memories of long-term recollections. Although Berger and his team want to tackle deteriorating neurological illness, For instance, an artificial hippocampus that would also be a big move forward towards the potential cyborg-machine mergers and epilepsy and other disabilities that affect the hippocampus (which would prohibit a human from maintaining new memories).

Berger developed mathematical neuronal activation models in a rat's hippocampus in order to construct an artificial hippocampus and built a chip (located outside the brain) for the imitation of the signal processing that took place in various sections of this hippocampus. Interestingly, by sending random pulses to the hippocampus, neuronal activation models were recorded in different locations to see how they were transformed and equations that represented the signals were then extracted. The chip containing the algorithms was also linked by Berger and colleagues with the rat brain via electron. Berger did not bring "into the brain human memories," but he was able to create memories in the brain with the implant.

The researchers conducted an analysis to see whether the chip could prothesize a weakened hippocampal area to see whether it might circumvent a core portion of the hippocampal pathways. In a basic memory-task qualified rat, Berger's team checked the system. The two levers were put in every rat (with the prosthesis). First the lever was presented on just one side of the chamber and the rat pushed it after a brief waiting time; the levers emerged on both sides of the chamber, and the rats would be rewarded with a drink of water if the rat pressed an opposite lever from the previous one. However, the rat needed to recall which height it moved originally to accomplish this mission successfully.

Berger and his team administered some of the rats with the medication that altered their natural memory to assess if the memory prosthesis was functioning as intended and then tested the animals in their lever experiment. The rats (with the prosthesis) were able to press the necessary lever to drink. They had the ability to form fresh memories, and they recalled the rats' brain implant. Noticeably, the researchers observed, even without the medication that had damaged their memory, that prosthesis would boost the memory of rats.

Robert Hampson and his colleagues at Wake Forest University School of Medicine recently demonstrated a hippocampal prosthesis on nonhuman primates as the phylogenetic size expanded. While the system is far from the fully implantable "chip" of the hippocampus, these trials from rats to monkeys show that the "evidence principle" is successful as a neural prosthethetic for an artificial hippocampus. Robert Hampson is preparing to launch human research. Although Hampson and Berger's work is a long way away from a brain hard disk, it's a step towards a "backup" or hack, and as soon as possible the next step will be a big step towards a cyborg future, moving the knowledge to artificial hippocampuses. Once an artificial hippocampus is inserted into the brain, a question explored in the chapter on cognitive liberty might, of necessity, be selected by third parties.

11.11 IMPLANTING FALSE MEMORIES

There are tremendous legal and regulatory ramifications of being able to control memory: will people even change memories because people can influence them? Could recollections of an individual be decoded towards and used in the tribunal as testimony, and could people remove recollections and substitute them for fresh ones? And can modifying your memory for artificially intelligent machines just change the lines of your code? Is this ethical; will a potential regulation to guard against this be required? When an artificially smart machine uses computer vision and algorithms to interpret the world, would modifying the program be like controlling a witness or digital lobotomy performance? And if you look at the possibility of enhancing your brain, certain people dealing with cognitive development ethics point to the risk of having two groups of people—some who have access to improved technology and those who would be bound to age-fading, static memory.

The idea of implanting synthetic memories in the brain may first appear fantastical, but this has already been accomplished, at least in the case of the laboratory mouse, which is often used as a test subject by scientists. Susumu Tonegawa, professor of biology and neuroscience at MIT, uses genetically modified mice, and Dr. Xu Liu has been using the optogenetics methods to access mouse brain neurons with his colleagues at the Riken-MIT Center for Neural Circuit Genetics. In essence, they inserted a fiber optic cable into the mouse to enable them to reactivate previously documented neuronal circuits. According to Dr. Liu, any time our memory improves, we will insert new knowledge into older perceptions and so a false memory can develop. Interestingly, the MIT team may falsely connect the mouse with an adverse encounter from a variety of surroundings by the implantation of a falsified memory. How they did this was to first respond to light by means of a network of neurons implanted in the mouse brain of a certain area. The scientists then put the mouse in a harmless red room. The next day, the mouse explored a blue wall chamber and at the same time they gave it a slight jolt and led the neuronal recollection of the red room. This was achieved in such a way that the mouse correlated the shock-free red room memory with fear of shock. The researchers needed to see on the third day whether this mistaken correlation had worked. They put the mouse in the red room but there was nothing horrible that happened to it. On the basis of the response from the terror, the MIT team hypothesized that there was a false recollection. Reporting to BBC News, cognitive researcher Neil Burgess from UNCL London said that the studio was a "impressive example" of producing an appalling response in a world where nothing frightening has existed. While it will not be feasible in the next few years to use an implant to prepare a memory in a human brain, a human memory should, in general, be separated and triggered, because challenging technological difficulties are solved. Psychologists have a well-known impact on law and regulation, the concept of mistaken memories. DNA evidence ultimately overturned the ruling, but it was proven to have been used against witnesses and suspects in several cases. In a step towards understanding the way these defective memories are made, this chapter has shown that false memories can be modelled and inserted in the brains of a mouse at the neuronal circuit stage. However, it is important to remember that neuroscientists have often discovered that all of these memory fragments are of the same type as the authentic memories. "Whether it is a fake or true recollection, the neuronal structure behind the memory alert in brain is the same," says MIT's Susumu Tonegawa.

It is not a new problem to concern ourselves with the potential implantation of fake memories in the brain only due to the emergence of brain-computer interfaces and a cybernetic future. For example, a Wisconsin jury gave $1 million to a couple alleging their daughter had experienced faulty childhood misconduct at the hands of clinicians. Indeed, many individuals have been paid millions in settlements in counterfeit therapist trials. Yet Dr. Charles Johnson and his wife Karen were the first to have neglectful therapy charges by the parents of the patient—over the patient's objections. The case was pursued for years until the Supreme Court in Wisconsin held that the treatment history of the daughter did not provide the therapeutic-patient right in 2005. The Johnsons used these documents to justify their argument that the clinicians used the notorious method of "recovered memories." What was to be done? The court found that there was a public interest exception to the clinical patient right and privilege of patient health reports where negligent treatment leads to parents unfairly accused with sexually or physically abusing their child. Will the brain-computer interfaces require a comparable privilege?

In a more general context, I cannot help but see the connection between the information and technologies of memory implantation, fake memories of the psychiatrist,

Cyborg Body

and an appeal heard by the European Union's Court of Justice that people have a solemn right to commit errors and then delete them, namely the right to be forgotten. The definition is drawn from an individual's ability to "determine his life's progress in an independent fashion, without continuously or occasionally being stigmatized as a result of a particular action taken previously." It is not only impossible to forget the past if the memories can be inscribed, but it is easy to stigmatize an individual by acts that have ever existed. There are fears among legal experts that the existence of a right to be forgotten contributes to censorship and a potential rewrite of history.

11.12 HACKING THE SKIN

Will we hack the skin, the biggest sensory organ, if we can hack the brain? The response is unexpectedly yes, but a digression into contemporary music first. A new survey by Pew reveals that almost 40% of Americans under forty have at least one tattoo, producing an industry of 1.65 billion dollars. However, the tattoo industry needs, like any industry, to develop and attract new customers. The transition to new technologies in an analog world is one way of innovating in Figure 11.2.

Instead of passive tattoos, digital tattoos work; they do stuff and they become clever. In these days the promise of digital tattoos is not just art or self-expression but also digital gadgets, even though they are laudable targets, as useful as the smartphone to track our health. It is not the far future; the infrastructure now exists to manufacture digital tattoos. For example, in a tattoo that reacts to electromagnetic fields, it is possible to employ a form of ink that provides several possibilities. Nokia has in fact patented just this invention, ferromagnetic ink, which can interfere with

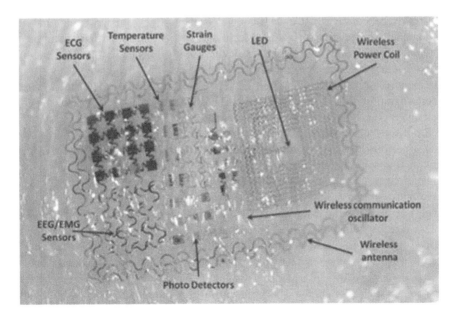

FIGURE 11.2 Electronic sensor tattoos.

a device. The basic principle is first to enrich tattoo ink with remodeled metallic compounds until the ink is implanted in a person's flesh (by exposing the metal to high temperatures).

Unlike passive tattoos, digital tattoos work; they do stuff and they become clever. Rogers is now working on how to connect electronics with other gadgets such as smartphones so that applications can be started. The idea for a digital tattoo containing numerous sensors and gauges, including strain gauges for the tracking of strains (as the individual is flexing), ECG, EMG (electrical impulses in the skeletal or nerve structure), ECG (cardiac activities), and temperature, has been patented by the Advanced Technology and Ventures Division of the company. Google is not too late in its production of digital tattoos.

Like other implantable technologies, it is more likely that the skin will be hacked for health purposes. By 2016, 100 million accessible wearable medical devices will potentially be used by individuals. For instance, the previously mentioned digital tattoo of Roger tracks health of the individual and measures healing close to the skin surface. Then a wearable tracker tells you when it's time to have a beer. Researcher Ronen Polsky from Sandia National Laboratories has created a prototype of a fluid's microneedle chip, capable of detecting and analyzing electrolytes in fluids around skin cells. The instrument consists of a series of micro-adhesives on the underside of the device, which protrudes through the skin of a human to test interstitial fluid levels, water within the cells of a human. Anyone who goes below a certain limit will be alerted to this number. Introducing sensors on the skin is an innovative concept with tremendous potential to track an individual's health, but sensors on the skin will also be able to perform other tasks, such as detecting environmentally interesting details and linking a person with billions of things that will in the future be linked wirelessly.

However, the skin is not only hacked in order to track our wellbeing, and artists uniquely merge skin surface and technology. Consider Moon Ribas, a modern Catalan choreographer and cofounder of the Cyborg Foundation, an international group that advocates "cyborgism" as a social and cultural phenomenon. It's also an international organization. Through an interesting application of technology and cinematography, she linked to her elbow a seismic sensor that enables her to sense earthquakes due to vibrations caused by earthquakes.

Professor Stelarc developed an amazing ore on the skin of his left arm in a more intense performing arts case, by a series of operations. The theory of Stelarc for body hacking is to apply technologies to improve the physical capacities of the body, where a human can do something they have been unable to do previously because of physical disabilities—that is achieved by adding a microchip and an artificial ear microphone. To develop the artificial ear, an embedded expander of skin in the forearm produced excess skin. A kidney shaped silicone implant extended the skin by injecting the saline solution into a subcutaneous port, producing a surplus skin pocket which was used to surgically create the ear. In a second operation, the skin was sucked in and a Medpor scaffold was inserted. The implant of Medpor was stitched together and formed in a variety of parts to shape the ear. A miniature microphone was mounted inside the ear during the second treatment. But the microphone implanted had to be removed due to bacteria, though testing was successful. If the hacking is done,

Cyborg Body

Stelarc will be able to hear what his artificial ear "hears" from afar using the necessary equipment. In overview Stelarc's theory of body hacking, he remarks that, as technology proliferates and micro-materializes, it is bio compliant in size and in substance, and it can therefore be absorbed as a part of the body.

11.13 HACKING THE EYES

Technology is the most important thing for many people, and the absence of real life is so terrible that a great deal of work is being poured into improving technologies for the eyes of those who have lost it. In the particular information, too, body hackers are involved. Meet Neil Harbisson, born under an unusual condition (achromatopsia, black and white) that helps him to see only in shades of gray (Figure. 11.3). Following a talk about cybernetics in the spirit of a hacker in 2003, Neil wondered if he would transform color to the sound, on the grounds that the sound wave might be equal to a particular frequency of light. When Neil first learned of the theory, he did not realize that in 2014 experiments would ultimately reveal that auditory input detected by the ears is processed by the visual cortex. Neil had a sound conduction chip inside his head and a flexible shaft with a robotic camera on it connected directly to his skull, in order to become a cyborg (the Eyeborg). Neil claims that he is able to hear ultraviolet, infrarot, and telephone calls and has a Bluetooth connection that lets him reconnect Eyeborg with the internet with the new tech update. Of course, the addition to his passport photo of the Eyeborg prompted everybody to dub Neil the first formally recognized cyborg by a sovereign state. Neil had to persuade a medical team to perform the surgery as a prelude to potential bioethical conversations on cyborg technology: that is, he had to inject the chip into his brain. Since medicine was intended to get the body back to its natural form, Neil must have convinced the doctors that his system could enable those who missed it to regain function and not only because he desired to have a sixth sense by perceiving things in his visual field with the sound of the bone. The second ethical dilemma comes up here. If Neil is to have an implant that

FIGURE 11.3 Neil Harbison hearing color.

helps him to sense outside of the normal range of human seeing and human hearing, again, it is not usually a justification to get an implant (hearing from a bone allows a person to detect a broader range of sounds from infrasounds to ultrasounds). But since, for example, the same electronic eye implant can be used to interpret or sense obstacles by a person who converts words into sound or lengths into sound, in principle, it may be used to restore functions to those who lose them.

The surgical team was reassured by this reason that the implant would restore function and the operation was performed. This example poses, however, a significant policy question: which forms of cybernetic implants are ethical and legal for society, and which are not? Let the conversation proceed.

Amazing vision disability technology produces a brand-new cyborg population, some of whom never felt they would ever see again. Different kinds of eye prosthesis in research laboratories are actually beginning to appear. In America, for example, for for those who have the symptoms of retinal pigmentosa, the FDA has approved a retinal implant. The implant does not fully reinstate vision but is intended for people who will need correction of degenerative conditions of the eye to partly regain vision.

Photoreceptors (rods and cones) of the retina turn the light of a healthy eye into brief electrical and chemical signals that are sent as pictures by the optic nerve and into the brain. If, due to factors like retinal pigmentation, photoreceptors are no longer functioning properly, the first stage in this phase is interrupted and the visual system is unable to turn light into photos. Enter the 2011 European Argus II Retinal Prosthesis System (Argus II) and 2013 U.S. system. This prosthesis is supposed to totally circumvent the weakened photoreceptors. A film monitor, which is located inside the patient's lenses, takes on a scene with the prosthesis and transfers the scene video to a small device worn by patients to be interpreted and translated into commands that can be returned through the cable to the lenses. These commands are then wirelessly sent to the antenna in the implant and delivered to an electrode array in the retina. The tiny electric pulses are engineered to circumvent weakened photoreceivers and to activate the remainder of the retinal cells that relay the sensory information to the brain through the optic nerve. The aim of this method is to establish an understanding of light patterns to be perceived by patients as visual patterns.

Can we boost vision if we can start to restore vision? How about the telephoto vision concept (Figure. 11.4)? For the 20–25 million or so people worldwide with advanced age associated macular degeneration (AMD), a condition affecting the primary, comprehensive vision area of the retina, it provided a comparatively recent product, mainly an implantable telescope, that offers the greatest cause of irreversible vision loss and legal blindness in people over 65.

In 2010 the FDA approved the IMT, which functions like a camera's telephoto lens. The IMT technology minimizes the effect of the middle-vision blind spots because of the AMD process and projects the patient's artifacts into the stable region of the non-degenerated, light-sensing retina. The Implantable Miniature Telescope (IMT) device eliminates the symptoms of an AMD end-stage central vision spot and projects items the patient examines onto the stable region of the non-degenerative, light detecting retina.

The operation includes extracting the normal lens of the eye and replacing it with the IMT, as with the cataract surgery. Behind the iris, the vivid muscular ring around

Cyborg Body

FIGURE 11.4 Implantable telescopic eye. Picture provided courtesy of VisionCare Ophthalmic Developments.

the pupil, is an embedded small telescope. Although the telephoto eyes are not entering an auditory department soon, this is an encouraging step forward. The legislation is, as well, part of that equation as the techniques for equipping a person with a rear camera eye are subject to tort law in the event of injury, to the regulations on product liability if an implant fails and the FDA controls implant technology as a medical product.

11.14 HACKING THE BODY WITH SENSORS

Anthony Antonellis, a body hacker, has added onto early work by Professor Warwick involving an embedded sensor in his skin and has an RFID chip inside his hand that can be accessed through a smartphone wirelessly, which allows Antonellis to reach and display an animate GIF on the phone stored on the implant while the chip only carries around 1–2 kB of data. Antonellis can swap 1 KB of files at will because the RFID chip can pass and receive data. Antonellis sees the implant as a "net art tattoo," which is typically used for quick response codes (QR, or barcode matrix). Rather than the QR code, the RFID chip would make quick change of art through increase in the storage capacity of the chip; further it will be of interest for a body-hacking movement for the convenience of a subdermal wireless hard disk. Likewise, Karl Marc, a Parisian tattoo artist, created an animated tattoo using a QR code and smartphones. The coding effectively unlocks the program on the phone that pushes the tattoo in the camera.

Moreover, for different purposes, some have inserted RFID chips. Dr. John Halamaka, for example, of the Harvard Medical School, preferred to use the 2004 medical details for RFID chips. Its implant file contains information that can guide anyone to a page with patient information with the right reader. In cases where patients enter the hospital unconsciously or unresponsive, he assumes chips like these can be appreciated. Meghan Trainor, another individual with an RFID implant, is even less functional, but highly imaginative compared those who obtained them. Trainor had the implant done in order for the NYU Immersive Telecoms Program as part of her master's degree thesis. Your implant is part of an immersive show of sculpture. RFID tags are incorporated into sculptures that can be manipulated in an audio library to replicate sounds. Trainers will use the electrode in their arm to control the sounds further. Brian Litt from the University of Pennsylvania, a neurologist and biotechnologist, inserts lead panels under the skin for the purpose of physical surveillance, in view of a wireless tattoo designed for medical monitoring purposes. These tattoos contain silicone electronics with a thickness of less than 250 nm, built on biocompatible, water-solvent silk substrate. As it is injected into the surrounding tissue with saline, the silk substrates conform and gradually fully disappear, only preserving the silicone circuit. The electronics can be used as photonic tattoos for powering LEDs. Litt is trying to perfect a version of this technology to create wearable medical devices, such as a tattoo to provide knowledge about diabetics' levels of blood sugar.

However, Tim Cannon, a grinder, has inserted a Circadia 1.0 biometric tracker under his arm and leg skin for the follow up of variations in weather, in what I consider to be an exceptional endeavor. The sensor/computer will wirelessly connect to an Android smartphone, allow temperature adjustments readings and, in the event of fever, give Cannon a text message. To insert the machine, cuts were made above an existing tattoo on Cannon's forearm. His skin has been raised and stripped from his tissue and the implant placed in the pocket until it was sealed. The LEDs function as "status lights" to lighten an arm tattoo under which the sensor is installed. The first iteration of the sensor recognizes variations in temperature, but later iterations in principle will be used to detect other critical signs and body changes. Some opponents argued that the equipment Tim implanted doesn't measure the body's temperature realistically—but exploring this topic doesn't serve this chapter except as an introduction to that dispute. How the legislation has addressed problems such as these and with faulty implantable equipment in general is a topic for other chapters in this book.

11.15 SENSORY SUBSTITUTION AND A SIXTH SENSE

Will a new meaning be developed when it comes to hacking the body? If by "new sense" one wishes to raise a characteristic impedance such that sensory input can be sensed beyond the range of the sensory receiver(s), then indeed. In reality, swapping one feeling with another is a well-intended subject and another means of hacking the body and creating a cybernetic future. The range of our sensations can be growing and/or expanded because of the frequency of vision and hearing, which enables the eyes and ears to perceive only input from the sensory receptors in a certain distance. In the future, when hacking the body via X-rays, or images and increasing olfactory sensibility, it may be necessary to include gustatory or haptic information.

Neuro-scientist Miguel Nicolelis from Duke University and his team say that, through a brain implant, infrarot light is being sensed by laboratory rats; they create a "sixth sense." Even if the infrarot light cannot be detected, rats will sense the light in the portion of the brain that is responsible for the sensation of touch in the rodent, by electrodes. Duke investigators inserted electrodes into rat brains, which were connected to an infrarot sensor, to give the rats their "sixth sense." The electrodes are then connected to the brains of the animals that handle contact results. The rats immediately start to detect the "touch" source and switch to the signal.

The research of Nicolelis and his team is a sixth sense, or not, in my opinion and leads to the future of a cyborg in introduction of brain machine technologies into the human body. Eric Thomson, who participated in the project, said previous experiments in the brain machine were intended to restore the function to weakened brain regions, not to build it. "It is the first research used to increase the function of a neuroprosthematic system, which literally helps a typical animal to grow a sixth sense. In addition to these intriguing observations, the Duke scientists found that constructing the sixth infrarouge sensor could never deter rats from processing contact signals, even though the electrodes were positioned in the tactile cortex (providing feedback for the infrarouge detection system).

11.16 HACKING THE EAR

Grinders expressed some interest in manipulating the ear as well. For starters, Rich Lee, a self-proclamed grinder, implanted magnets to expand his auditive sense in his ears. Part of the mechanism involves a spin that produces an electric field that vibrates sound-producing magnets implanted in its ears. Lee claims the sound level is like a pair of cheap of earbuds. The working phenomenon is known as thermionic emission, so that energy can be produced by mechanical motion (e.g., generators) and that the sound transforms into electric currents (i.e., microphones). One of the worries of Rick is the use of the implant to display the real hacker attitude. One of the desires of Rick is to use the implant to attach to other gadgets in order to increase his own sensations and talents in order to show his own spirit. However, Rick loses sight of a key part of his interest in hacking his auditory sense. Rich could offset lack of vision by learning to echolocate and to perceive his environment structure and proportions by responding to the conveyed sonic waves passed on to the embedded magnets through his headphone implants and other technologies. In addition to listening to music or podcasts, Lee wants to use earlobe implants for his phone GPS so that instructions can be sent right through his heart. Lee wants to connect the earphones up to a longitudinal microphone to listen to conversations; this obviously has privacy concerns.

Hackers weren't the only one who needed to restore or boost hearing. In 2012 about 324,200 individuals were implanted with cochlear implants to enhance the hearing in the country, according to the FDA. A cochlear implant definitely gives a person who is intensely hard of hearing a sense of hearing. The implant does not regain natural hearing; it will give a deaf person a helpful picture of sounds and make them learn their expression. The standard implant is: the outer part behind the ear and the second part, which is worked on a skin with a microphone, which collects sound and selects and arranges sounds from the microphone; the transmitter and the receiver/stimulator, which receive and transform signals from the speech processor. A cochlear implant varies considerably from an auditory aid that amplifies sound to be detected by injured ears. Instead, cochlear implants circumvent affected ear parts and activate the auditory nerve directly

In the United States, the FDA complies with laws specific to the manufacturing and delivery of cochlear implants as such products known as medical instruments. The most prominent federal law is the FDA auditory aid rule, which mandates the practitioner, ideally trained in ear conditions, to inform the patient prior to the sale of a hearing aid that it is in his best interest to see a physician before the purchase of an auditory aid. Like other implant technologies, there will be complications with cochlear implants and timely court action. In Kentucky, for instance, a verdict has reportedly given $7.24 million to a young woman after her speech and language implant failed. Advanced bionics, an organization of surgical devices, was found to be responsible for deliberately distributing faulty cochlear implants.

11.17 SENSING ELECTROMAGNETIC FIELDS

Grinders are now hacking their bodies by inserting magnets under their fingertips, as they extend the work of Miguel Nicolelis (feeling light) and University of Reading's Kevin Karrick (implantable chip) to detect energy sources that go beyond ordinary perceptions, electromagnetic fields. Likewise, robots are designed with the ability to feel objects before they are touched by magnetic fields that imitate the sensory experience of sharks in their situation.

Whether we notice them or not, electromagnetic fields are surrounding us. In reality, whatever a transformer or direct current uses (like domestic appliances) generates an electromagnetic field. Stuff like cable transmission lines, microwave ovens, and desktop fans were visible for a cyborg using implantable magnets. Each target has a special field with a higher magnification and "feel" according to grinders with the embedded magnets. Body hacker Tim Cannon has a lot of expertise with magnetic field analysis through his magnetic implant neodymium. Especially when Tim obtained his implant, he reports that the invisible region in the cash register can actually be felt, the frequency of the pulse differs from where his embedded magnet has his finger with respect to the unit. However, Tim has not been completely optimistic about magnetic field experiences. In the end, DNA evidence proved that the decision had been utilized illegally against witnesses and suspects in other instances.

Cannon experimented on an implantable device named Bottlenose, which is an echo locating unit that gives a human a sonar sense in an expansion of their corporal hacking attempts to generate a magnet sensation. An apparatus of approximately 50 mm is slid over the finger of a human. It's called the electromagnetic impulse after the echolocation used by dolphins, which calculates the time it takes for it to bounce back. The implant will refer to the sonar information if a person is fitted with a finger magnet to provide information from a distance. The embedded magnets in tattooist Dan Hurban's arm would be a final example of hacking of the skin to keep an iPod Nano in his body. While a harness would have done it properly, none the less, this is a fascinating case of confirmation of conception.

11.18 CYBERSECURITY AND THE CYBORG NETWORK

The body becomes a local area network involving a dedicated spectrum as individuals are fitted with wireless sensors and implants. And it can be compromised much like any wireless network. In the United States, the Federal Communication Committee determines the number of technologies used or incorporated into a human body to interact outside of the body by using a cellular communication channel connected to the medical body area network or MBAN. The range assigned to MBANs is intended primarily to calculate and record physiological parameters and other information for patients or to carry out diagnostic or therapeutic functions mainly in the health care field. MBANs are subject to a range of restrictions—they may only be used for medical and medicinal reasons; a patient should be provided one only under the supervision of an accredited health care professional; a body-worn MBAN device should not touch a body-worn MBAN device straightaway; and maximum emission bandwidths of five megahertz are permitted. With cellular medical implants, the network has many vulnerabilities, including risks described as "unauthorized access to a computer," or hacking, that may result from unintended signal interference. A host of possible flaws exhibits medical equipment, including untested firmware and applications and unencrypted wireless communication. It can happen in many ways: battery life limitations, vulnerabilities in remote access, interruptive wireless signals, unencrypted transmission of data, interruption vulnerability, malicious alert systems, dependency on old and redundant technology, and inability to download protection patches. These vulnerabilities may contribute to system configurations being tampered with by hackers, key device functions being disabled without user awareness, confidential patient data being obtained or device failure being induced.

11.19 CONCLUSION

At a time in which technology helps individuals to increase their bodies, human progress appears to be a step ahead in transforming humanity's very existence without the constraints of evolution by natural selection. Body hacking, rise, cyberization, anything you want, drive towards implanting body technology is a mature foundation for regulation although new innovations are being made in the fields of sensors, prostheses, and brain-computer interfaces. Law and science must work together when we reach this modern phase of human progress. The "capable" individual is the normal living adult, with all original sections intact, according to existing regulations, and no extension or alteration to decide that anyone is impaired is relevant in the eyes of the law. But this theory doesn't work in a cybernetic era where genetics and robotics lead to a revised distinction between people with a disability, being able to and "best capable," as people with both disabilities and healthy people are rapidly technologically extending their bodies and even minds. The use of interfaces between brain and machine would go way beyond restoration and enhancements and actually bring new functionality to the mind.

REFERENCES

[1] Penley, C., & Ross, A. (1991). Cyborgs at large: Interview with Donna Haraway. In C. Penley & A. Ross (Eds.), *Technoculture* (pp. 1–2). Minneapolis: University of Minnesota Press.
[2] Plato. (1982). *Timaeus*. (R. B. Bury, Trans.). Cambridge, MA: Harvard University Press.

[3] Poster, M. (1990). *The mode of information: Poststructuralism and social context*. Chicago: University of Chicago Press.

[4] Poster, M. (1995). Postmodern virtualities. In *The second media age* (pp. 23–42). Cambridge, UK: Polity Press. Available on-line at www.hnet.uci.edu/mposter/writings/internet.html.

[5] Porush, D. (1994). The rise of cyborg culture or the bomb was a cyborg. *Surfaces*, 4(205), 1–32.

[6] Reid, E. M. (1996). Text-based virtual realities: Identity and the cyborg body. In P. Ludlow (Ed.), *High noon on the electronic frontier: Conceptual issues in cyberspace* (pp. 327–346). Cambridge, MA: MIT Press.

[7] Rorvik, D. M. (1971). *As man becomes machine: The evolution of the cyborg*. New York: Doubleday; Rushing, J. H., & Frentz, T. S. (1995). *Projecting the shadow: The cyborg hero in American film*. Chicago: University of Chicago Press.

[8] Sandoval, C. (1995). New sciences: Cyborg feminism and the methodology of the oppressed. In C. H. Gray (Ed.), *The cyborg handbook* (pp. 407–422). New York: Routledge.

[9] Shannon, C. E., & Weaver, W. (1963). *The mathematical theory of communication*. Urbana: University of Illinois Press.

12 Cyborg Futures on AI Robotics

12.1 CYBORG AS AN INTELLIGENT MACHINE

Consideration over advances in humanoid robot production in mass culture, reiterated by the robotics involved, begins with the questions: how are we humans to act in view of the "rise" of these human-like machines? This chapter suggests that our first answer should be the challenge, the recognition of the rhetorical sleight of hand behind this seeming call to consideration and intervention; see Figure 12.1. It suggests approval of the proposal that these inventions advance to mankind and that this progress is imminent. The inevitable consequence is founded on the assumption that robotic creation is like global climate change—a natural/cultural event. As with global warming, however, much human intervention will lead to the spread of robots; it now has its own dynamic [1].

However, AI and robots are very different types of civilization from global warming. Really, there are patterns unfolding until they are disrupted or remedied effectively. These dynamics, though, are much more fully human, less intertwined with the more human, and more ready to intervene politically. In certain areas (processing capacity, storing of data, algorithm complexity, and networking) technical initiatives are advancing; however, noticeable progress is being made in attempts to attain human capabilities. The prevalent misperception of the state of robotic arts and sciences, however, obscures these distinctions.

12.1.1 UNIVERSAL MACHINE INTELLIGENCE

In their respective representations of technology as either the out-of-control problem or the in-control solution for the human condition, these quotes reiterate the universalizing progress narrative that underwrites announcements of our supersession as humans by machines. It is that narrative that comprises "Singularity," an idea popularized by AI enthusiast Raymond Kurzweil, to reference the "tipping point" at which machine intelligence will exceed our own. This principle accumulates in the realms of science fiction and science fabulation and reinforces expressions of concern like the ones given by Hawking, which call us to respond [2].

And what is really concerning about the event horizon is not its future so much as the imagination of the developing person and his (using this pronoun) technological relationships as summarized in STS student Hélène Millet's *Analysis of Stephen Hawking himself* (2012), who refigures Hawking from a limited person to what Millet calls "a distributed-centered topic" and that provides a helpful response. Millet's analytical work dissipates Hawking's organization's borders from the actual genius to a group of people in diverse positions and technology, allowing their companies

FIGURE 12.1 Machines behave like humans.

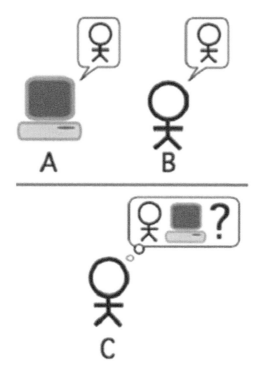

FIGURE 12.2 Structure of game playing puzzles.

(including their network constructing) together. We need to work on this relocation as a tool.

Among other items, redistribution includes returning cultural tales to robotic narratives, particularly identity politics. As a start, mathematician Alan Turing from 1950 could return to the paper that was the source, alternatively known as the imitation game, of so-called Turing tests for machines (1950). Figure 12.2 displays the most often seen diagram of the exam, where the questioner tries to distinguish between an unmarked "human" and a test machine. However, Turing explains the result differently:

> it has three men, a male (A), a woman (B), and a questioner (C) of any sex. The questioner is held in a separate room

12.1.2 Generating Robotic Progeny

Feminist critics of the AI project indicated that their goals entail a kind of male birth; the robotics imaginary also deploys the infant figure, the creation of individuals and of animals and the development of their organisms. A quotation from Turing's seminal paper shows he was often fascinated by the assumption that a computer system might be developed that had a human child's competitive strength: Turing's thinking shows the combination of Anglo-European colonialism and cognition, which is a characteristic of the AI, enacted by the "acceptable training" of the (unmarked/white/male) adult and infant brain and their normative training [3].

In modern futuristic renditions of the human computer, this fantasy is reiterated: The original thought machine HAL 9000 in the 2001 film *1968* was only one famous example: *A Space Odyssey* refigured as a near-perfect human replica a device by Steven Spielberg's *A.I. Artificial Intelligence* film, in 2001.

12.1.3 Framing of Robot Code

Whilst these examples activate articulations as a means to widen the time-historical context of AI/robot programs, we are also able to broaden the framework in space to recover relationships and networked interactions that are elusive for dominant renditions. One example is the first technical assembly that IBM's Deep Blue project has announced as a proof of the accomplishment of humane artificial intelligence. Due to its strong connection with a human genius, chess is the canonical place to implement AI and it is suitable for the specific calculation capabilities as far as it is a shut-down environment with a certain number of regulations generating an exceptionally wide range of possibilities. Most of the computers playing chess depend on "brutal force," i.e., pure strength and accuracy. The story goes, in a decisive game that led to widespread media declarations that AI had finally reached the holy grail of a machine intelligence which exceeds humanity, the IBM computer called Deep Blue defeated world chess champion Garry Kasparov.

12.1.4 Bodies in Relation

In the world of AI and robotics it is known that there are no embodied agencies and affective ability for chess-playing machinery. Rodney Brooks and MIT's computer science and artificial intelligence laboratory colleagues at the beginning of the 1990s focused on what Brooks called "Situated Robotics," an important position of incarnation in creating a smart organization [4].

Brooks developed this method for humanoid robotics projects in the autumn of 2001; I had my own meetings in the MIT AI laboratory with the robots Cog and Kismet. In the apologies, the graduate student who led our tour explained the Cog was inactive as seen in Figure 12.3. Since there were no scientists working continuously to improve the program, Cog suffered from what was widely called "bit rot" syndrome, which was the deterioration of the animated software of the robot without continuous repair and upgrade. Yet we saw Cog sitting in a corner of the laboratory. He was inanimate. Although still an intimidating figure, the rest of his "body" which was not clearly apparent in media pictures, most struck me. The basis of Cog's torso was a massive cabinet, which provided a large sheaf of cable connections, spinning like a centaur on a roof-top bank of processors that delivered the computer power needed to give life to Cog. The visibility of the robot "in the house" in the lab, located in this backdrop setting, offered the ability to see the extensive network of human labors and associated technology that Cog offers its organization.

FIGURE 12.3 Robots Cog and Kismet.

12.1.5 Slow Robots and Slippery Rhetoric

Dennis Hong, Virginia Tech Robotics Laboratory's Director, says to the interviewer in the BBC video coverage of the event:

> As a lot of people assume of robots, they watch too many science fiction movies and believe that robots can do anything they can. You can really see through this competition that this is not the reality. The robots are going to crash; it really is going to be sluggish.
>
> **(Paterson 2013)**

DARPA Director Arati Prabhaker also said: "I believe that robotics is a field in which our creativity has gone ahead of technologies, and this challenge is not science fiction but science reality" (ibid.). Though many aspects of the market dispute the separateness of fiction and reality (particularly the investment in robots as humanoids by its sponsors and competitors), there is still an important distinction (Figure 12.4).

However, in a whiplash-inducing moment at the conclusion of the BBC clip, these warning messages are contradicted as the Boston Dynamics Project Manager, Mr. Boundary, who mentions the first flight testing of the Wright brothers, reassuring us: iIf anything goes on, by 2015 we can imagine robots that can, you know, assist the firemen and our police, and help them. Even in comparison with the extraordinary flight history" (Paterson 2013). The premier league, Shaft (post purchased by Google) from the University of Tokyo, owes its distinction advantage to a new high-voltage liquid-cooled engine system using a condenser rather than a power battery.

12.1.6 Challenges of Robots

Although the invocation for the robotic rescues of the nuclear accidents begins with the question, what will be the hardening of robots against the effects of radiation, and the costs at which robots are competing (robots that have already taken up the challenge have cost up to several million dollars each)?

FIGURE 12.4 Helping robot.

FIGURE 12.5 Robot use in industries.

He states that the robot that won the Robotics Challenge in 2015 "considers that home assistance is the major business potential [for] humanoid robs" and that "these are the culmination of several years of Japanese science, motivated in major measure by fears about Japan's rapidly growing population," a statement reiterated in Figure 12.5. What may link these components of heavy equipment with home treatment is left to our interpretation, but Pratt remarkably indicates "that the problems facing the DARPA robots in hospital emergency rooms are very close to those confronting them."

12.1.7 The Labors of Violence

The "Super Aégi" and its kind are threatened not by the prospect of a "sensitive" bot or a humanoid cyborg but by a much more universal progression in the improving technology of missile defense systems: automation of identification of certain human categories (those that fit a defined and machine-readable profile) as legitimate goals for the designation of specific humans. The deterrence of robotic weapons systems from their human operators is the next logical move towards warfare automation—a course I took with my colleagues to work together to stop the warfare, "Stop Killer Robots" by Human Rights Watch.

Noel Sharkey points out that, instead of taking humanoid form, the deadly, independent weapons would most certainly look like conventional arms structures such as tanks, battle fighters, and jet combatants, as a robotic and founding member of the International Committee for Robot Weapons Control (ICRAC) [5].

He states that an international prohibition that prejudges delegation of "decisions" to destroy robots is the central concern for an autonomous weapons campaign. One key point here is that the "decision to kill" delegation relies on the assumption that algorithms for differential detection of a valid target are defined in a computationally tractable form. The latter is an adversary that is involved in war and faces an "imminent danger" according to the terms and conditions, international humanitarian law, and the Geneva conventions. We have abundant evidence that differentiation is becoming more unclear.

12.2 AUTONOMY OF AI

The movement voiced fears about the existential challenges posed to the "future of mankind" and the "future of civilization" arising in Figure 12.6. Philosopher Nick Bostrom, a key figure in the movement, describes a "existential danger" as "a risk that will cause the Earth's destruction, or that the capacity of a desirable further growth, will otherwise be permanently and dramatically destroyed." The idea that AI could pose such a danger comes in line with the need to create enormously potent computers consisting of completely independent operation. This ambition is related to the utopian concepts of how these robots can lead to "exponentially increasing" for the good of mankind, as the film *Transcendence* (2014) demonstrated [6].

12.2.1 EXISTENTIAL DANGER DEBATE IN ARTIFICIAL INTELLIGENCE

Unintentionally pointing out a tacit truth regarding the social structure: the work of "diligent hands" relies on human intelligence. However, they mask this dependency by separating the idea of knowledge and raising its importance with regard to physical practices and material systems, in defending innovation and engineering. Increasing the value of knowledge is the continuing function of Hawking and others' (2014) narrative on Enlightenment Development to exploit the additional assumption that AI is essential, as it "enhances" what the human condition can do. The existential danger debate in AI seen in Figure 12.7 thus reaffirms the philosophical schema of Enlightenment in which the notion of "intelligence" is defined as the agent "of all that civilizations have to give."

FIGURE 12.6 Save life from future computing.

FIGURE 12.7 Intelligent machine self-schema design.

12.2.2 AI AND THE ILLUSION OF TRANSCENDENT RATIONALITY

In 2014, *Transcendence*, a film by Hawking et al. (2014), revived its commitment to *Good's* (1966) and popularized speculative technology science dialogue on autonomous AI (2005). In the AI danger discourse, the film provides some insight into the paradoxical form of human subjectivity and the connection among rational thought and autonomy. *Transcendence* says how a dying computer scientist (Johnny Depp played Will Cartermind) is uploaded to a supercomputer to stay with his loving wife Evelyn and resume their science work. The experimental generates a subconscious personality (the will), which appears on computer screens in a spectral way. While Will as the AI system has the strength and capacity for technology advancement that scientists would wish for, he is regarded as a rogue agent who undermines social order. The dispute between the autonomous AI system and the state agencies that aim to regulate its conduct is eventually settled by the killing of the personality on the computer.

12.2.3 CAPITALIST CONTEXT OF RATIONALITY

With these images, the issue of what actually forces autonomous identification is not readily clarified by the identities and thoughts of individuals interested in technoscience research and contrary to common myths of valiant scientists struggling with "future" of "technological society." This restriction is in fact expressed by a criticism of the autonomy of the "transcendental question." Terms such as "economic society" or "technological society" are widely used to describe the definite "society" sense where automation and progress-related crises appear as thematically important subjects. The importance is in how the work, intelligence, machinery and natural resources of humans are arranged precisely as "powers of creation" as Adorno observed (2002). These concepts of "society" are correctly understood. However, it is the exclusive emphasis on business, technology, and change that distorts the contextual factors deciding awareness and agency of the subjects.

12.2.4 Knowledge of Transcendent Rationality

Although idealism will in some respect be the last to acknowledge it, the transcendental subject is more real, which is that it defines people's and society's true conduct even more than those psychological individuals, from whom the transcendent subject was abstract, who have nothing to do with the world. The living human, as obliged to behave and for whom he was also internally designed, is closer to the transcendent subject as a homo-economicus incarnate than the living person could automatically be.

12.2.5 Bostrom's Occult Motivations of AI Machines

Bostrom believes that the distinguishing feature of a class of rational beings, including biological species such as humans and artificial bodies, is that of rational behavior. A super rational being, nevertheless maintained in an organic (not biological) body, will have motives that are strange or not humanly understandable; (i.e., contra Will as AI-system, whose undiminished love for Evelyn was the romantic hook in *Transcendence*). In view of the human tendency to anthropomorphize, he maintains that the idea that the AI "inherently cares for something" that is concerned with "humanity" is unreasonable. He provides the example of an AI "whose only essential objective is the counting or maximization of the number of clips in the potential light cone of the grain boundaries on the Bora Cay." Furthermore, he assumes, that various "species" of rational individuals could inhabit the same room, citing a Hobbesian scenario of uncertain future in which some of the "automates" have rational forces well outside the standard and motives that make them responsible for others. But in order to provide such a concern, I propose that Bostrom should waive the assumption that the people of this space were "useful to maximize logical agents."

12.2.6 Colonization of Cyborg

However, instead of pondering the treatment of the consequences of a "destroying gift" for our world, he implied that since we are a colonization species with a history of resource depletion causing crisis, it is time to cross over the Earth in search of another planet. Although it is an unquestionably collective objective to defend citizens from anthropogenic risks, this surprisingly fatalistic response to the global ecological crisis shows that the AI danger debate does not have such security in its final analysis. Why does it happen? My response is, the value of "humanity" applies not to humans but to the value of accumulation of wealth. The distinction among "us" and "them" is thus to be discerned in the sense of 'humanity' that Bostrom described as "intelligent life of earth origin." Humanity's keyframe animation spirit is inspired by "what we have justification to love" rather than "knowledge" (Bostrom 2013). There is no distinction, however, in the context of the qualifier "what we have cause for worth." At first glance, it seems to Bostrom that freedom to decide its own mode of incarnation—biological or mechanical—is such a value. He does not argue, however, that this freedom itself is an end. Rather, the value dilemma is divided between today's bodies and forms "we" will later select. By postponing "what" we are now—a postponement encrypted in the claim of "the future of" humanity—the universal "we" appears to circumvent the finitude of bodies, individuals, and the world [7].

12.3 VISIONS OF SWARMING ROBOTS

Evaluation, assumptions, and suggestions on the potential production and implementation of deadly automated robotic systems as found in reporting, surveys, and demonstrations of what is still well-known as a weapons industry are shown in Figure 12.8. This word is also appropriate as a more or less overt collection of ties among think-tank networks, lobby groups, defense agencies and government agencies, policymakers, scholars, and university-funded research ventures as well as their interconnections with the private sector and private equity and military "clients" continue to describe it. What are both end customers about the new technology initiatives and beta-testers? I would like to focus particularly on promoting artificial swarming intelligence in science and strategic scale research that promotes the development of this robotic future of warfare—if it also has to be named if artificially intelligent systems manage combat missions.

FIGURE 12.8 Swarming robot use in defense.

12.3.1 AUTONOMOUS ROBOTIC SYSTEMS

The effort to create and incorporate autonomous robotic systems in real government and police procedures is an example of what Stiegler calls "noetic" work. Noetic came from the ancient Greek noesis—intellect or interpretation that comes from the English word "nous." The thesis of Stiegler on the essence of the work of the human mind, which I need to present briefly, is based on two similar arguments. First, noetic work is the understanding of the human mind's capacity, which is necessary and often takes place when it comes to mutual knowledge. The unmanifested person sees, contemplates, encapsulates, and thinks of his experience and experiences. As she explains, she speculates (more or less explicitly) in dialogue, building on existing experience. Second, because of the difference between ourselves and the intellect of other people, human intelligence is a phenomenon emerging from a reflection on and exchange of experience, which is similarly contingent on the accumulated understanding of the shared experience of national memories. The substrate of this information acquired is art factual, i.e., technical. We inherit prior insights and knowledge from what remains beyond the previous ones. The community's social, cultural information banks are filtered through the material legacy of previous encounters and observations and plans. The nematode spirit may feed and replicate itself but does not experience or travel across the earth. It takes care of the fundamentals of living, draws food from the ecosystem, and reproduces it. The receptive (or perceptive) soul is a part of the life forms in which animals hide, flee, wander, wait for prey, or meet their reproductive partner in the environment. For Aristotle only the noetic (smart, thinking) soul belongs to entities who can think, reflect, use words, and have access to logos, reasoning, and justification [8].

12.3.2 SWARMING AS NATURAL INSPIRATION FOR MILITARY FUTURES

The numerous CNAS studies and publications, including the two-volume "Nanotechnology on the Battleground" study initiative, represent a much stronger defense of the roles autonomous systems play in future U.S. military activities (Scharre 2014b, c). In this respect, the second volume, *The Coming Swarm*, has an interesting effect because it consolidates the conceptual and functional reasons for the use of military power by swarming robotic components. In *The Coming Swarm*, Scharre proposes the "disruptive" paradigmatic shift from "direct individual monitoring" to "human monitoring" at the required times and "to Maneuver and execute operations on their own inhabitable systems" (Scharre 2014c, 6). For the future, Scharre predicts that advances in AI will go beyond the existing "networked" doctrine and achieve "real swarming—cooperative behavior between dispersed elements that will lead to the creation of a coherent smart whole" (10). The study encourages the ability of independent structures to provide the U.S. military powers with a definitive (and cost-effective) means of sustaining their dominance and global reach in a developing environment.

The forthcoming military swarm of self-sufficient machinery will travel, interact, and "think" quicker, even more quickly than human combatants. Scharre mentions the "ground breaking monograph," entitled *Swarming and the Future of Conflict*, of

RAND's academics John Arquilla and David Ronfeldt (2000) in a segment on the roots of the applicable principles for a militant military mobilization of swarming robots. The *Coming Swarm* adopts the RAND thesis to incorporate the biological idea of swarming into the formulation of military strategy in a very rigid, misleading, and inconsistent way. Arquilla and Ronfeldt note at the start of their discussion of *Swarming in the Nature* that "Military swarming cannot be closely modeled after swarming in the human species. But it can draw some helpful lessons and viewpoints" (Arquilla and Ronfeldt 2000, 25). There are no grounds for not closely modeling military swarming on freshwater ecosystems [9].

12.3.3 Emergent Stupidity

This outlook on the near future of U.S. military confrontation is typical inside institutions and is wholly only a manifestation of an intentional blindness—a type of willing ignorance. This image (which is also the imagination) works, like all fiction, to strategically choose (and exclude) the elements to be used in this scenario view. Traditional ships are rounded up in one series by a protective screen of robotic vehicles that respond in a swarming assault to the arrival of incoming missiles. It is not known whether the unfriendly vessels are piloted humans or are autonomous. Another series involves the creation of a wave of terrestrial robots inland, which eliminates enemy forces before the icons that represent traditional armored personnel carriers. The nucleus of controversies and discussions that have been waged over the last decade or so in the United States' military and security operation of semi-autonomous and remotely controlled systems is conspicuous by their lack of any visualizations of future war: no non-combatants, no villages, no temples, no town square meeting, no farmers, no fishers, no bus trips to marriages or soccer matches, no "other" presumed terrorist fighters' appearance (and even this is a weak and unnecessary implication).

In these photos, the disappearance of the inhabitants of the disputed territories is expressed in *The Coming Swarm*, which provides no substantive consideration of the question of discrimination between the fighters and the non-combatants. This brings a lot of civic protest and advocacy in recent decades to the forefront of U.S. military politics and hides the use of drones by the CIA to track and target the massacre in Afghanistan, Pakistan, Yemen, and Somalia's skies. The operation of drones over populated areas, recognized as war zones, has been criticized by human rights organizations. For example, the resident population in Waziristan in northern Pakistan is constantly threatened by the constant overflowing of drones. There is a second explanation I named a dumb projection in this study about the future of swarming armies, one relating to the idea of human intelligence as a complex and transient tendency in particular. This silliness is most obviously evident here not as a flaw in logic or collecting specific facts but as an approach that embraces the reduction of current warfare and, with it, the sensory perceptions ability of people who commit genocide [10].

12.4 BUSINESS OF ETHICS, ROBO, AND AI

Robot ethics and artificial intelligence (AI), driven by states, firms, researchers, and campaigners, is a widespread explosion in the public arena. Global technology firms such as Facebook, Google (Alphabet), IBM, and Microsoft are actively campaigning ethicalizing robotics and IA and have developed research units in their companies that are dedicated to the ethics of emerging technologies as seen in Figure 12.9. Robot and AI ethics also contribute to technical advances created by robots and AI, financed by the European Commission and Governments in China, the United States, Russia, France, and Britain, which had meetings with its membership of stakeholders and established its own ethical standards. In European and U.S. technical, nongovernmental, and private organizations, ethics as a model to address potential risks posed by robots and AI has become prominent. Ethics has been a major business, to put it lightly. I am writing this chapter, based on my experience researching robotic and AI ethics in the two ventures sponsored by the European Commission, developing robotic enhanced autism therapy (DREAM) and responsible robotics ethical learning 7 (REELER) and participating in ongoing debates on the value of ethics in Robotics and AI within a lively society made up of researchers, businesses, the government, and members of the public.

12.4.1 Cyborg Ethics?

Any analysis of ethics would prove that it is not a unifying set of thoughts and that there is no ethic model for the "true life." In reality, ethics can differentiate between good and theoretical concepts: Immanuel Kant's work (1978), for example, concentrates on rationality, the main subject of ethics, vs. David Hume's (1940) concentration on emotions. Although certain nations are under armed forces and sanctions, they purchase and sell destructive weapons and they destroy. Ethics universals such as "thou shalt not kill" may be believed. We might think that "do not rob" is a basic maxim, but what is robbery then? Is a legitimate threat to a high-interest bank or a sort of legal robbery? How would a virtue booster be if the ethics were excellent or wrong, like Aristotle, who talked about how grown men might comprehend their merit or virtue while justifying slavery and a subordinate woman? When we

FIGURE 12.9 Robot involved in business.

explore ethics, we see a collection of ideas that are not impartial or indifferent, but ideas formed under particular circumstances that are marked by biases and hostility towards females. There is also a perception that ethical reasoning should be removed from its morally objectionable sections, such that the virtue ethics scan of Aristotle can be removed from its reasons for slavery and misogyny, while other parts are kept, such as a well-known definition.

12.4.2 Why Ethics Now?

Representative democracy changes around emerging technology and its effects on humans; the climate and nature are accompanying ethical discussions. Moral arguments sometimes have limited the advancement of such technology in companies and universities for innovations such as genetically modified organisms (GMOs). In the 1990s, for example, in response to complaints by environmental activists regarding the impact of GMOs on the environment, the European Union heavily controlled and limited the use of GMOs. This has resulted in the use of the "cautionary theory," a principle used to provide precautionary advice for emerging technology (Beck 1992). GMOs are legitimate and permissible in the U.S. and countries in Africa. Businesses and scientific stakeholders have criticized the EU to argue that legislation restricts industrial entrepreneurship and sustainability.

New ethical principles and ways of legislation include robotic and AI technologies. Robots can convert working conditions as virtual robots. Historically, tools, machinery, and robotics have replaced human labor and have enabled the processing of products or services to be efficient. Implementation of expanded job automation raises unemployment issues. "Flexibility" has been the central phrase for Fordism (Valles 1999), which is reflected in the increase in IT technology, the reduction in conventional workers and syndical participation, the transition to outsourcing, and the "feminization" of workers. The consequences of Post-Fordism were evident in the United States and the United Kingdom in the 1970s and have changed the working world for a century.

12.4.3 Why Robots and AI and Ethics?

I continue to pursue the requirement for robotic and AI ethics. I advocate the division of this inquiry into two sections:

1. These are artifacts (robots and IA programs) operating in the environment that generate results without always understanding the behavior or effects of their creators (through the use of big data using machine learning or deep learning algorithms).
2. Robots and AI can become so sophisticated that they are irrelevant to people.

There is no sense of the distinction of a person and a machine. Company and university study organizations build robots and AI. Robotic systems and cognitive technologies are theoretically useful as increased automation, on the one hand, can minimize the strain of stressful or repetitive work; however, on the other hand, it can also increase unemployment and lower national taxes (Frey and Osborne 2013). Are the advantages and disadvantages of robotics and AIs really conceivable, without mentioning capitalism as an economic system? The actual possibilities of understanding how the robots and AI impact humanity are still neglected by ethicists in the capitalist system. Historically, tools, machinery, and robotics have displaced human labor and have enabled the processing of products or services to be efficient. Implementation of expanded job automation raises unemployment issues. Fordism's fundamental word, "flexibility," has been mirrored in the rise of IT technology, the decline of traditional workers' and syndical engagement, the shift toward outsourcing, and the "feminization" of the workforce (Valles 1999). The effects of Post-Fordism were felt in the 1970s in the United States and the United Kingdom, and its repercussions will be felt in the workplace for the next thousand years.

12.4.4 CORPORATE ROBOT

The drum and bass command economy is absorbed by patriarchal ethics, which promote philosophical investigation into the existence of matter (corporate anthropomorphized creatures). This is the logical end of a society that gives the few dominance above everything and that doesn't even claim to maintain the myth of humanism as a project of equality before the law. Transhumanism popularizes the rejection of the humanist project (Bostrom 2005). Bostrom is free to begin its logical end with the liberal topic of Enlightenment, enabling the uncontrolled commodification of man. The name of the game is alienation and consumer goods are refurbished in human forms: as compañeros, living beings, and slaves. Hardaway (1991), who dismissed the mythic humanist topic as a self-contained being, is indeed a project of abandoning man. Instead, the cyborg was declared a political entity of the biomedical and military-industrial complex; see Figure 12.10.

FIGURE 12.10 Cyborg in military industry.

12.4.5 Why We Need a Feminist Ethics of Robots and AI

As described, the implementation of emerging technology presents several serious problems, some of which I describe in this chapter. Feminist ethics is marginalized as an alternative to the mainstream of classical male ethical doctrines (deontology, utilitarianism, or virtue to name three dominant modes). What I would like to say is a feminist ethic, not an alternative but a transformational ethic that addresses political, financial, and social exclusion. Feminist ethics is based on the interrelationship in you that is essential but also conscious of age, sex, class, and race power imbalances. Feminist scholarship concentrates on interaction and empathy to deal with ethical practice, which celebrates isolation, displacement, manipulation, and cultural relativism in businesses.

12.5 FICTION MEETS SCIENCE

For humanitarian motives, Aaronson tells *Common Science*, many thinkers have deep interpretations of consciousness. After all, if the giant mind game approximates a thinking or a feeling in this field (or C-3PO or Data, or HAL 9000), "who are we to say that the awareness is less real than ours?" Scott Aaronson is an informatics scientist, so what has "this area" to do with *Star Wars*, *Star Trek* or *2001: A Space Odyssey*? As a literary scientist, in analyses of artificial intelligence and robotics, I am always struck by the multiple references to literature, so how is this fiction mobilized? Why do fictional robots fail in the robotics industry, for example? And how do robotics and artificial intelligence picture science and fantasy in a different manner? Even though there are few transdisciplinary discussions among fiction students, this chapter also seeks to promote this conversation and question the argument that science fiction is consistent with scientific facts and to restore the divide between science and fiction.

The newspapers of the 21st century are full of fictional claims: "The Craziest Science Fictional Fantasies This Year," *Wired* reveals. "8 science fantasy forecasts that really come," reads *Huffington Post*, "Things in scientific fiction that exist anymore now," claims *Buzz Feed*, many of which deal with robots and AI.

It is also not rare for local museums to draw comparisons between technology and industry by designing their exhibitions on this topic.

Built at the Carnegie Science Center in 2009, Robot World is using a timeline that matches the actual robots. "The first robots have been inventions of creativity rather than engineering. This permanent display tells the visitors: "the reproduction robotics that lined the 'famous walk' are a testament to the creativity of those who built real following robots." The 2017 show *Robots: the 500-year-long quest to make human machines* at the Technology Museum in New York often combined imaginary robots with business robots and their title played a fictional trope with an old history: robotic humans who seek human status.

Robot engineers, programmers, and manufacturers often quote literature and movies as a source of motivation: "My Star Trek robot becomes a thing, and much stronger than I ever thought." In a similar spirit, Jeff Bezos said at 2016's code conference, "It was a fantasy from early sciences to speak to a robot, and it's real,"

Cyborg Futures on AI Robotics 311

revealed Amit Singhol, a former engineer and vice-president of Google (Bergen 2016). Cynthia Breazeal, director of the MIT Media Laboratory personal robot community, often frequently lends *Star Wars* encouragement to her robots with military funds: "all of those droids have become full-fledged characters; they have been caring for humans. I believe that was what caused my creativity to ignite" (Greenfield 2014). An interview of the new Jibo design by Breazeal, introduced as the very first family robot, begins with the following comment: why, however, are robots collapsed in such a way as to be imaginary in the world? The flopsy bunnies of Beatrix Potter, for example, are never conflated with true rabbits or Disney's the Lion King for a live lion. Fiction is packed with references to weird objects, from spacecraft to benevolent dragons. Robots in *Star Wars* are not less fictitious, but business robots read through the prism of fantasy frequently portrayed in their fictional equivalents as interchangeable and constant. This confusion presents some problems, for example, a commercial for Jibo features clips of R2D2 from *Star Wars*, *Lost in Space*'s nameless robot, Johnny Five from *Short Circuit*, Rosie from *The Jetsons* and WALL-E from a 2008 postapocalyptic film of the same name. "We've dreamed of him for years now and then it is actually there," the narrator says, as if fantasy is involved in a teleology that inevitably leads to actual humanoid robots being materialized. These clips animate this new creation, like other parallels to fantasy in the promotion of this technology. Chain decaying of this fictitious heritage used to market the product appears like a black miracle lantern; it has a three-axis engine equipped with stereo, gesture, and speech sensors and a touch-screen operating platform.

If the Jibo advertising uses fictitious information to make a plastic shell containing wires and chips "part of the kin," it overlooks the allegory, similarity, metaphor, metonymy, and the social or political meaning of its origins of literature. Jibo is introduced as a friendly mechanical worker as a story illustration that comes into being when it sells people's imagination. However, the embedded photographs tell another story. The happy family shown on the ad embraces the bourgeois consumer's fantasy. *The Jetsons* is a lively TV show set in 2062 that started in the early 60s. George Jetson worked for an hour a day in the show. Joan Jetson was a homemaker who had been shopping for clothing and modern appliances, and Rosie the robot did the household chores and the family flew by. This image of Jibo reminds us of an entertainment center, which is always in danger in the promotion of "future" technology: the enduring hope that robots can support and even make life easier to afford to those who can accommodate them, thereby freeing us from boring work.

Unsurprisingly, the announcement for this potential invention evokes a feeling of longing for the imagination of the whole American family—healthy, laughing, blond, white children in household happiness, living in a single-family home with a garage and car in a tree-lined suburb with a woman baking comfortably and having a big family meal. However, in a much more unstable America, Jibo is being released with housing forfeitures, precarity, and the shrinking middle class. In addition, the Jetsons' leisure fantasy turned sour decades back. Not only do individuals in our "still linked" society work longer hours, but declining infrastructure and excessive toxic deposits are causing Jibo to be more near to the nervous WALL-E world, the last clip in which fat people float in their room with mindless consumerism enslaved by machines and separated from one another, despite having dismantled the planet.

So, what is the "he" of whom we "dreamed" in the ads? More than a nightmare is the line between Rosie and Wall-E of fictional robots.

While science has made fictional statements, it is often careful which version of fiction is mentioned and cautious of dystopian versions: "fear is probably not going to stop experimental study, but different concerns are about to delay the process." The preface to *Robot Ethics* argues the need for mechanisms to differentiate between actual threats and imagination and hysteria, fed by science fiction. A *Guardian* headline makes it sound the same way: "Brave New World? "Make Policy Decisions of AI," warns the specialist. Although science acts as a "true" or "hype" paradigm for fiction, by definition, fiction is fiction and it is resistant to this literal method. So, how were robotic and artificial intelligence imagined by scientists and literary critics?

Robots are also compared with people and robots in the fields of robotics and AI. For instance, the Good AI website, by a Prague research institute, quotes that "Our future artificial intelligences can experience incentives in the same way a person does—by seeing, sensing, communicating, and learning—and use them to generate actions, carry out tasks and react to the motivations of human mentors." The prevalence of this founding trope is explained in a brief scan of an MIT technology topic. "We do not expect a robot to learn from scratch alone, as we would expect the human being to do," claims Ashutosh Saxens, the director of the RoboBrain Project. "Imagine a child without access to textbooks, literature, or the Internet." . . . A robot wants all of the information a human need in one position, just as a person does (2016, 15). Stefanie Tellex, a professor of computer science at Brown University, references to childhood as a "massive method of data collection." in another report on robots training each other. Another paper again wonders whether AARON and the Full-on Idiots in the World of Modeling are artistic, innovative, and sensitive to emotions and computer programs creating "machine art." The machine produces photograph comparable with Rubens or Van Gogh. AlphaGO, Deep Mind Industries' AI software, beat the South Korean grandmaster in the old Chinese GO game (which was purchased by Google). The algorithm uses "deep learning," which is rather grossly and predominantly based on biological neurons, far more complex living cells. However, the software "learns as human beings—when you look at the environment, consume data, and analyzed thematically and laws" (Lee 2016). Geoffrey Hinton, the Google Brain lead scientist, said the winners of the program were "a lot of intuition" (Lee 2016). I questioned one of Hinton's former students (now a professor of computer science) in a conversation on using this phrase. In the sense of machine learning, that's what Hinton meant by "intuition," that is, how a machine can circumvent rationality and behave based on emotions. He replied in a message that Hinton considers every "mental task" or perception—regardless of whether it's thought, dreaming, reasoning, feeling, etc. That "intuition" is indeed much more of an electrical/computing mechanism than we consider it to be, "if any of it is neuronal activity." Hinton claims that once computers are able to model the brain accurately by identifying patterns and learning from errors, perception and "consciousness" are just an issue of physics and that computer algorithms that mimic neurons present the brain as nothing more than a processor.

12.6 RACING OF ROBOTICS

The Black British theorist, critic, and suspected Afro-futurist Kodwo Eshun brings together two unique figures that were singularly influential in the past from different intellectual traditions: American African Activist and the Polymath W.E.B. Du Bois and American scientist and philosopher Norbert Weiner, Father of Sun, in his now classic and soon to be reissued *More Brilliant Than the Sun: Adventures in Sonic Fiction* (1998). In this regard, Eshun links these dramatically divergent thinkers to the classic works of Weiner's *The Human Treatment of Human Beings* (1950), identifying them as Du Bois' epochal work of African American identity and relationship with American races. "Enhanced for the analog Era" the souls of Black Folk.

Eshun explored this link, but his contribution—where two distinct modes of civilization, one technical and one ethnic, are compared—allows me to argue that the literary, critical, and cultural capital of the African Diaspora provide great gleanings in the fields of technology studied. Such documents may or may not be directly focused on technology but, as a complement to traditional history of science, robotic and IA, they are focused on race, slavery, colonization, and global systems of dominance and sex—not only in the creative forms made possible. Afro-futurism, or potentially complex efforts to delineate race as a science and not merely anti-racism, is often reductive, by seeing technology as a simple replacement for whiteness or insisting that Black people are or have always been simply subject to their colonial control. It clearly takes a close look at the material history of race and technology that enables such creative speculations and surprising conjunctions.

The first truth is that technology never was lacking, foreign, or different from racial discourses and vice versa. The research is a study of this kind. I would say that between these fields, there has long been an affinity that feeds on those eerie experiences that empower the genre of science fiction (SF). This intimacy is precisely what allows Eshun to propose that African Americans' history (slaves and former slaves, known as the "double consciousness" of a distinctly racialized kind of uncanny) bear some connection with the SF, or that the fight for Black mankind against aggressive and violent powers shares common narrative features of cybernetics. This analogy should be familiar to those acquainted with the basic tropes of Afro-futurism because it evokes the renowned equivalence of Greg Tate between slavery and science fiction. His impact on the work of Eshun is uncertain: critic Mark Dery will, often famously, refer to "Frofuturism" in a "psycho-grouping" of "African American concerns within the context of twenty-first century technoculture—and, in general, African-American significance that appropriates technological images to a practically "improved future" rather than the multi-faceted political and artistic/intellectual step (Dery 1993, 180). The range of African American, Caribbean, and Black British culture and subcultures from literature to music and the widening complex of book-based comics and paralysis are profoundly inferred or believed to be the parallels among race and science, cybernetics, and slavery. But its propensity to live mostly in epiphanoms is one of the shortcomings of Afro-futurism. The wealth of material culture is so ignored that they remain incredible and reveal suggestions and comparisons and are essentially empowered by imagination.

From the aforementioned quotes one might think of my interest simply as a reinterpretation of African-American perspectives in science fiction, along with critics such as Mark Dery, the late Mark Fisher, and, of course, Greg Tate and Kodwo Eshun. Such an observation will be welcome but, like Tate's and Dery's, the manner in which he orders intelligence is historically wrong. Science fiction, I contend, is instead, as a fundamental part of our technological history, a simple extrapolation into and through slavery and colonization of Black material interactions as well as other disadvantaged classes. SF does not permit race or slavery. Eshun's membership in Du Bois and Weiner is not possible only due to the convenience of the metaphor but also because, at least a hundred years early, Hugo Gernsbach will also be naming the genre for the evolution of robotics and cybernetics explicitly taken from and pictured in the rhetoric, assumptions, and social positions of Black people. Indeed, Norbert Weiner himself was so mindful of the parallel between race and technology, Black slaves and machinery that it may be argued that cybernetics was hunted by race when it was first expressed in the 1950s. The prevalence of racism and Black protests strongly reflect Weiner's thoughts as he writes this epochal novel, and the effect of these issues is his thinking of how people use other people (or deprive them their humanity). If Eshun knew in *More Brilliant Than the Sun*, it is not obvious, yet human use and numerous lectures and publications published in Weiner's epoch made clear reference, in order to identify the limitations of machines, cybernetic entities (the word cyborg was, of course, invented six years later by scientists Manfred Clynes and Nathan Kline, as well as artificial intelligencia), to African Americans and to slavery.

The initial reply of Weiner to African Americans and slaves and slavery is simple: why? This was not Isaac Asimov's hyperconsciousness, who had already tried to reverse what had become a weak cliché: robots as symbols for minority people, or, more generally, for racial or social differences. This was also not an overt effort to use this cliché that had been naturalized by SF as a "golden age," but can be traced back to the 19th century and into the dawn of those imperial texts. I shall be discussing these texts later, but to underline the material bases of my remarks, Weiner recognized his own reliance on the metaphor or "analogies between living beings and machinery" as Marinetti and the Futurists had already done some decades earlier (who were already obsessed by the technology of Black and Colonial Africa) and the optimistic impact on culture of emerging technology on the dystopian (Weiner 1950b, 48). But, in a solely scientific sense, the most critical point regarding Weiner's slaves is not only its interruptive regularity but that they have been employed to create a leading morality of his modern research and its eventual technical goods.

And there was an ethic that was deeply rooted in a racial consciousness. Cybernetics was deliberately presented in the fields of American ethnic politics, mastery, and slavery as a science of connectivity and power. Of course, the rhetoric is known: industrial life makes us all machines. In truth, Thomas Carlyle's insight into the term robot is Victorian. But, contemporaneous with Weiner's technological revolution, this insight is an American cliché, which was made American by the rhetorical inevitability of slavery. Caribbean pensioners like C.L.R. James to Sylvia Wynter and Antonio Benitez Rojo contend that the government and the subject formations of corporate capitalists are the presidents of a system.

12.7 CONCLUSION

Several separate cybernetic improvements and artificial intelligences have been examined in this chapter. Experimental cases have been published to show how human beings and/or animals will thus combine in a technology that raises a wide range of societal, ethical, and technological concerns. In each case details have been given on real practical experiments and not just a theoretical term.

The historical achievements and failures of AI can give us a lot of valuable insights. A reasonable and harmonious relationship between application-specific initiatives and bold research concepts is necessary to maintain the advancement of AI. In addition to the previously unheard of enthusiasm for AI, there are also concerns about how the technology will affect our society. To guarantee that society as a whole will benefit from the development of AI and that its possible negative impacts are reduced from the start, a clear approach is needed to take into account the related ethical and legal concerns. Such apprehensions shouldn't impede AI's advancement but rather spur the creation of a methodical foundation for it to grow. The most crucial of all is that science fiction must be distinguished from actuality. AI has the potential to drastically change our society's future, including our way of life, our living conditions, and our economy, with consistent financing and proper investment.

REFERENCES

[1] Shanti, A. (1993). Cyborgs in the n-dimension: The heretical descent of non-Euclidean geometry. *Constructions*, 8, 57–82.
[2] Shapiro, M. J. (1993). "Manning" the frontiers: The politics of (human) nature in Blade runner. In J. Bennett & W. Chaloupka (Eds.), *In the nature of things: Language, politics, and the environment* (pp. 65–84). Minneapolis: University of Minnesota Press.
[3] Siivonen, T. (1996). Cyborgs and generic oxymorons: The body and technology in William Gibson's cyberspace trilogy. *Science Fiction Studies*, 23, 227–244.
[4] Springer, C. (1996). *Electronic eros: Bodies and desire in the postindustrial age*. Austin: University of Texas Press.
[5] Stabile, C. A. (1994). *Feminism and the technological fix*. Manchester, UK: Manchester University Press.
[6] Stone, A. R. (1995). *The war of desire and technology at the close of the mechanical age*. Cambridge, MA: MIT Press.
[7] Taylor, M. (1993). *Nots*. Chicago: University of Chicago Press.
[8] Taylor, M., & Saarinen, E. (1994). *Imagologies: Media philosophy*. New York: Routledge.
[9] Thibaut, J. W., & Kelly, H. H. (1959). *The social psychology of groups*. New York: Wiley.
[10] Trinh, Minh-ha T. (1989). *Woman, native, other: Writing, postcoloniality and feminism*. Bloomington: Indiana University Press.
[11] Miguel S. Valles. (1999). *Técnicas cualitativas de investigación social. Reflexión metodológica y práctica profesional*. Editorial Síntesis, Madrid. ISBN: 84-7738-449-5.
[12] Bergen, L., Levy, R., & Goodman, N. (2016). *Pragmatic reasoning through semantic inference. Semantics & Pragmatics*, 9(20). Available on-line at https://semprag.org/article/view/sp.9.20.
[13] Aber Bergen (TV series). (2016–2018). *IMDb*. Available on-line at https://www.imdb.com/title/tt5759648/.

Index

A

aesthetics, 54
AI, 279
algorithms, 94
anatomy brain, 62
androids, 232
animated emotions, 95
anthropological, 34
artificially intelligent brains, 189
artificial memories, 220
artificial organism, 120
avatar, 236, 238, 242, 244, 246

B

behavioral therapy, 48, 50
biological enhancements, 111
biological evolution, 279
bodily integrity, 171
brain architectures, 194
brain-computer interfaces, 110
brain nets, 167

C

cantor of attraction, 101
cardiac diseases, 277
categorical imperatives, 47
chlorosis, 95
cognitive liberty, 213
colonization, 303
combative violence, 96
copyright law, 247
copyright protection, 247
cyberization, 7
cyber-objectivism, 11
cyber-subjectivism, 44
cyborg body control, 1
cyborg disputes, 118, 119
cyborg network, 292
cyborg reproduction, 88

D

digital world, 27, 28, 45, 105
divergence, 278

DNA, 247
doctrine, 121, 179, 180

E

effeminate, 100
electro- communication, 275
emergent stupidity, 306
enlightenment cyborg, 62

F

facial recognition, 236
FDA rules, 120
female fortification, 96
female vanity, 95
Freudo-Lacanian, 48

G

genetic material, 277
genuine innovation, 41
GPS bracelet, 276
GPS tracker, 276
gratification, 36, 49
grotesque, 97

H

hacking machines, 279
hacking memory, 282
hacking the ear, 291
hacking the eyes, 287
hacking the skin, 285
hallucinatory, 46
humanoid, 108
hypothetical, 12, 25, 50, 54, 87, 107, 133, 220, 250

I

illegitimate gratification, 99
imagination subjectivized, 59
implantable technologies, 276, 286
inflammation, 276
information age, 44
innovative cyborg, 28
intellect, 42

intellectual female cyborg, 84
intellectual property law, 153

L

law and brains, 187
law and policy, 161
lie detection, 217
Luddites, 127

M

machine intelligence, 295
machine learning, 194
man-machine, 77
media eradication, 44
media pleasure, 57
mediator, 34
mental development, 57
metaphysical, 95
misappropriation, 236
misogynistic, 96
mucous membrane, 55

N

narrative visual, 58
natural forces, 41
natural inspiration, 305
network of sensors, 141
neural interface, 274
neuromodulation, 276
neuroprosthesis, 205

O

occult, 303

P

pancreatic, 278
pedagogical evolution, 11
personalities, 5, 19, 68, 86, 122, 245
phobia, 50
Platonic, 27, 32, 34, 44
posthuman, 7, 15, 17, 44
preception vision, 31
priesthood, 99
proof-of-concept, 274
prosthesis limbs, 109
prosthetics, 105
prosthetizing devices, 106
psychoanalytic, 46

R

real annihilation, 56
regulating sensors, 149
reliance on sensors, 139
reminiscent longings, 30
remotely sensed, 151
retroaction, 40
RFID, 275
rheumatoid arthritis, 276
roadblocks, 211
robot code, 297
robotic progeny, 297
robotic revolution, 120

S

scarification, 273
semiotics, 45
sensory substitution, 291
singularity, 27, 39, 43, 104, 114, 116, 123, 158, 160, 174
skill literacy, 14
slippery rhetoric, 299
soul communication, 65
surgical malpractice, 275
surveillance, 155
swarming robots, 304

T

Taildoes, 97
technological belief, 58
technologization, 36
telepathy, 167
telepresence and sensors, 143
textual transition, 9
theologians, 95
transcendent rationality, 302
transcending human, 28

U

Ugly Laws, 240
uncanny valley, 238

V

virtualization, 39, 43
virtual reality, 116
virtual technology, 36
vital disavowal, 58
vivid mind, 95

W

water-engine, 104
woman cyborg, 97
woman-machine, 93, 95, 98, 99
womb geometry, 102
world of sensors, 136